LIVERPOOL INSTITUTE
OF HIGHER EDUCATION

LIBRARY

WOOLTON ROAD,
LIVERPOOL, L16 8ND

£15.99

INDUSTRIAL CHANGE
AND
REGIONAL ECONOMIC TRANSFORMATION

LIVERPOOL INSTITUTE
OF HIGHER EDUCATION

Order No.
L710

Accession No.
134796

Class No.
914 ROD

Control No.
ISBN

Catal.
8 MAY 1992

INDUSTRIAL CHANGE AND REGIONAL ECONOMIC TRANSFORMATION

The experience of Western Europe

EDITED BY
LLOYD RODWIN and HIDEHIKO SAZANAMI

HarperCollins*Academic*
An imprint of HarperCollins*Publishers*

© United Nations 1991
Industrial change and regional economic transformation
is published for and on behalf of the United Nations.

This book is copyright under the Berne Convention.
No reproduction without permission. All rights reserved.

The authors assert the moral right to be
identified as the author of this work.

Published by
HarperCollins*Academic*
77–85 Fulham Palace Road
Hammersmith
London W6 8JB
UK

First published in 1991

Library of Congress Cataloging in Publication Data

Industrial change and regional economic transformation: the
  experience of western Europe / edited by Lloyd Rodwin and
  Hidehiko Sazanami.
    p. cm.
  Includes bibliographical references and index.
  ISBN 0–04–445882–7 (HB). — ISBN 0–04–445883–5 (PB)
  1. Europe—Economic conditions—1945– —Regional
disparities—Case studies. 2. Europe—Industries—Case studies.
I. Rodwin, Lloyd. II. Sazanami, Hidehiko.
HC240.I525   1990
338.094—dc20   90–41046
                                                              CIP

British Library Cataloguing in Publication Data

Industrial change and regional economic transformation: the
  experience of Western Europe.
  1. Western Europe. Economic development
  I. Rodwin, Lloyd II. Sazanami, Hidehiko
  330.940559

ISBN 0–04–445882–7
     0–04–445883–5

Typeset in 10/12 pt Melior by Nene Phototypesetters Ltd
and printed in Great Britain by
Mackays of Chatham PLC, Chatham, Kent

Contents

Preface   xi
Lloyd Rodwin & Hidehiko Sazanami

I OVERVIEW

1 European industrial change and regional economic transformation: an overview of recent experience   3
   Lloyd Rodwin

   Decentralization, deindustrialization and regional economic change   3
   The country studies: synopsis   6
   The perspective studies   19
   Concluding observations and inferences   25

II CASE STUDIES

2 Structural transformation in the regions of the United Kingdom   39
   Peter Hall

   The anatomy of regional change   40
   The role of high-technology manufacturing   46
   The role of producer services   49
   Producer services: two case studies   52
   Toward explanation   54
   Regional case studies   58
   Postscript: the economic revival of the 1980s   65
   Conclusion: the search for policy   67

3 Deindustrialization and regional development in the Federal
   Republic of Germany   70
   Franz-Josef Bade & Klaus R. Kunzmann

   Introduction   70
   National trends in structural change   75
   Spatial differences in the shift to services   80
   Some key factors explaining regional differences in deindustrialization
   in the Federal Republic   93
   Regional policy responses to deindustrialization   100

4 Spatial impacts of deindustrialization in France   105
   Jean-Paul de Gaudemar & Rémy Prud'homme

   Introduction   105
   Deindustrialization in France   107
   Regional deindustrialization patterns   113
   A tale of three regions   121
   The case of the automobile industry   130
   Conclusion   133

5 Regional deindustrialization and revitalization processes in
   Italy   137
   Roberto P. Camagni

   The deindustrialization condition   137
   Regional performance in Italy, 1970–84   145
   The deindustrialization and revitalization process   151
   Some case studies of revitalization strategies   158
   Conclusions   164

6 Structural changes in the Spanish economy: their regional
   effects   168
   Juan R. Cuadrado Roura

   Recent developments and changes in the Spanish economy:
   an overview   168
   The regional impact of the industrial crisis   179
   Reconstruction of the industrial structure and areas of attraction 188
   The contrast between two macro-regions   195
   Some final remarks   199

7	Regional perspectives on the deindustrialization of Sweden *Folke Snickars*	202
	The issue of regional deindustrialization	202
	Structural change in the Swedish economy	206
	Twenty years of regional change in Sweden	215
	Changing policy perspectives in Sweden	224

III PERSPECTIVE STUDIES

8	Problems of regional transformation and deindustrialization in the European Community *Paul Cheshire*	237
	Some definitional issues	239
	Measuring deindustrialization	248
	Measures of regional deindustrialization in the European Community	251
	Regional transformation as a source of adjustment problems	258
	Some concluding remarks	264
9	1957 to 1992: moving toward a Europe of regions and regional policy *Paul Cheshire, Roberto P. Camagni, Jean-Paul de Gaudemar & Juan R. Cuadrado Roura*	268
	Introduction	268
	Major features of the European Community budget	271
	Community regional policy: the first moves	278
	The political and statutory framework of industrial restructuring	283
	Community programmes for regional restructuring	288
	The 1988 reform	291
	The regional impact of economic and monetary union	296
10	Europe's regional–urban futures: conclusions, inferences and surmises *Roberto P. Camagni, Paul Cheshire, Jean-Paul de Gaudemar, Peter Hall, Lloyd Rodwin & Folke Snickars*	301
	Anticipated effects of 1992	301
	Some specific spatial effects	307
	Europe and Japan: a comparison	312
	The Greater Europe of the 1990s	313

IV POSTSCRIPT

Structural transformation in Japan: issues and prospects for regional
development in the coming years   319
Hidehiko Sazanami

Backdrop of global economic change   319
Regional economic transformation in Japan   324
Implications of Tokyo's phenomenal growth   332
Challenges of balanced growth   338
Options for regional development in the coming years   341

Chapter appendices   347
Chapter notes   369
Notes on contributors   388
Acknowledgements   392
List of tables   393
List of figures   396
Index   399

Preface

THIS BOOK is a sequel to *Deindustrialization and regional economic transformation: the experience of the United States*, which was published in December 1988 under the auspices of the United Nations Centre for Regional Development (UNCRD). The focus here is on the experience of Western Europe. Both studies are part of a much broader set of studies sponsored by UNCRD dealing with the regional aspects of the vast changes now occurring in the world economy, and their implications for Third World countries.

The book on the United States had two parts. The first part consisted of case studies of five regions, examining the characteristics and trends of the particular regional economy, the leading economic sectors of the economy, the private and public policies to deal with the problems of industrial changes, and such evidence as was available on the outcomes of these efforts. The second part suggested some additional ways of interpreting these experiences.

The five regions were: the midwest industrial heartland, the region which has sustained since the late 1970s the greatest losses of all the US regions in manufacturing jobs and employment; two former problem regions, the South and the Massachusetts region, which have experienced economic resurgence; and the two prosperous regions of California, identified in particular with high-tech activities and the burgeoning service economy, and the New York Metropolitan regions, an example, par excellence, of the growth of financial and business services. We were well aware of the radical differences between these regions – differences in size, economic base, culture, history and in range of problems. However, we felt that these very differences might illustrate how comparative analysis could be applied effectively to quite diverse economic experiences.

There were also five perspective studies. One examined the broad intrametropolitan spatial readjustments of industrial activities in response both to shifts in industrial activities and to the advantages today of metropolitan location and services. Another analysed growth prospects of the service industries and why their further expansion at anywhere near the same rate is unlikely, while just the reverse might be anticipated for manufacturing. A third reviewed regional theory and practice in light of the changes and adjustments in the different regions in relation to both the US and the international economy. The fourth dwelt on how little we know as planners and the obstacles we can expect in the way of effective implementation as indicated by how regional development theory and policy has changed over the past generation. An overview explained how the studies related to each other; it also provided some of the historical and ideological background before reviewing the findings and unsolved puzzles.

In this European study we examine the experiences of industrial change and regional economic transformation in Britain, Germany, France, Italy, Spain and Sweden. None of the nations of Eastern Europe were included because of the great differences in the economic systems and trade relationships. This situation is likely to alter significantly with the dramatic political and economic changes which began in East Germany, Poland, Hungary, Czechoslovakia and other countries in Eastern Europe in 1989. Each of the six countries studied is at a different level of industrial development. The countries differ in location, size, political complexion, administrative organization, economic base and economic history. Five are in the European Community, one is not. Each has sustained losses in manufacturing activities – mainly but not exclusively – in their prosperous as well as their older declining resource-based industrial regions. Each has experienced varying degrees of frustration and success with efforts to modernise or revitalise industry.

Three perspective studies supplement these country case studies. The first focuses on the patterns and problems of regional transformation viewed from the vantage point of Western Europe as a single, integrated economic community. The second traces the evolution of the regional policy from 1957 to 1992. The third articulates some of our expectations regarding Europe's regional–urban futures. This is based on the likely regional, urban and spatial effects of further integration of the European Community in the context of an internationalising economy, an informational revolution and the liberation of Eastern Europe.

Not surprisingly, the European regional studies were far more difficult to produce than the US studies. This was because of the greater scope and range: the analyses involved several regions rather than one, and European as well as international and national contexts. Also, there was a modest effort to reflect on the effects and implications of our findings for the rest of the world, including newly industrialising countries (NICs).

The original concern of the European studies, too, was on regional 'deindustrialization', but the literature in the field and general usage yields very little agreement on how to define the term. Most of us are inclined to share Professor Roberto Camagni's view that the most virulent problems involve situations where there are simultaneous loss of competitiveness and productivity. Be that as it may, in retrospect, it is accurate to say that we have studied regional economic transformation in the United States and Europe, with a special focus on the varieties of industrial change as well as other factors responsible for these transformations. We trust, too, that our studies have been conducted in a way that will facilitate comparative analysis of these experiences as well as exploration of the implications for both the Third World and more developed countries.

This enterprise involved more than the commissioning of papers and their preparation by the contributors. There were five meetings of the group: the first in Paris on October 17, 1988 to discuss the structure of the study; the second in Brussels on January 13–14, 1989 to review the first drafts; the third in Paris on June 5, 1989 to review the second drafts; the fourth in Nagoya and Tokyo on September 18–22, 1989, where the papers were formally presented at the International Conference on Industrial Transformation and Regional Development: Regional Economies in a Borderless Age, sponsored by UNCRD and the Organizing Committee for the International Conference on Comparative Regional Development Studies. There was also a final meeting in Paris, on December 16, 1989 to re-examine the papers in light of the discussions in Japan, before final submission of the manuscript to the publisher.

Finally, a word about comparative analysis. It is an ambitious aim, and most of us pay lip service to this fashionable idea. It resonates so well with the current interest in our global economy, and it is another way of overcoming parochial perspectives. But to practise comparative analysis is far more difficult than one might suppose. The best comparative analyses, like the best interdisciplinary studies, often take place within one skull. Few people have the necessary background, particularly the knowledge and understanding of the different cultures and values and the way things work in these varying contexts. Most of us can master a subject like regional economic change and the way it is manifested in a particular region or country. We can and often do venture a lot beyond that range, but those of us who are sensible do so with caution. The more knowledgeable we become, the more we are aware of what we don't know. That is why we try to work with multidisciplinary teams and promote interdisciplinary professional meetings as one of the better – perhaps even indispensable – ways of becoming more adept in these efforts.

Lloyd Rodwin
Ford International Professor, Emeritus

Hidehiko Sazanami
Director, United Nations Centre for Regional Development

I
Overview

1

European industrial change and regional economic transformation: an overview of recent experience

LLOYD RODWIN

Decentralisation, deindustrialisation and regional economic change

AT THE OUTSET, this book was to have been an investigation of the deindustrialisation experience in Western European countries from a regional perspective. Now we regard it as a study of industrial change and regional economic transformation in the contemporary world economy. The shift of emphasis reflects how we have chosen to approach these studies.

The economic transformations we are examining are contemporary. They involve the experiences – national and Western European – associated with regional economic changes over the past two or three decades rather than the past two or three centuries. These changes are occurring in what is now the largest and, at the same time, the smallest world economy in history: the largest geographically when compared with the Phoenician, Hellenic, Roman, Chinese, Ottoman and European economies of the past[1] and the smallest in terms of the shrinkage of space and time created by modern communications and swift travel between the major centres of the world.

Three powerful interacting forces are accelerating these changes: the intensity of international competition, the cumulative impact of research and innovation and the enhanced importance of amenity. The impact of these forces may be seen throughout each national economy in astonishing advances in transportation, communications and the handling of information; in the extension of the span of management capabilities; in the increased ability of both large and small business organisations to function efficiently; in the shift not only in the composition of goods and services in favour of services but also in the blurring of the distinction between goods and services; in the capacity of firms to locate flexibly their component functions; and, not least, in the reordering of entire production technologies to facilitate

> computerised factory schedules and inventory controls to cut costs; intelligent production machinery that can shift processes in the middle of an assembly line ... and [an] increasing ability to produce local products adapted to local markets, but reap world economies of scale in research and development, raw materials sourcing, and production balancing.[2]

In the wake of these changes there have been many losers as well as winners, and enormous economic, social and human costs: blighted communities and regions, loss of infrastructure, disappearance of jobs, erosion of skills, increasing inequality of income and sheer human misery. Fear, hostility and concern have been aroused in many quarters and many persons are convinced that a better way to handle the incidence of costs and benefits must be found. It should come as no surprise that, as in past phases of the industrial revolution, we are now seeing

> the battle of ideologies around the inexorably growing economy, some blindly opposing, some seeking to retard its more ruthless thrust into the social fabric, some single-mindedly or simple-mindedly hailing its every advance. We witness the rearguard action of the champions of the older order, the impotent discomfiture of the upholders of [tradition], the easy triumph of the orthodox economists who neatly explain it all. But the advancing front leaves ruin in its train, and the hastily built defenses crumble before it. We see how with a new liberation went a new servitude, and we [must] measure the challenge that now faces our age.[3]

Specific aims and approach

As in all such enterprises, resources were limited, and all the more so given our aims. We wanted to examine several countries, not one, and several regions within those countries. We planned analyses in the context of national as well as European trends, but we also intended to probe the problems of Europe as a single integrated economic community. More specifically, the chapters that follow have five objectives. These are to:

(a) describe the course of industrial change and regional economic transformation in each of the countries: the United Kingdom, the German Federal Republic, France, Italy, Spain, and Sweden (Fig. 1.1) and explain why the changes are occurring;

Figure 1.1 Europe.

(b) point up the cases or regional experiences (successes as well as problems) that particularly illuminate the trends and policies;
(c) depict the macro-European experience with regard to industrial change and/or subnational regional economic transformation;
(d) raise questions or test hypotheses about various issues, e.g. the growth prospects of the industrial and service sectors; the nature of deindustrialisation; the role of such factors as changes in international markets, corporate organisation and management; the current effectiveness of local, regional and national policies; and the parallelism with and implications for experience elsewhere; and finally
(e) report what we could say about the effects and implications of these findings for Europe, Japan and many newly industrializing countries.

Now what in summary are the main findings?

The country studies: synopsis

The United Kingdom

The first case study is of the United Kingdom the country where industrialisation began. The United Kingdom's problems in the 'North' (its poor regions of northern England, Scotland, Wales and Northern Ireland) led to major industrial assistance and relocation programmes some 50 years ago. Regional differences reflected the contrast between the traditional heavy industries in the peripheral north and the consumer goods industries of the industrial heartland – particularly the Midlands – and the south (Fig. 1.2).

In Chapter 2 Peter Hall notes that since the mid-1960s the United Kingdom has experienced massive losses in manufacturing in inner London and in

> the non-standard, non-assembly line, non-automated production in old northern industrial [heartland] cities that are ... disinvesting [The differences in the period and the technology of these activities help to explain] why industrial decline is no longer concentrated in the traditional heavy industrial sectors, as in the inter-war period, but has now spread into the consumer manufacturing industries that spearheaded the post-war recovery, especially the industrial heartland of the West Midlands.[4]

This decline sharply contrasts with the exceptional growth of high-tech activities and the higher management control, and business and producer

Figure 1.2 Regions of the United Kingdom.

service activities in London and the South East. Especially important in this region is the 'Western Crescent' around London, i.e. the area from Hampshire through Berkshire and Buckinghamshire to Hertfordshire and Cambridgeshire.

Put somewhat differently:

> beginning around 1974, but intensifying after 1979 ... those regions that were most manufacturing dependent – the West Midlands, the North West, Yorkshire–Humberside, the North, Wales, Scotland and Northern Ireland – have suffered most from manufacturing decline and gained least from new private sector service jobs. Half the population of the United Kingdom live in these regions, and by the mid-1980s unemployment rates in all of them exceeded 15 per cent, producing two nations separated by widening economic and social disparities.[5]

The north–south disparity, Hall reminds us, arising in part from the concentration of services – particularly financial and commercial – in the south, may be traced to the early nineteenth century and perhaps even earlier;[6] and the close spatial and functional association between high-tech manufacturing and producer or business services has evolved from this initial advantage. High-tech manufacturing and service firms in Britain have favoured locations where producer services are also concentrated[7] and avoided those having a high proportion of traditional economic activities. This preference partly reflects the image of the declining areas, the lack of appropriate local skills and the likely difficulty of attracting technological and scientific staff to such locations.[8]

Still another marked trend is the regional concentration of producer services and their deconcentration among regions. If these particular trends continue, Hall expects that by the early 1990s there will be more commercial office floor space in the South East (excluding London) than in central London itself.[9]

The growth of the economy since 1981 has further sharpened the differences in performance between the south and the 'old' and 'new' features of the north. Hall thinks the forces behind the trends lend credence to two rival and not altogether inconsistent interpretations of the trends. One is the neo-Schumpeterian stress on the role of innovation, entrepreneurialism and emergence of new firms shaping the contemporary south. The other is the neo-Marxist view that the failure of the northern regions to shift from process to product innovations, from industry to services and from goods-handling to information-handling is because basic management and decisions occur in the south.[10]

Regional planning policies over the past two decades have been designed to mitigate the effects of the changes. These include the elimination of the restrictions on industrial development in London and other regions; the shrinkage in the size of areas receiving assistance; and the shift of attention from inter-regional to intra-regional policy. The main focus of the latter has been to recycle or promote smaller, derelict zones or areas potentially responsive to redevelopment measures. In retrospect, Hall now ruefully concedes that although the earlier policies may have

> worked moderately well in the boom years of the 1960s, when there were plenty of new industrial jobs to be diverted from prosperous to less prosperous regions [they became more or less obsolete in the transition from an industrial to an informational economy] apart from a residue made necessary to enjoy access to European Community regional funds. In its place [has] come a mixture of policies, some spatially specific, some not, to try to regenerate the worst-hit pockets in each region, as appropriate.[11]

The Federal Republic of Germany

The Federal Republic of Germany now has one of the strongest economies of Europe and ranks among the most advanced countries in high tech manufacturing activities. Franz-Josef Bade and Klaus Kunzmann report that on the basis of value added, the Federal Republic had the same growth rate for manufacturing as did the United States between 1973 and 1978, but that the loss of employment in the Federal Republic was much more severe.[12] (The Federal Republic actually lost jobs between 1973 and 1978, while in the United States net industrial employment expanded.) At the same time, growth and decline trends in services often paralleled the ups and downs of manufacturing, unlike the experiences of the United Kingdom and the United States. The expansion of employment in the services sector did not offset the downward trends in manufacturing. In fact, the relative share of the Federal Republic's service-sector employment ranked among the lowest of the industrial nations.[13]

Like Britain, the Federal Republic, Bade and Kunzmann note, has a north–south disparity. There has been much more growth in business services and high-tech activities in agglomerations in the south (notably München, Stuttgart and Frankfurt, and their surrounding cities and regions). There are also more universities in these regions and a more attractive social and cultural image. Baden Württemberg has become a centre of mechanical and electrical engineering as well as of the automobile industry; and there have been successful efforts in the south, in places like Ulm, Karlsruhe and Heidelberg to encourage science and research-based enterprises and to promote all forms of cultural activities. Also, over the past decade or so, the new modern research, development and production centres of the defence industries have located in München and Baden Württemberg.

In contrast, other areas, particularly the older, resource-based industrial heartland such as Nordrhein-Westfallen (Fig. 1.3) (with its coal, energy, steel, chemical, mechanical engineering, textile and consumer goods industries) have seen much less growth or even decline. Employment from 1961 to 1987 dropped in the industrial sector from about 63 per cent to only 49 per cent whereas employment in the service sector rose from about 35 per cent to almost 50 per cent involving a loss of 1.5 million jobs

Figure 1.3 States of the Federal Republic of Germany.

in the traditional sector and a gain of only 420,000 jobs in the service sector.[14] However, Bade and Kunzmann think that efforts now underway to modernise the infrastructure, improve the environment, increase the number of industries, encourage small and medium-sized innovative industries and transform the qualifications of the labour force – and even the forbidding image of the region – have met with modest success, as exemplified by cities like Dortmund.

France

Despite new regional authorities with powers to raise revenue and spend it,[15] France is still a more highly centralised and administered economy than the Federal Republic of Germany or the United Kingdom. It has a higher rate of industrial loss in the Ile-de-France (Paris) region and the northeastern industrial areas like l'oc-de-Calais and Lorraine, and lower rates in the south and west (Fig. 1.4). But Jean-Paul de Gaudemar and Rémy Prud'homme indicate that services have not offset the loss of industrial output or employment: in fact, they have fared better where industries did best because some of them were complementary and still others reinforced or spurred industrial development.

In the past, designers of public sector industrial policies sought to slow down the rate of industrial growth in the Ile-de-France and to increase industrialisation in the poorer regions of the west, the south and selected areas of the southeast. The effort was not inconsistent with the trends; and it proved moderately successful, according to de Gaudemar and Prud'homme. In part, this was because of developments in the private sector, particularly in some important branches like the automobile industry, e.g. increased merging of firms coupled with the decentralisation of production facilities, greater centralisation of headquarters and management, and increased subcontracting for components and services.

However, problems afflicted the industrial northeast, France's formerly prosperous heartland.[16] With the possible exception of Lille, this region lacks significant urban centres. Levels of education are surprisingly lower than elsewhere in the country. The regional economy, in addition to the resource-based activities of coal and iron ore mining, is still dominated by large oligopolistic activities in steel, machinery, automobiles and basic chemicals. In the economic squeeze between 1976 and 1986, the region lost 30 per cent of its total employment (about 400,000 jobs). Assisted by public policies – the nationalisation of steel and the protection of the automobile industry – the large producers managed to improve productivity and earn some profits even in those low-growth sectors. The region is also favourably located in relation to the main population and economic centres of the European Community. Despite these hopeful signs, de

Figure 1.4 Regions of France.

Gaudemar and Prud'homme still doubt whether efforts to attract other activities through technical assistance, research, vocational training and infrastructure investments (especially for transportation) can offset the fundamental sources of the economic malaise.

The south, they say, has fared much better despite a low initial level of industrialisation, the predominance of small and medium-sized firms and relatively large losses of industrial jobs.[17] These losses were compensated for by relative stability in the equipment industries and significant gains in the electronics, computer and aircraft industries and other technology-intensive areas. This economic resurgence, de Gaudemar and Prud'homme suggest, was made possible or further reinforced by higher demand and higher levels of education, a dynamic business climate and the propulsive role played by actors in the financial, industrial and commercial services supported by subsidies from the European Community.

In terms of industrial job losses and increases of service employment, the most industrialised, tertiarised, diversified and urbanised part of the nation – the Ile-de-France – did less well than the rest of the country.[18] These economic trends were reinforced by deliberate government decentralisation policies. Nevertheless, the area experienced less unemployment and attracted (as in the case of the decentralised production facilities of the automobile industry) high-level, well-paid managerial and technical jobs. It also grew economically, despite deconcentration, because of its favourable location as one of the dominant metropolises of the European Community and its triple role as France's leading political, economic and cultural centre.

Italy

One conventional image of Italy projects a prosperous north – run by governments of varying political orientations – pursuing continual efforts, of varying degrees of seriousness and effectiveness, to industrialise or otherwise transform the poorer, backward Mezzogiorno. These efforts have ranged from investments in infrastructure and the creation of growth poles to the establishment of branch plants plus diffused industrialisation based on small firms and other miscellaneous decentralisation strategies (Fig. 1.5). The reality, according to Roberto Camagni, is more complex. In fact, aside from the Mezzogiorno, there have been deindustrialisation problems in the north – in both new and old industrial regions (Veneto, Emilia Romagna, Liguria, Piemonte, Lombardia, and Toscana) – following the strong industrialisation of the 1960s and 1970s.[19]

However, in Italy's Northwest, unlike the United Kingdom and France, and to a lesser extent the Federal Republic of Germany, a market-led reindustrialisation occurred in Lombardia and Piemonte. This 'rejuvenated' the region and contributed to the closer integration of industry and services, with the former leading the latter.[20] It also dispelled the onset, or at least the fear of the onset, of long-term structural deindustrialisation, at least at this time. Several factors made the reversal possible. The more important, Camagni suggests, were critical high-tech innovations in production and management processes plus direct or indirect support from national sectoral or macroeconomic policies. Both combined to reduce costs, improve industry's competitive position and avoid the most virulent form of deindustrialisation: the simultaneous losses of jobs and competitiveness.

However, Camagni observes no structural change in the Mezzogiorno, and in particular, the poorer southern regions of Abruzzi, Campania,

Figure 1.5 Regions of Italy.

Puglia and Sicilia, despite some state assistance and increased investment and employment. Crises arose in the so-called growth poles, especially in the cities of Napoli and Palermo, and the further growth of the north appeared likely only to sharpen traditional north-south differences.[21]

Spain

Because in many ways Spain has only recently begun to transform and modernise its economic structure, it is of special interest to newly industrialised countries (Fig. 1.6). With the growth of the world economy in the decade and a half following 1960, Spain experienced a tremendous economic spurt and industrialisation. Then, between 1975 and 1985, in the course of an international economic recession, Spain suffered a period of inflation and a serious economic setback. Juan R. Cuadrado Roura indicates that during this period, and even earlier, international competition intensified and patterns of demand changed with damaging effects on particular sectors and regions.[22]

The main drop in investment and losses of jobs occurred in agriculture and mining and in industries with low growth prospects (e.g. shipbuilding, steel, automobiles, machine tools and chemicals). Two other problems, Cuadrado Roura adds, exacerbated the situation: the failure to adapt or diversify significantly the existing economic activities or to innovate new ones; and the increasing deterioration of physical and economic infrastructure, especially business services. A formerly prosperous area, the Cantabrian Cornice (i.e. the Basque country, Cantabria and Asturias), appeared to be not only ailing but on the verge of a severe secular decline. Madrid, Barcelona and a few other isolated nuclei, such as Ferrol, Vigo and Cádiz, were also adversely affected.

Figure 1.6 Regions of Spain.

Not all the developments were negative. Since output rose somewhat during this period, there were some gains in productivity. Services also increased, albeit mainly in the public sector. Most astonishing, however, were the political and institutional changes as Spain was transformed from a highly centralised dictatorship into a constitutional monarchy with representative institutions and very substantial political decentralisation.

Eventually, from mid-1985 on, general economic conditions improved, partly in response to the renewed growth of the world economy and access to an enlarged market following Spain's entry in January 1986 into the European Community. Also helpful was the Community's financial assistance (for poorer backward regions – i.e. for the economy of Spain generally in relation to the rest of Europe – and for declining or threatened industrial regions). The recovery was further reinforced by the favourable business climate created – to almost everyone's surprise – by the leadership of the new socialist government.

In the process of recovery and reindustrialisation,[23] Cuadrado Roura indicates, much of the territory from Cataluña to the south, and the Ebro Valley emerged – in addition to Madrid – as dynamic regions of major economic growth, international investment and the development of high-tech activities as well as business and consumer services. The Cantabrian Cornice also benefited somewhat, although its distance from the large European urban centres, poor transportation facilities and political turbulence in the Basque country posed serious handicaps to economic progress.

Sweden

A geographically peripheral European country (Figs. 1.1 and 1.7), Sweden is not a member of the European Community. Its domestic market is too small to permit economies of scale and it is committed to very advanced social policies and welfare programmes. In fact, the main increase in services from 1960 to 1990, according to Folke Snickars, has been in the public sector. Sweden has a regional policy based on 'the principle of equal access to services and work opportunities in all regions'.[24] Subsidies have been used – not very successfully – to offset regional differences in production and, more recently, in service costs.

This combination of factors, Snickars suggests, makes Swedish industry an obvious candidate for deindustrialisation, however defined. Manufacturing employment, a key indicator, declined about 45 per cent between 1950 and 1990.[25] Nevertheless, Snickars says, Swedish industry is managing to cope with these disadvantages in several ways. Its industry transforms domestic natural resources into intermediate products used by

Figure 1.7 Counties of Sweden.

the nations of Western Europe. More recently, Swedish industry has been competing – with surprising effectiveness – in specialised dynamic, high-value milieus in foreign countries on the basis of product quality rather than price. Swedish firms have also emphasised R&D and creating a highly skilled labour force.

Snickars thinks that aspects of public policy have helped in several ways. One has been the investment in infrastructure (initially mainly in roads and later in telecommunications); another, perhaps inadvertently, has been the imposition and upgrading of quality and environmental standards. Still more recently, the government has provided decentralised and linked educational facilities, particularly in different regions, anticipating that continuing education is likely to be a 'fundamental part of work and social life ... in the knowledge ... society'.[26]

Swedish industry recognises the importance of markets, particularly international markets, and therefore now focuses either on locating production units near the markets or on improving the logistic systems to facilitate access to markets without moving production capacity.[27] It also uses international co-operation and mergers 'to attain economies of scale in knowledge production, innovation and product development'. The aim is to achieve the comparative advantage of a highly skilled management and labour force in addition to the exploitation of more traditional factors such as location or resource endowment.

Spatially,[28] Snickars notes, the net effect of these trends has been an increase of jobs and population in the large metropolitan regions and a decrease elsewhere, i.e. in the steel belt of middle Sweden, in the shipyard cities on the west and south coasts and in the northern mining regions. In several parts of the country the industrial production capacity was obsolete in communities with only single plants or firms; and quite often the decision making in these plants was controlled by head offices located in Stockholm or outside the region.[29]

Stockholm's growth in the past decade, Snickars reports, was exceptional: almost double the growth of other metropolitan regions, a growth rate expected to continue for the next decade. This is due in the main to its role as a market, and as a financial, cultural and intellectual centre. However, as in the case of London, Paris and New York, Stockholm has been experiencing a great loss of manufacturing activities.[30] The manufacturing activities that remain require fewer manual and more service jobs. Also, most of the relative growth of the business-oriented service jobs has been higher outside the metropolitan regions. Over the past decade and half, there 'has been a steady decline in the Stockholm region's share of national employment in occupations with a distinct R&D character'.[31]

These trends, Snickars emphasises, were not without cost.[32] One is that there is now less social pressure to achieve an important goal of the past, i.e. the easing of the problems of the people in the poorer regions or in regions directly afflicted by structural change. But another goal – attaining regional balance – is still stressed, perhaps because of the sustained political pressure made possible by area representation. The goal of regional balance has, in fact, raised two questions: is traditional metropolitan organisation becoming obsolete, given the economic transformation away from the industrial sector? Are more decentralised multiple core network cities and educational opportunities possibly more efficient for the new high-tech information society?

The perspective studies

Deindustrialisation, regional transformation and the European Community

The emphasis in the studies noted above is on regional economic transformations in several national contexts. How does the view change if we examine the transformations from the overall perspective of the European Community? Paul Cheshire reminds us that it is anything but easy to provide answers based on reliable data. The available data differ in quality, definitions, underlying concepts and time horizons. For his comparative analyses, Cheshire relies on Functional Urban Regions (FURs). These 'consist of a core, defined in terms of a concentration of jobs, and a hinterland or ring, defined in terms of commuting relationships'.[33] However, a variety of 'adjustments' had to be made when some of the necessary data were not available. In addition, he had to distinguish (a) differences between the loss of firms and employment within and from a region; (b) cyclical and trend (structural) changes as well as (c) relative and absolute employment and output (or value added) measures of deindustrialisation; and (d) corrections necessary to account for the decline in industrial employment as per capita income rose above certain levels.[34]

When these and other factors are taken into account, Cheshire manages to sharpen, qualify and even raise controversial questions about a number of views. For example, some of the studies show:

> There is no correlation ... between change in industrial output and industrial employment over the period 1983–7, and the country that did worst on the employment measure, Ireland, did best on the output measure. The country that did best in the output measure did worst on the simple productivity measure.[35]

Furthermore, his view of the European Community as an economic unit discloses the same gross component features as the country studies: i.e. decentralizing and probably recentralizing service-oriented metropolitan regions, a range of high-tech growth centres, threatened industrial heartlands, declining, older, resource-based industrial regions and poor peasant-dominated backward regions. While Cheshire refers often to the phenomenon of deindustrialisation and draws attention – with reservations – to its frequent association with increasing income inequality, he is sceptical (as is Camagni) about the deindustrialisation theme. To 'focus on deindustrialisation', he concludes, 'is at best misleading; at worst dangerous'.[36] Citing Camagni, he notes that deindustrialisation is often only one facet of a transformation having many hidden aspects – intra-regional and inter-regional changes as well as organisational and sectoral dimensions.[37] Cheshire believes that just as we have placed less emphasis on de-agriculturalisation in the period of the Industrial Revolution, so must we focus less on deindustrialisation and not ignore the important organisational, sectoral and spatial aspects of industrial change and regional economic transformation.[38]

He stresses that if unemployment is a serious concern of policy, the evidence is that the older industrial regions, based on coalfields or ports, and the poor, backward rural regions have the most severe problems of unemployment. He notes also that heartland industrial regions like Birmingham, Stuttgart and Torino where 'specialisation was skill-based rather than resource-based ... have adapted to capital–labour substitution and to loss of employment in their traditional industrial sectors more readily than the old, resource-based industrial regions' and have been more successful in introducing innovations, transforming institutions and enhancing productivity.[39]

Cheshire's analysis also throws some doubt on the widely held view that the weakest and most vulnerable manufacturing plants and firms in the Community are always in the peripheral regions. This may have been true, he says, for the transplanting of heavy industry to Italy's Mezzogiorno, but elsewhere 'many peripheral regions – the southwestern regions of France, the southeastern regions of the Federal Republic of Germany, Ireland and South West England – exhibited some of the most rapid rates of manufacturing employment growth'.[40]

Since the problems of regions experiencing agricultural transformation appear to be exceptionally severe, particularly in so far as unemployment, migration and low income are concerned, the European Community has given high priority to the problems of poor, backward, rural regions of Southern Europe. However, even within these regions, Cheshire observes, the worst problems are in the urban areas, where the uprooted peasant population is concentrated. As a consequence, an issue becoming increasingly critical on grounds of equity and efficiency is whether most Community aid ought to be directed to these centres.[41]

Evolution of regional policy

But serious questions may be raised about what can be expected from such a policy change based on the analysis of the evolution of European Community regional policy made by Cheshire, Camagni, de Gaudemar and Cuadrado Roura.[42] The total revenues of the Community only approximate 1 per cent of Community GDP compared with 10 per cent to 15 per cent for national revenues.[43] Of this total, about three-quarters is earmarked for agriculture, mainly for price support.[44] For the most part, also, Community aid is often a substitute for, instead of a supplement (as intended) to the assistance provided by existing national programmes. Therefore, to date, the actual sums involved for regional 'structural' changes have been small and often inadequate.[45]

However, it has turned out to be difficult to ignore the initial and persistent aims of the 'structural' programmes. One was designed to redress the problems of regional imbalance and inequality resulting from the preponderance of agricultural and industrial change and structural unemployment. The other was to counter, when possible, the negative effects of European integration on particular regions. The authors noted constant reminders of these aims, all the more evident because they helped to offset, if not correct, some blatant inequalities. For example, one of the persistent spurs for these 'structural' efforts was the recognition by countries like Italy and the United Kingdom that the regional programmes might offset, at least in part, the disproportionate per capita agricultural subsidies going to fairly prosperous agricultural countries such as France and the Netherlands. Over time, the authors note, support for these measures by more urbanised countries seems to be increasing. Thus, with the successive reforms in 1979, 1984 and 1988,

> regional policy has been strengthened, funding has been increased, co-ordination (both different regional policy actions and regional policy with other Community policies) has been improved, and, perhaps most significantly, a measure of responsibility for the initiation of action and for the formulation of policy objectives has moved from a national to a Community level.[46]

Also, not surprisingly, sectoral pressures and policies of the Community, albeit indirect, have had even more influence on regions, both favourable and unfavourable. The most well-known examples include the crises in the steel, shipbuilding and textile industries. These created the need and the pressure, the authors note, for the Community to exercise control over national aid, and to provide incentives (mainly loans but also grants) for restructuring and modernisation efforts through organisations like the European Investment Bank, the European Fund for Regional Development and the Social Fund.[47] The rehabilitation programmes increased productivity and accelerated deindustrialisation but cut employment drastically and indirectly benefited the regions that had more efficient plants.

The Community and the impetus for economic integration also exerted at least a catalytic effect on plans for the high-speed rail networks linking most of the major urban regions of Europe: particularly the London, Paris, Bruxelles, and Amsterdam network and the 'corridor' from Barcelona, Marseille, Nice, Genoa, and Milano. These investments will surely

> confer great competitive advantages on those regions that are well served ... and penalise those that are not, such as southern Italy, Portugal and Greece. It will benefit particular urban regions that integrate the rail system with other transport systems, especially air transport. Here Paris and Amsterdam have the most advanced plans. In general, it will reinforce existing tendencies to recentralisation in the larger metropolitan regions served by the system.[48]

Despite their great promise, these and other efforts of the Community still provoke a sense of unease and uncertainty for the authors of Chapter 9. They observe that the outlays are marginal in comparison with the Community agricultural funds, and also marginal in comparison with national programmes such as the Federal Republic's efforts to reconvert the Ruhr. The authors recall, too, the many research programmes in nuclear engineering, electronics and telecommunications that have been constrained by what they deem undue caution and timidity by European Community decision makers. But they also acknowledge the potential value of Community-sponsored programmes, even at a low level of funding:

> There are real resources involved and the move to Integrated and Community programmes has also had a demonstration effect. National and regional authorities have been compelled to co-ordinate their efforts and to consider the total effects of sectoral deindustrialisation and transformation on a region. These programmes may have positive results and contribute to redevelopment in ways that exceed the value of their resources.[49]

Europe's regional–urban future

Chapter 10, by Camagni, Cheshire, de Gaudemar, Hall, Rodwin and Snickars, considers Europe's regional urban future.[50] Forecasting is a hazardous undertaking under most circumstances, and the present attempt is no exception. The authors' predictions fall into two sets. The first appear to be reasonably likely and quite obvious, although they may confirm the worst expectations of some readers about the basic conservatism and limited vision of the group. The second set of predictions are also quite likely, but less obvious.

The first set of Anticipations

As European integration proceeds, the authors suggest that most existing trends are likely to be reinforced. They expect the odds to favour the well endowed; but they acknowledge exceptions since the advantage even of exceptional endowments may change over time. The former prosperity of industrial heartlands provides one example; the former advantages of access to ports and minerals for many industrial cities is another. However, in general, they do not quarrel with the view that 'to him that hath is given'. They expect the expansion of the market to enhance many opportunities for innovation, economic growth and institutional change; and they think it will provide more options for social and individual choice both for the great metropolises and regional centres and their residents, and for those medium-sized and smaller cities, within an hour or more commuting range, or with particularly attractive amenities and surroundings. The big cities' dynamics, markets, wealth, scale, variety, labour force, infrastructure, universities and amenities clearly offer very tangible benefits. These magnets attract or hold the high-tech firms and the specialised financial and business service activities. The authors note these assets are critical, too, even for metropolitan areas trying to cope with the loss of traditional manufacturing activities and to adapt or transform the economic base of their economy.

The second set of Anticipations

As the technical and economic transformation proceeds, the authors expect a number of spatial changes. One is the recognition of the changes already emerging in what they call multinational regions, i.e. regions where 'both individuals and firms are internationally oriented and that will make decisions that will impinge on the rest of Europe and the world'.[51]

Over time, too, they expect the national urban hierarchies – ranging from the relatively flat ones of Germany and Italy to the pyramidal patterns of France – 'to merge into one common urban system',[52] and they anticipate even more intensive rivalry for status between the European world cities. In terms of size and financial influence London is still at the top because of its high-level functions: its two international airports and 'its unique position at the intersection of three major trading economies, the Anglo-American, the Commonwealth and the European'.[53] But they expect London's position in the European hierarchy to weaken. There are intense pressures and competition, for example, the vigorous pace of Paris as the second greatest European economic, political and cultural centre; the changing role of Bruxelles as the Washington of the growing European Community; and the growing financial and economic activities of Frankfurt (and, in the future, perhaps Berlin).

Also, the continued growth of the world economy, these authors say, will reinforce the headquarter and other hegemonic functions of a very few world cities. Telltale signs are already in evidence, for example, 'London's employment began to increase from 1986 after continuous decline since 1939 [and] the trend of migration has steadily changed from the early 1970s. Exactly the same pattern ... was observed in Copenhagen.'[54] At the same time, they expect growth to nourish the decentralised concentrations of lower-level functions within and outside the macro regions. Evidence of these trends may be seen in the metropolitan fringe 100 kilometers or more southeast of London, in the Copenhagen and south Sweden regions and also in various parts of Southern Europe, e.g. the Toscana region in Italy, the Toulouse region in France, the Cataluña region in northern Spain, and the region between Frankfurt and München in the Federal Republic.[55] Another continuing trend is the spread of metropolitan clusters along axial corridors. Megapolis, England, the Dutch–German megapolis and the Mediterranean corridor are examples. New high-speed rail transportation systems are likely to reinforce these patterns just as they did after the opening of the Tokkaido Shinkanzen.

Finally, the authors suggest that three questions may provoke lively discussion in the future.[56]

One involves the differences in the urban systems and policies of Europe, Japan and the United States, and the relative emphases of their economic systems on growth and development at the expense of rising standards for housing, community facilities and space. The contradictions are increasingly evident in reappraisals of the urban infrastructure investment programmes being pursued by these countries, particularly in transport and communications.[57] The second involves the reversal of those European attitudes and policies that in the 1960s and 1970s encouraged the immigration of workers from Third World countries to fill jobs in Europe. The reversal has already reinforced the pressures on firms to relocate manufacturing to Third World economies, thus benefiting them.⁵⁸ The third involves the likely strengthening of the urban systems of Berlin and Wien because of the transformation of the economies of Eastern Europe and the Soviet Union. An increase in investments in these areas is likely to reduce the investment funds available for other, poorer regions of the world.⁵⁹

Concluding observations and inferences

What are some of the common denominators of these trends? How do they compare with those in the United States? What other inferences might be drawn from them?

The composite image: a comparative perspective

What is particularly striking is that most of the regions in the different countries examined are losing manufacturing jobs. That is happening in thriving as well as declining economies and in quite diverse circumstances. Examples include regions with economic activities afflicted by sudden appreciations of exchange rates; those involving a decline in demand for goods and services from regions specializing in these sectors; and areas with mature, diversified economies hobbled by one or more grave problems, e.g. high and relatively inflexible costs, bureaucratic ossification, and/or unappealing environmental conditions and amenities in comparison with other areas free of these problems and of equal interest to new growth activities.

However, the most intractable difficulties are those confronting both the poorer backward regions and the regions with secular trends involving severe productivity and job losses, and limited competitive and response capabilities.⁶⁰

It is also evident that almost all of these countries have pronounced north-south disparities. That is to say, they have important threatened or declining regional economic cores, located in what might be characterised metaphorically as their new north. The areas experiencing these difficulties are different from the poorer and often peripheral rural, resource-based or older industrial regions of the past, e.g. the south of Italy, the west of France, Appalachia or the old south of the United States and, to a lesser extent, the north of the United Kingdom.⁶¹ These formerly prosperous industrial areas now face combined losses of job productivity and competitiveness. Generally, they are production centres (based on some combination of metals, machinery, electrical equipment and consumer durable goods) that are now grappling with a common array of problems: declining investments, shut-down plants, considerable unemployment and outmigration, declining revenues, and ageing or obsolete infrastructure stemming from the different circumstances noted above. Nationwide, large segments of the economic interiors of the six countries examined here are hollowing – losing capital, jobs and people. Hall suggests that innovations and entrepreneurship are less apt to flourish in these threatened or declining regions for they are, or are likely to be, subject to outside control and dependency.[62]

However, there is evidence, at least in France, that dependent regions may turn out to be 'the most dynamic'.[63] There are also older, skill-based regions, as Cheshire, Camagni, Bade and Kunzmann and others have shown, where the crisis has spurred significant innovations and restructuring. Such 'come-backs' or transformations required a combination of favourable circumstances.[64] These have included entrepreneurial traditions; managerial competence; one or more universities well grounded in science and technology; training and research facilities; an ample supply of skilled and semi-skilled labour; and minimal infrastructure. Typically, there are several efficient modes of transportation affording access to markets and appealing recreational and cultural hinterlands. Supportive national and regional public policies and programmes in these regions have ranged from good-quality schools and vocational education to a reasonable supply of housing and community facilities, business and consumer services, and workable development incentives.

Finally, there are the 'boom regions' or 'hot spots'. These are the areas of exceptionally rapid growth, few in number (from one to four or five) often including 'sunbelt regions', relatively well endowed with infrastructure, climatically attractive and possessing considerable amenities. Often concentrated near, or somewhere along, the perimeter of countries, they possess modern transportation, telecommunications and financial services and facilities: these provide effective access to large internal and external markets, such as the growing regions of the European Community core like Paris, Bruxelles, Frankfurt and München.

In all – or at least in most of the thriving and either come-back or transformed areas – there is a less easy to define, yet critical difference: an exceptional motivation and response capability that infuses the population and leadership with innovative spirits, energy and dynamism.[65] McClelland tried to explain aspects of these traits in his investigations of 'the achieving society'.[66] In the language of Schumpeter, such characteristics contribute to the 'creative response' in economic history.[67]

At least five identifiable forces have interacted and shaped these and related changes. First are the innovations – in concepts, in tools and methodology, and in technological changes in transportation, information processing and telecommunications – that continue to mushroom. Second, firms have been able to arrange for increasingly efficient, autonomous, decentralised offices and contracting or subcontracting arrangements – both to manage or to procure important components or supply services. Third, small and medium-sized firms have thrived by exploiting the new technologies. However, through mergers, acquisitions and other means, hierarchical control and influence of large firms is also continuing or expanding through mergers.[68] Fourth, there are incessant pressures within the international economy to maintain or increase productivity, to identify secure niches and to hold on to or increase market shares. Fifth, in many of the countries (especially the United States, the United Kingdom, the Federal Republic of Germany and France) defence programmes and contracts have spurred the innovations, investments and reinforcement of the non-spatial alignments now in evidence.

In combination, these forces, coupled with the new or enhanced energies and capabilities, have enabled firms to tap global resources, to function more efficiently in situ or to relocate and manage branch offices and plants in regions or countries with less costly labour and a more favourable image and physical and business climate.

A historical perspective: from decentralisation to deindustrialisation

The relocation of industrial manufacturing activities is not a particularly new process or phenomenon. Intra-regional and inter-regional shifts of industry occurred throughout the first half of the twentieth century and through much of the nineteenth century and even earlier.[69] In these previous periods, manufacturing continued to grow mainly in the big cities. Since the 1930s the accelerated shift of industry to outer areas of metropolises has caused urban reformers to press for legislation combining penalties and incentives to achieve several aims. These were to reduce big cities' rates of growth; curb congestion; encourage land use controls, development plans and other measures to promote orderly decentralisation (or, when feasible, shifts to lagging regions); and to improve the physical environment in different metropolitan areas.

Those aims shaped key policies in Western countries for several decades when planners felt they could tame the behemoth. But now the emphasis has shifted largely from three-dimensional spatial issues to the entire biophysical environment. One of the most important concerns of metropolitan planners today is to stem the loss of firms and jobs from the central city to other areas, whether to the outlying rings of the metropolis, outside the region, or to offshore locations in other parts of the world. Contributing to that change is the recognition that most of the new growth activities, particularly the high-tech and higher income service firms, are likely to be lured to only a few physically attractive, well-located areas within or reasonably near existing metropolitan complexes. These are the locations with the necessary labour force, infrastructure, services and cultural facilities, or at least the potential to achieve these minimum conditions and amenities.

In short, in less than three or four decades there has been a reversal of attitudes to the big city and job promotion strategies. Many of the aims and measures designed to control city growth now appear out of date. In coping with competitive pressures on an international scale, there is a far more sensitive appreciation of the advantages of both urban size and amenity. The range and adequacy of business and consumer services – coupled with physical and cultural amenities – are increasingly shaping the attitudes and choices of the managers and staff of the new growth industries. Only a metropolitan scale makes this combination possible. Planners today are less certain about the extent to which these trends may be shaped, but they are more confident of their ability to exploit or rationalise them.

Alternative industrial images of the future

In contrasting post-, de-, meta- and other industrial theses, we veer back and forth from optimistic to less confident positions.

Most of us are familiar with the post-industrial view that as technology improves and per capita gross domestic product and income increase in an advancing economy, services and high-tech activities also increase and there is a drop in the sectoral share of manufacturing in total national jobs – and even of output. Two factors shape these trends: first, the greater rates of growth of capital investment and of productivity in manufacturing in relation to services (as well as to agriculture); and, second, the greater income elasticity of demand for services in comparison with manufacturing.

Knowledgeable critics do not quite disagree with these general propositions but they do mock the complacent assumptions of ideologists who ignore social costs: the loss of jobs and income, declining tax revenues, inadequate and deteriorating public services,[70] and neglected infrastructure investments and maintenance. They also cite increasing income inequality, and caution that too many firms, households and whole regions are now on a downward skid.[71]

The debate about the loss of industry is likely to simmer for a long time because one side wants to have public policy slow down the pace of change and the other does not. The resulting policy issues surface in various ways. For example, government might discourage firms, by tax and other measures, from moving plants to less expensive or more satisfactory locations. It might provide protective barriers against competition, decide which industries to support, be the arbiter of what unprofitable goods and services are useful, or require compensation for the social costs of business decisions.[72]

The controversies, of course, are also concerned with the facts. Industrial losses measured by employment are quite different from those measured by some version of productivity, output, competitive capability and market share. The numbers also change depending on whether firms decide to buy a service or perform it directly. More often than we care to say, the very choices of the analysts' measures reflect a hidden agenda: the desire to reinforce deep faith or scepticism in the workings of a particular alignment of the market or the public sector.[73]

It is most intriguing to observe how the insights and the direction of the discussion of industry and services change as the angle of vision shifts. For example, Hall, looking at services as one of the basic innovational industries and frontiers of southern England, suggests that the great problem facing Britain's north, new and old, is the failure to make the switch from industry to services and from goods handling to information handling.[74]

In the case of the Federal Republic of Germany, Bade and Kunzmann also view services as a kind of safety valve, a potential job reserve that is currently underutilised. The United States, they note, has added many more lower-quality and lower-income jobs in the services at the expense of productivity while the Federal Republic over the past two decades appears to have focused on efficiency and therefore suffers from a higher rate of unemployment.[75] However, analysts do not agree that these differences between the two countries reflect only a trade-off of employment and efficiency. These divergent outcomes may also be shaped by differences in culture and behaviour. Camagni and Cuadrado Roura, in turn, see industry and services today as closely interrelated and integrated, not as separate or opposed to each other. The right name for our period, Camagni suggests, is not post-industrial or de-industrial but meta-industrial.[76]

The other national studies underline the complementary relationships between industry and services, with one faring well when the other does, and vice-versa. However, the future growth prospects of both are not systematically examined in any of the studies.

For this reason, Lester Thurow's perspectives are especially arresting.[77] He tries to gauge future prospects of services and industry in the United States by examining the reasons for the growth of various services of the past, and the likelihood that the same forces would prevail in the future. He suggests that the major growth in services in the United States is now over because two-thirds of the new service jobs occurred in trade, hotels and restaurants; real estate and business services; and financial and insurance activities. The first two occurred in large measure because full-time working women require round-the-clock and weekend shopping, and these changes have now reached the point, at least in the United States, where a slow-down is likely. Financial services are also unlikely to grow faster than GNP in the next decade. This is because recent growth in the United States was due to deregulation, the development of world capital markets and volatility. He suggests that the first two factors were probably one-time phenomena and that the degree of volatility is now so high that governments are likely to curb any increases.

Thurow is persuasive because 'general propositions do not decide concrete cases'.[78] He has empirically tested the assumptions and propositions about the growth of the services in a particular country and set of circumstances. His approach was especially challenging because he raised parallel questions about the growth prospects in the United States for manufacturing. His answer, once again, is surprising. He argues that there will be inexorable pressures for growth in some activities, sustained by long-term demand, as national policies promote the creation of a surplus to reduce its debts and pay for imports. Thurow concludes that since neither agriculture, nor loans, nor financial services are likely candidates, only manufacturing (not necessarily US-owned) is capable of becoming a leading growth sector.

It is possible that, compared to Europe, the United States is at a different stage and we should not expect to see similar trends in Europe. But Prud'homme and Cuadrado Roura suggest that there is already evidence of somewhat similar trends in France and Spain.[79]

Outcomes

Size of firms

These industrial changes and, in particular, the decline of manufacturing employment, have reinforced the view in some quarters that large firms with centralised management and 'Fordist' mass production operations have become inefficient dinosaurs.[80] This perspective saw them being replaced either by small firms or at least by firms with more autonomous units and/or much more flexible subcontracting arrangements. In the late 1960s and 1970s policy makers and scholars were intrigued with the burgeoning of small high-tech firms or plants in such diverse places as Boston and Route 128, Silicon Valley, the Veneto region and much of the area between Milano, Firenze and Venezia, with their potential for innovation, entrepreneurship and job creation.

However, current evidence suggests that neither of these views adequately anticipated the actual trends. Small firms seem to be creating fewer jobs than anticipated. While continuing to grow, they are also merging or being acquired. They often suffer serious technological gaps in comparison with larger firms. Meanwhile, big firms, faced with sharply declining profits in the 1970s, went through a prolonged period of retrenchment. This included short-term pecuniary-oriented mergers and acquisitions, and even the wholesale 'hollowing' of their companies. During this period the big firms have also been paring down their core operations and are contracting out (within and between regions and countries) many of their traditional activities. However, even as retrenchment proceeds, the big firms continue to exert centralised control. It is the conviction of Bennett Harrison that 'even when certain specialised suppliers provide high tech research and development or design services to the big firms, the truly strategic choices about the production (or service delivery) process, product mix, site location and relations with governments are still made at the top – or centre'.[81]

The larger firms are also making 'strategic alliances' with other corporations in different regions and countries, thus enlarging their resources, technology and access to regulations and procurement officers. These new linkages and relationships are blurring even more the distinction between manufacturing and services.

Income effects
The longer-term regional as well as national income effects of these regional transformation trends are somewhat elusive. There are several reasons, including time constraints and the limitations of the data available for the subnational area classifications. But, for the short term, the evidence as indicated earlier suggests that income inequality has increased both for Europe and the United States, especially the latter. In the United States, the majority of the new jobs in many services, particularly wholesale and retail trade, have involved lower wages and lower fringe benefits. Also, most of the new jobs in manufacturing over the past decade have been for managers, salaried professionals, technicians and secretaries while the number of production workers has declined.[82] These trends indicate some of the reasons why income inequality has increased nationally and within most regions. However, inter-regional inequalities have been reduced, because of the decline of a more prosperous region, the industrial Midwest, coupled with the limited but very real economic resurgence of two less prosperous regions: the South (gaining in part from the shifts of economic activity from other regions) and New England (benefiting from the boom in high-tech, business service, construction and defence activities).[83]

As for Western Europe, Cheshire concludes that – even allowing for the inadequacy of the data – deindustrialisation and tertiarisation appear to be 'accompanied by income polarisation and inequality',[84] especially in the United Kingdom, 'where the processes are most advanced'.[85] One wonders whether inter-regional inequalities may have been reduced, or might be reduced, in some (not all) European countries for somewhat the same reasons that this equalisation has occurred in the United States? Britain's policy, Cheshire and Hall emphasise, has shifted attention from inter-regional to intra-regional disparities. The change occurred for two reasons. Unlike the decade of the 1960s and earlier, there were far fewer industrial jobs available for diversion to lagging regions; and the major problems in the cities of the region – inequality, blight and arrested development – demanded attention. This may be a harbinger; however, the European Community still focuses on inter-regional differences, particularly those between poor, backward rural regions and the prosperous regions of much of Northern Europe. This focus may be sharpened by the entry into the Community of Greece, Portugal and Spain.

Policy-shaping capabilities

Technically, the structural cyclical issue on regional economic transformation is whether the losses of manufacturing firms and jobs over the past two or three decades reflect much more than the shake-out and the flexible adjustments accompanying the severe downturn of the economic cycle. If so, the job losses, and most of the other related problems should disappear for most, if not all, the ailing core industrial regions when the cycle turns up again.

Structural problems, however, are less subject to correction, assuming correction is possible at all. They involve factors such as changes in taste and demand, technological change, decline of productivity coupled with new and more efficient competitors, etc. The prospects for formerly prosperous regions and sectors are less promising unless the existing activities can be transformed, new activities can be initiated, or the pressures of economic change can be significantly modified or redirected. If not, significant economic sectors and even regions in much of Western Europe and the United States may be destined to shift to the status of depressed or assisted areas.

Our regional case studies for the United States and Europe suggest that the changes are more structural than cyclical. We see a discrepancy between relatively effective responses of at least some segments of the private sector and the increasingly limited capabilities of the public sectors, national as well as local and regional. The responses of industry to fierce competitive pressures have

> included experiments in work organisation and labour–management

relations, wage freezes or cuts, plant reductions or relocations to areas of lowest cost and/or more receptive attitudes to innovation, corporate restructuring, and the shifts of capital out of the less productive sectors of the economy. As in Japan, companies engaged in hollowing, so that production activities which formerly took place inside the company were [outsourced].[86]

Public sector programmes ranged from upgrading regions, metropolitan areas and particular sectors or industries to providing infrastructure investments, agency relocations and incentives for new or existing enterprises. However, these efforts proved relatively ineffective, except when accompanied not only by complementary local, regional, or national programmes, but also by felicitous circumstances in which these efforts reinforced existing currents. Almost none of these policies and programmes held back the tide.

In the past, planners needed to be aware of the great influence on regional development of the unintended effects of other factors or programmes, such as new technology, infrastructure programmes, defence investments, and the impact of financial, hierarchical and other forms of control. This is even more the case today, as other serious constraints loom. Regional and local authorities must be ready not simply to cope with but to respond with vigour and imagination to decisions made by firms with markets or headquarters outside the boundaries of the nation as well as the region. Another factor beyond the planners' control is the increasing influence of international forces and agreements on macroeconomic policies that modify exchange rates and trade policy.

The European Community and Third World countries

True or not, many persons think that the French dislike the British, have qualms about the Germans, distrust the Italians and Spaniards, etc; and that these sentiments are in varying degrees reciprocated. It is not surprising that the European Community was often looked upon originally – indeed, is still looked upon in some quarters – as a somewhat naive, if not Utopian enterprise. There were many other reasons for distrust: the uncertainties of the incidence of costs and benefits, fear of loss of power and, of course, the previous history of suspicion, hostility and bloody conflict between the nations and interest groups in Europe.

Given these obstacles, the promise and some of the unexpected achievements of economic integration have produced an understandable euphoria. However, reinforcing the extraordinary leadership that made the Community a world region was the recognition of two hard realities: that war between the great powers was futile and disastrous, and that the prior economic structure of Europe was essentially obsolete. The national markets, resources and powers of most European nations were simply too small for them to pursue economic policy and trade negotiations or to mount the R&D programmes on the pragmatic scale their firms required to compete with comparable firms and organisations in nations as rich and powerful as the United States in the past and the United States and Japan today.

The European Community was a creative response. Most tariffs have already been eliminated. By 1992 the aim is to harmonise standards, excise and other taxes; eliminate non-tariff barriers; and remove restrictions on traded services and factor flows (e.g. mutual recognition of qualifications, eliminating controls on capital movements, etc.). Over the past decade or so, European nations and firms have been co-operating on huge R&D projects. These efforts are likely to make Europe a formidable competitor with the United States and Japan in many areas including computer chips, consumer electronics, telecommunications, communication equipment, high-definition television, atom smasher projects, space shuttles, missiles and aircraft production, to name only the more well-known European co-operative enterprises in manufacturing and research.

The financing arrangements, although as yet modest, are often quite pragmatic. For example, on many of these projects, a country or firm putting up 25 per cent of the money expects roughly 25 per cent of the jobs created.[87] In still other co-operative multi-project programmes, countries can choose to help finance projects in which they are interested.

The co-operative arrangements involve not only some giant national firms but small innovative companies such as those abounding in Silicon Valley. The aim is to provide an attractive environment for companies able to create new technologies and jobs. Private organisations have arisen specifically to facilitate links between small and big companies. They are trying to offset inbreeding by deliberately mixing staff from different cultures, nations and organisations.

Of special interest is the Community's regional policy. The resources involved, as indicated earlier, are quite low (proportionate to the scale of the total Community budget) in relation to needs, to national expenditure, or the scale of largesse available for agricultural assistance. Nevertheless, the programme does involve international financial transfers. The aims are to promote growth and reduce inequality by providing financial aid and development assistance to lagging regions, to areas of threatened or actual industrial decline and to rural areas. Under some of the criteria currently employed, e.g. per capita income, entire countries, such as Greece, Ireland and Portugal, can be and are defined as less-developed or lagging, while the poorest provinces of rich countries may fail to qualify. Assistance is offered on a matching basis: i.e. if member countries are willing to make outlays for some approved purpose, then Community matching funds will be provided. In principle, the Community cannot merely offset spending by national and local authorities, but in fact it often does.[88] European Community, national and local authorities must all co-operate and contribute to the programme and also monitor performance. Understandably, the whole process is, and no doubt will remain, both shaped and constrained politically.

There are already strong pressures to modify the criteria in favour of smaller targetted areas. The total sums that may be transferred are not negligible (ECU 9,000 million in 1992), and to date, a large portion (96 per cent of the funds for lagging regions) is now allocated to Spain, Italy, Portugal, Greece and Ireland. More than three-fourths of the funds for declining industrial areas will go to the United Kingdom, Spain and France.[89] With such a shift in emphasis, the more prosperous nations with poor regions would receive a larger share of the funds.

At present, the poorest regions of the Community are Portugal, Spain, southwestern France, southern Italy and Greece. With the development in 1986 of the Integrated Economic Development Programme, the Community appears to be trying both to allay fears and to test hopes: i.e. to allay fears that economic integration will disadvantage the peripheral parts of the Community; and to test the possibility that financial assistance will help to curb unwelcome migratory flows by promoting in these areas promising opportunities and new economic frontiers.[90]

To be sure, many snags will become evident: the size of the budget, language differences, the spread of projects over many countries, hoarding of top managers and research specialists and disagreements about what should be financed and how to proceed. At the same time there is an incentive to work out solutions. If any country or company objects too much, the work may well proceed without it and to its disadvantage. For all of these reasons, the general conviction is that 'the European experience in international collaboration is one that nations in other world regions will sooner or later try'.[91] Integration carries clout and is, therefore, an increasingly indispensable means of coping with representatives of the protectionist markets of regional blocs.

The United States, in one of its more generous and imaginative periods, encouraged the formation of the European Community despite the strong competition and likely conflicts it could expect if the Community succeeded. Surely one of the most impressive indicators of the degree of success achieved is that East Germany, Poland, Hungary, Czechoslovakia and the other newly 'liberated' nations of Eastern Europe (including possibly the Soviet Union) want to become associated with and possibly even members of the Community in one form or another. An intriguing question for the future is whether the United States, Japan and Europe and organisations within these entities will encourage similar types of regional co-operation for Southeast Asia, Latin America and much of Africa. While the present outlook for such efforts is anything but promising, it was no more encouraging for Europe when the initial efforts began. With knowledge, technology, research and production facilities spreading so swiftly to all parts of the world, surely the issue is *when* rather than *whether*, these world regions will try to replicate this remarkable precedent.

II
Case studies

2

Structural transformation in the regions of the United Kingdom

PETER HALL

THERE WAS a watershed in British economic history in 1966. That year, total industrial employment peaked at 11.5 million in mining, construction and manufacturing. It then fell dramatically to 7 million in 1984, with manufacturing down from 8.7 million to 5.4 million. In the mid-1950s the United Kingdom had been perhaps more industrialized than any other country in history, with more workers in industry than in all services; yet by 1983 there were almost two service workers for every industrial worker. The drops during 1966–73 and 1978–83 were among the greatest in any industrial country. Between mid-1979 and mid-1981 alone, manufacturing employment in the United Kingdom declined by no less than 14 per cent, from 7.1 to 6.1 million, and manufacturing output by 20 per cent. Total employment fell by 1.25 million and unemployment more than doubled to 2.3 million, just under 10 per cent.

This chapter attempts to analyse and explain the regional consequences of this transformation. First, it details the varying responses of the regions to the fundamental challenge of the time: the replacement of lost goods-handling jobs with service jobs, especially in information-handling. It then looks more specifically at the regional location of two growth sectors: high-tech manufacturing and producer services, with a detailed look at two of the latter, computer services and accounting. This is followed by explanation, the conclusion of which is that both Schumpeterian and Marxist explanations have something to offer. Following the schema of other contributions to this volume, the chapter then considers specific regional case studies: the South East, the United Kingdom's consistently most dynamic region; the neighbouring South West, also a strong performer; the industrial heartland of the West Midlands; and central Scotland. A postscript traces the regional pattern of recovery from recession in the mid-1980s, and conclusions are summarized.

The anatomy of regional change

The regional impact of industrial decline was far wider than during the Great Depression of 1929–32, including not only the peripheral industrial regions (the Assisted Areas) worst hit the last time around, but also the regions of the industrial heartland: Yorkshire–Humberside, the North West and the West Midlands. Manufacturing employment did not bottom out until 1984. Thus in just five years the manufacturing employment base fell by 1.7 million or 24 per cent, a reduction equivalent to more than half the total decline since 1966.[1] This emerges strikingly from Figure 2.1, which indicates that the positions of both the heartland regions (in the

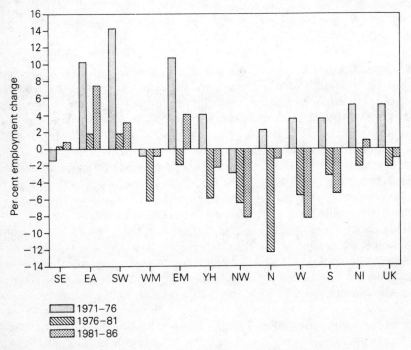

Figure 2.1 Employment changes by region, 1971–8, 1978–81, 1981–8.

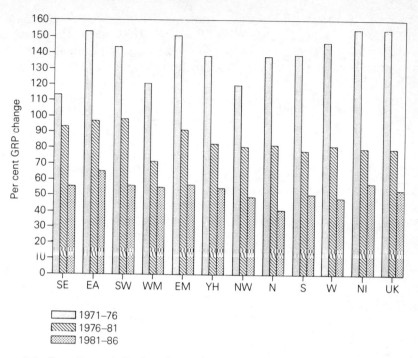

Figure 2.2 Gross Domestic Product changes by region, 1971–8, 1978–81, 1981–8.

middle of the figure) and the peripheral regions (on the right) deteriorated after 1981.

Since productivity was rising on average 3 per cent per annum after 1966, GDP would have needed to grow that fast to maintain employment; but it did not do so in the nation as a whole or in most regions, and in none after 1978. Figure 2.2, which shows GDP in current values, demonstrates that growth fell sharply in the quinquennium after 1976 and again after 1981, and in the last period nowhere exceeded 70 per cent. The regions exhibiting slowest GDP growth during 1976–81 were the West Midlands and Wales; during 1981–6, the North West, North, Wales and Scotland. The best-performing regions in both periods were East Anglia, the South West and the East Midlands (Fig. 2.3). Although *real* GDP in the United Kingdom was almost the same in 1981 as in 1974, the industrial heartland regions had declined 5.1 per cent, Assisted Areas had grown 1.7 per cent, and the South East region had grown over 3 per cent.[2]

The outcome is that an old theme in British history – regional economic divergence – has reasserted itself. However, the problem area is now much larger, including not only the traditionally problematic peripheral regions of Scotland, the North, Wales and Northern Ireland, but also the industrial heartland of the West Midlands, North West and Yorkshire–Humberside. These are regions with large problematic manufacturing sectors. Thus, beginning around 1974 but intensifying after 1979, a big gap has appeared between the south and east of the country and the industrial heartland.[3]

Right across the regions, in fact, a clear inverse relationship has

Figure 2.3 Standard regions of the United Kingdom.

developed between deindustrialization and tertiarization: those regions that were most manufacturing-dependent – the West Midlands, the North West, Yorkshire–Humberside, the North, Wales, Scotland and Northern Ireland – have suffered most from manufacturing decline and gained least from new private sector service jobs.[4] Half the population of the United Kingdom live in these regions, and by the mid-1980s unemployment rates in all of them exceeded 15 per cent, producing two nations separated by widening economic and social disparities. Their economic restructuring constitutes a major economic policy challenge.[5]

These trends may be viewed more closely in Figures 2.4 and 2.5, which group the three-digit categories of the Standard Industrial Classification to bring out the distinction between the services and production sectors, and the information-handling and goods-handling sectors, summarized in the form of replacement ratios. These represent the degree to which increases in service jobs offset losses in production jobs (Fig. 2.4), or, alternatively, the growth of information-processing jobs offset losses in goods-handling jobs (Fig. 2.5). (Zero entries on these figures indicate that both trends were moving in the same direction: in other words, both were increasing or both were declining, so that no ratio can be calculated.) In most regions,

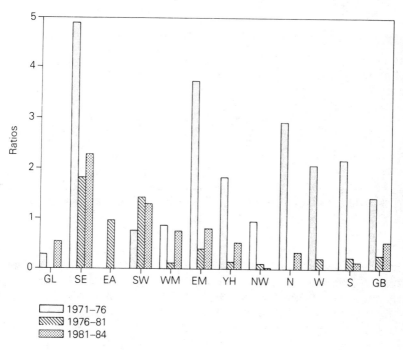

Figure 2.4 Ratios of services to production, 1971–8, 1978–81, 1981–4.

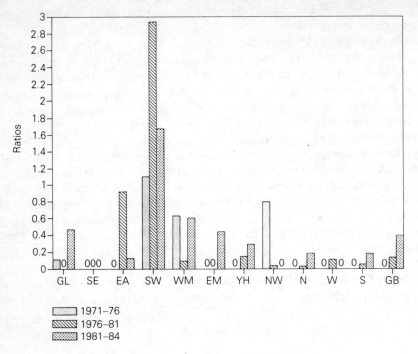

Figure 2.5 Ratios of information to goods, 1971–8, 1978–81, 1981–4.

both ratios were far from adequate after 1976; but the distinction between the three southern regions (except for London) and the rest of the country is extremely striking.

An alternative, more precise attempt to measure the informational economy has been made by Goddard and Gillespie, using an occupational classification. They estimate that in 1981 some 45 per cent of the workforce could be described as informational workers, but with very marked regional variations.[6] Table 2.1, derived from their work, shows that London, where 58 per cent of the employed population hold information-related jobs, and the South East, with 47 per cent, were actually the only places recording above the national average. Viewed more closely, the areas and places that have most successfully adapted their structures are concentrated in a diagonal belt, the 'Golden Belt' or 'Sunbelt'. This includes the outer fringes of the South East and adjacent parts of the South West, West Midlands, East Midlands and East Anglia standard regions. According to Martin: 'It is in southern England, and especially on the "sunbelt" corridor stretching from Cambridge through Berkshire to Bristol, that Britain's high technology industries and services, and more importantly the core functions of these activities, are concentrating.⁷ Both high-tech manufacturing and producer services are heavily concentrated in the South East, which has 41 per cent of high-tech engineering, 55 per cent of R&D, and 50 per cent of producer service employment.'⁸

Champion and Green's recent study on the quality of life in the contemporary United Kingdom uses a wide range of data, consisting of five 'static' indices (unemployment rate, unemployment duration, employment in high-tech and producer services and mean house price) and five 'change' indices (unemployment rate, employment, employment in high-tech and producer-service industries, population and house price). Champion and Green show that the top values on the Amalgamated Index are all in the southern part of the country while the bottom places are all in the North, Midlands, Wales and Scotland.⁹ The top 35 places fell into a single crescent-shaped zone extending around the western side of London from places like Chichester, Crawley and Winchester in the south to Cambridge, Newmarket and Thetford in the north; the sole exception is Stratford-upon-Avon.¹⁰ This zone corresponds closely to the 'Western Crescent' identified by Hall et al. as the major concentration of high-tech industry in the country.¹¹

These differences have long historic roots. Recent research has shown that there was a north–south divide in the nineteenth century, based on the concentration of services in the south; the gap widened from the 1850s. This was due not to transfers from the north but to a different economic structure in the south dominated by services, especially in

Table 2.1 Informational occupations as a percentage of total employment

	Percentage
London	57.8
South East	47.3
Great Britain	45.0
North West	43.8
South West	43.7
West Midlands	42.0
East Anglia	41.5
East Midlands	40.9
Yorkshire–Humberside	40.3
Scotland	40.0
Wales	39.4
North	38.8

Source: Goddard, J. B. & A. E. Gillespie. 1987. Advanced telecommunications and regional economic development. In B. Robson (ed.) (1987) *Managing the city: the aims and impacts of Urban Policy*. London: Croom Helm. p.88.

trade, that predated the Industrial Revolution, leading to a growth of financial services and aided by the concentration of government there. However, during the long post-war boom of 1945–70, unemployment everywhere sunk to unprecedentedly low levels, and differentials narrowed. The peripheral northern regions (North, Yorkshire–Humberside and North West) had unemployment that was 90 per cent greater than the three southern regions (South East, East Anglia and South West) in 1931, but this fell to 44 per cent in 1951 and 37 per cent in 1961; the gap then increased again, to 42 per cent by 1971, 54 per cent in 1981 and 59 per cent in 1984.[12]

According to the most fashionable analysis, this represents the transition from the 'modern' or immediate post-war period – characterized by the relatively even spread of development opportunities, convergence in income and unemployment between classes and regions, large-scale production and collective consumption, and considerable use of semi-skilled, sometimes feminized operator grades – to a 'post-modern' economy characterized by rationalization, especially in the big cities, and by flexible production systems, including subcontracting, which affects services as much as manufacturing. Such flexible specialization is characteristic of new industrial areas based on small towns, previously not highly industrialized. They are mainly scattered, but there is one major concentration in the corridor along the M4 motorway. Symbolic of these changes are the new high-tech complexes of southern England, dependent on defence procurement. They are found in small towns on London's metropolitan fringe, such as Aldershot, Basingstoke, Bracknell, Chelmsford, Crawley, Guildford, Harlow, Reading and Southampton.[13]

The role of high-technology manufacturing

In these areas of the 'Sunbelt', two kinds of fast-growing activity are found in close spatial and functional association: high-tech manufacturing, and producer or business services. Analysing the regional distribution of high-tech manufacturing and services, Begg and Cameron note that the 'striking feature ... is the concentration of high technology industry in the South of the country, especially in the outer South-East, which exceeds the national share by 53% in 1984'.[14] The South West and East Anglia also show up strongly. In high-tech manufactures the pattern is less skewed, partly because of the under-representation of London, while in services it is over-represented.

At an urban level, the top 25 high-tech locations contain a high incidence of new towns, mainly but not exclusively in the south. The bottom 25 are old industrial towns, nearly all in the north. Both high-tech

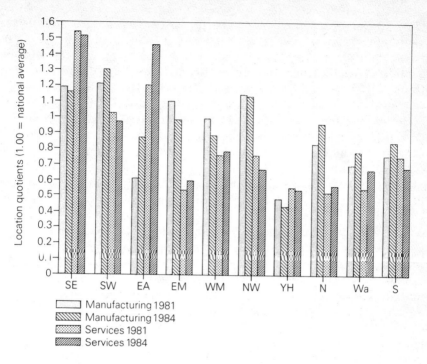

Figure 2.6 High-technology industry, location quotients, 1981 and 1984.
Source: Begg and Cameron, 1988, p. 366.

manufacturing and services reject locations with a high proportion of traditional economic activities; they are more likely to be found where producer services are also concentrated. This is due to the 'image' of an area, the inappropriateness of local skills and the likely difficulty of attracting technological and scientific staff to such locations. New infrastructure, a 'clean' image, a vigorous labour force and flexible attitudes adapted by development corporations to land and property requirements for growth combine to attract a markedly above-average share of high technology in almost any region.[15]

Between 1981 and 1984, East Anglia showed the best record for attracting high-tech manufacturing and services. The North, Wales and Scotland also gained; the East Midlands, West Midlands and Yorkshire–Humberside declined (Fig. 2.6).[16] High-tech manufacturing was not however a source of job gain at the national level. On the contrary; in sharp contrast to the United States, employment in high-tech industry in the United Kingdom actually declined in the 1970s. Comparable definitions show a gain of over a million jobs in the United States between 1972 and 1981, and of 728,000 in the brief period 1977–81, but a loss of 366,000 jobs, 10 per cent, in the United Kingdom between 1971 and 1981.[17] Within this framework, Greater London was losing high-tech jobs, 17,000 or 15.7 per cent over 1975–81 alone, while a 'Western Crescent' was gaining. Overall, the Western Crescent, stretching from Hampshire through Berkshire and Buckinghamshire to Hertfordshire and Cambridgeshire, is by far the outstanding high-tech area of Britain (Table 2.2).

Underlying these distributions, however, is a more subtle differentiation. The M4 corridor is distinguished from the high-tech complex in Central Scotland and from high-tech outgrowths in South Wales by 'its elite occupational structure and its dense decision-making network of activities which have the potential for spawning (and sustaining) new firm foundation to an extent not readily apparent elsewhere'.[18] It is also strong in sub-sectors of electronics that are still experiencing job growth: capital equipment, computers and computer services. Central Scotland, although quantitatively unimportant in comparison with the M4 corridor, is not a pure branch-plant economy. It has been fashioned largely through inland investment from the United States and other foreign countries, but these firms do far more than routine assembly; US firms are often in development work, although not leading edge R&D.[19] South Wales, in contrast, is a real branch-plant economy, without R&D or an indigenous sector, and its attraction is that it is at the subsidized end of the M4

Table 2.2 Major high-tech concentrations, 1975–81

	Employment 1981	Location quotient	Change 1975–81 thousands	percentage
London & Western Crescent				
Greater London	91,400	0.85	−17,000	−15.7
Hertfordshire	45,100	3.60	+5,914	+15.1
Hampshire	33,800	2.00	+2,179	+6.8
Berkshire	19,700	2.04	+7,600	+62.6
Buckinghamshire	6,900	1.15	+1,811	+11.1
Surrey	18,200	1.79	+1,811	+11.1
Total	215,100			
North West				
Greater Manchester	27,300	0.86	−2,980	−9.8
Lancashire	30,000	2.01	+1,392	+4.8
Total	57,300			
Silicon Glen (Central Scotland)				
Strathclyde	23,600	0.88	−3,847	−14.0
Lothians	9,000	0.93	+1,461	+19.3
Total	32,600			
Great Britain	640,900			

Source: Hall, P. 1987. The geography of high-technology industry: an Anglo-American comparison. In: Brotchie, J., Hall, P., Newton, P. (ed.). *The Spacial Impact of Technological Change*. London: Croom Helm. p. 145.

corridor. It has more of a 'headless' occupational structure, with relatively few technically skilled or managerial employees.[20]

The role of producer services

The other crucial component is producer service employment which, depending on definition, includes between 4.5 million and 5 million jobs or 22–24 per cent of total employment.[21] The share is growing for two possible reasons: first, an increasing level of intermediate service inputs for any level of output; second, lower productivity growth in the office-based service sectors than in other sectors of the economy.[22]

Although services as a whole, and consumer services in particular, are distributed broadly in line with population, producer services have deviations that more closely parallel manufacturing. Producer services are particularly concentrated in the South East, with 40.3 per cent; subsidiary concentrations are in the North West (11.3 per cent) and the West Midlands (8.8 per cent). There is an urban bias, with 45 per cent in Greater London and the former metropolitan counties.[23] The highest Location Quotients for producer service employment are found in and around London, especially to its west; in East Anglia; in the Bristol area and in the Exeter–Plymouth corridor; and (scattered) in the East Midlands and North West.

Changes in producer service employment occur at two scales; nationally, and regionally within the urban hierarchy, as defined in terms of Local Labour Market Areas (LLMAs). At the national scale, during 1971–81 the South East's producer service employment grew by 11.4 per cent, less than the 13.4 per cent national average; but the South West grew 42.6 per cent, East Anglia 41.3 per cent, and the East Midlands 34.1 per cent. Numbers actually fell in the North West, while growth was below the national average in Scotland, Northern Ireland, the North and Wales. At a finer geographical scale, at each level in the urban hierarchy, metropolitan areas in the southern half of the country performed more impressively in job generation than did their northern and western counterparts. However, the intra-regional pattern is more dramatic everywhere: in the cases of both London and the major provincial cities there is very marked deconcentration from cores to smaller places around them. The evidence is set out in Table 2.3. Producer service employment is concentrating regionally but deconcentrating intra-regionally.[24]

This is most evident in the South East. Much of this growth has occurred outside Greater London, especially in the Outer Metropolitan Area. Only banking, insurance, finance and business services and miscellaneous services increased in London, but in both cases the absolute

Table 2.3 Changes in producer service location by local labour market areas, 1971–81

	Percentage change Manufacturing	Percentage change Producer services	Change in location quotient Producer services
London	−33.8	+14.7	−0.19
Conurbation dominants	−33.2	+24.2	+0.03
Provincial dominants	−26.9	+44.3	+0.05
London subdominant cities	−16.1	+59.4	+0.04
London subdominant towns	−14.9	+61.2	+0.06
Conurbation subdominant cities	−32.4	+60.4	+0.14
Conurbation subdominant towns	−30.4	+70.7	+0.08

Source: Gillespie, A. E., & A. E. Green, 1987. The changing geography of producer services employment in Britain. *Regional Studies* **21**. p. 405.

increases outside were greater. All other categories declined in London, counteracted by large gains elsewhere in the South East. Producer service employment has been part of these trends, but to a lesser degree than consumer services, which have followed population. A measure of this change, given present trends, is that by the early 1990s there will be more commercial office floor space in the South East outside central London than in central London.[25]

The growth of employment in producer services has, however, provided only a limited and partial compensation for employment losses in manufacturing. Nationally, employment in manufacturing declined by 1,918,000, while in producer services it grew by only 472,000. Only 4 of the 19 urban groups had a positive balance: the southern and northern Rural Areas, and the southern Commercial and Service Towns. The limited degree of compensation is particularly evident in the Conurbation Dominant and Subdominant LLMAs: the Dominants recorded 85,000 extra producer service jobs but lost 686,000 manufacturing jobs. In general the 'compensatory effect' is most evident in the southern labour markets outside London, which also have less severe manufacturing job losses than their northern equivalents. Thus the differential effect of producer service growth can be seen to reinforce rather than reduce the relative job gap between north and south, and reflect the growth in extensions of head office functions which, despite improvements in technology, remain tied to the greater South East.[26]

A particular sub-group – business, information, or office-based services – is even more concentrated close to the capital than are all producer services (Table 2.4). It is heavily over-represented in the capital and in the larger cities within its immediate sphere of metropolitan influence. The London Local Labour Market alone possesses over 40 per cent of total national employment.[27] Analysis of the distribution of particular types of business service offices reinforces the picture of strong concentration in the South East (Table 2.5). Banking, insurance and financial services increased everywhere during the 1970s, but the lion's share, 49 per cent, went to the South East.[28] All services, apart from accountancy, are more represented in the South East than would be expected on the basis of total employment. Market-related services such as marketing and market research are most concentrated, with more than 84 per cent of national offices in the South East and more than 50 per cent in Greater London. Virtually all provincial regions have fewer offices than expected for every service, except for accountancy and, to a lesser extent, management consultancy.[29]

There is also evidence that it is the higher-level services that are concentrated in the South East. Provincial areas, especially the major cities, tend to have more branch or subordinate offices belonging to multi-site firms, while the South East has more main offices. This pattern – the marked concentration near the capital, the local decentralization from London and major cities, the slower growth in northern areas – cannot be understood simply in terms of site rentals or changes in communication costs, nor of manufacturing demand. Rather, a corporate urban hierarchy exists, with national headquarters and many administrative, research and technical functions in the south and east, and branch offices carrying out production activities or serving local markets elsewhere. Banking and insurance have decentralized away from London, but have gone to smaller labour markets in the South East and in the

Table 2.4 Business and producer services: location quotients, 1981

	All services	Producer services	Business services
London	1.19	1.61	1.85
Conurbation dominants	1.02	1.06	1.07
Provincial dominants	1.00	1.01	0.99
London subdominant cities	1.07	1.04	1.21
London subdominant towns	0.99	1.03	1.05

Source: Marshall, J. N. et al. 1988. *Services and uneven development*. Oxford: Oxford University Press. p. 72.

Table 2.5 Types of business service: percentages of national offices, 1981

	Management consultancy	Accountancy	Market research	Marketing	Advertising	Computer services
Greater London	23.0	14.2	62.7	59.4	44.7	31.9
South East	25.9	15.7	21.9	25.5	18.9	24.1
North West	11.5	12.8	2.0	2.5	8.8	9.3

Source: Marshall, J. N. et al. 1988. *Services and uneven development*. Oxford: Oxford University Press. p. 72.

secondary financial market of the North West. However, the growth of international trade in financial, engineering, oil and consultancy services has encouraged growth in and close to London, especially near international airports.[30]

A cricital role has been played by the location of corporate headquarters. Between 1971 and 1981, the outer South East's share of the top 1,000 companies' headquarters rose from 7 per cent to 15 per cent, while the share of the provincial regions fell from 40 per cent to 31 per cent. The implication is that the expansion in the rest of the South East has been at the expense of provincial cities rather than of central London, which has retained its level of activity. The presence of a headquarters is important because there is clear evidence that they generate a demand for locally generated producer services. Branch offices of large multi-locational firms get their services via headquarters. These in turn either generate them internally or obtain them locally, i.e. within London and the South East. The North West and West Midlands have done better in this regard than Yorkshire–Humberside because their provincial capitals are stronger; Manchester in particular is used by firms in the West Midlands and Yorkshire–Humberside. This indicates that producer service activities can perform a basic role in regional development.[31]

Increasing globalization enhances London's role as a major corporate complex; the city provides good international communications both with the parent countries of corporations and with potential markets, as well as access to specialist services, clients and occupational skills. Dunning and Norman's research shows that in 1976, 95 per cent of 261 US-based business service firms were in London and 82 per cent were in central London. They observe that the 'trend towards internalisation of producer-service provision as a consequence of the growth of large multisite organisations, together with the national and international location behaviour of business and other services is clearly helping to exacerbate the distinction between the core and the periphery within the British economy'.[32]

Producer services: two case studies

Computer services

There is a close association between the location of high-tech manufacturing and that of producer services, especially those services that depend on information technology. Howells's study of the computer services industry shows that areas that have both specialized in and benefited from computer service activity are found disproportionately in southern England, especially 'in a major band running from the west of London down through to Portsmouth (IBM), Cambridge and the Severn Valley', together with a node in the North West. In these regions, Howells finds evidence of a 'dual economy', with both an important set of elite multinational computer firms and a high rate of indigenous new firm formation. Wales and the North each had less than 2 per cent of employment. These disadvantaged regions have high rates of new firm formation but also high failure rates, a phenomenon that cannot alleviate the economic problems of these regions.[33]

Computer services provide a crucial indicator of the degree to which regions and places are moving toward the informational economy. Information is being increasingly commodified, and also increasingly concentrated in the hands of the transnational corporations, which can best exploit its characteristics of high-speed digital telecommunications. Goddard and Gillespie write: 'Rather than encouraging greater efficiency through more competition, then, the new technologies are facilitating the development of oligopolistic rivalry amongst the biggest companies'.[34] As already noted, these are highly concentrated in London.

Another illustration of the trend to spatial concentration is provided by value-added networks. Run by private companies, and providing specialized services to subscribers, they are a new feature of the competitive telecommunications market in Britain. Of the 164 registered down to October 1985, 125 were in London and the Home Counties, 15 in the South West, 4 in the West Midlands, 3 in the East Midlands, 3 in Yorkshire–Humberside, none in the North East, 9 in the North-West, 4 in Scotland, none in Wales, and 1 in Northern Ireland.[35]

Even with regard to traditional services, there is clear evidence of marked regional disparities of uptake within most European Community countries. In the United Kingdom, telephone subscriber take-up varied in 1980/1 from 42 per 100 people in London to 23 in the Northern Region; in France, from 40 in Paris to 23 in Franche-Comté and Lorraine; in Germany, from 53 in West Berlin to 24 in Regensburg. A specialized service like Prestel, which is particularly suited to the small user, shows comparable big variations in the United Kingdom. Similarly, international calls are a small proportion of total calls in the North compared with the South East, even though subscribers in the latter can reach nearly one-third of total business phones in the country at the local rate compared with 1 per cent in Newcastle.[36]

Accountancy

A producer service that has shown exceptionally fast growth over the last two decades is accountancy. Again, it is disproportionately concentrated

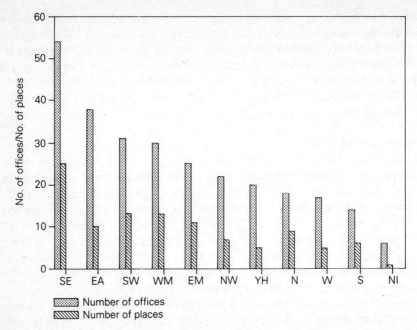

Figure 2.7 Top ten United Kingdom accountancy firms: regional distribution.
Source: Morris, J. L. 1988. Producer services and the regions: the case of large accountancy firms. *Environment and Planning* **A20** 750.

Table 2.6 Top ten United Kingdom accountancy firms: major locations

Number	Location represented
10	Aberdeen, Birmingham, Bristol, Cardiff, Edinburgh, Glasgow, Leeds, London, Manchester, Newcastle
8	Liverpool, Southampton
7	Leicester, Nottingham, Reading
6	Belfast
5	Sheffield

Source: Morris, J. L. 1988. Producer services and the regions: the case of large accountancy firms. *Environment and Planning* **A20** 750.

in the South East (Fig. 2.7 and Table 2.6). In a survey made by Morris, all firms confirmed that the major growth had been in the London office. They also quoted growth in the outer South East, because of decentralization of back office functions and the growth of business there.[37]

Toward explanation

There are two warring theoretical explanations for these trends. One, which can be called neo-Schumpeterian, stresses innovation, entrepreneurialism and new small firm foundation. The other is neo-Marxist and stresses a new national (and international) division of labour that is no longer based on the industrial sector but rather on hierarchical function, with increasingly centralized corporate control in huge transnational corporations.

Neo-Schumpeterian explanations

Goddard and his colleagues have used data from the Science Policy Research Unit to study regional innovation rates. They find a considerably greater rate of product innovation in the South East (88 per cent) than in intermediate (82 per cent) or development areas (73 per cent). Furthermore, the variations are greater for single plant innovations where the origin is more clearly local (85, 76 and 55 per cent, respectively). Only 13 per cent of establishments in the South East had external sources of product innovation compared with 15 per cent in intermediate areas and no less than 33 per cent in development areas.[38] They conclude that 'the South East region is the primary source of external technical information. In general, plants in development areas mentioned relatively few local useful technical contacts. They appear to prefer or are forced to use technical sources at a greater distance than are firms located elsewhere.'[39]

Worse, innovations in the development areas mainly take the form of process innovations that save labour. Table 2.7 shows that the North gains fewer jobs through product innovation but loses more jobs through process innovation than the South East, with the most significant losses occurring in northern multi-plant firms.[40] Harris has extended this work to show major variations in innovation in manufacturing industry over a long period. Figure 2.8 shows that the South East outperformed all other regions, mainly because of a strong structural effect overcoming a large and negative residual effect. However, industrial structure appears to have played a particularly detrimental role in the North, South West and Scotland, while Northern Ireland suffered from a very large negative

Table 2.7 Employment effect of innovation

| | Mean employment effect (first year) | | | |
| | Product | | Process | |
	North	South East	North	South East
Single plant	+0.25	+1.25	−0.50	+0.20
Multi-plant	+6.00	+7.45	−6.53	−0.92
Total	+4.30	+4.92	−3.81	−0.49

Source: Goddard, J. B., A. Thwaites & D. Gibbs. 1986. The regional dimension to technological change in Great Britain. In J. Amin & J. B. Goddard (eds.). Technological Change, Industrial Restructuring and Regional Development. London: Allen & Unwin, p. 151.

residual.[41] Just as theory suggests, the peripheral regions have produced more process innovations, and over time the concentration of innovations in larger externally owned plants has resulted in an ever greater relative importance being placed on process innovations in these regions. At the same time, the concentration of innovations in externally owned plants, coupled with the growth of the branch-plant economy, has meant a larger proportion of UK innovations emanating from the South East, where these larger firms are located, and fewer from the Development Areas.[42]

Significant differences exist in new firm formation rates: they are higher in the prosperous areas and lower in the non-prosperous areas; and these prosperous regions are currently the most entrepreneurial.[43] Shift–share analysis demonstrates that once agriculture is excluded, the structure of employment in each region is less important in explaining these differences than non-structural factors (Table 2.8). Storey and Johnson have generated a composite entrepreneurship index based on six groups of factors shown to be associated with entrepreneurship either positively or negatively.[44] This index is found to correlate well with composite take-up

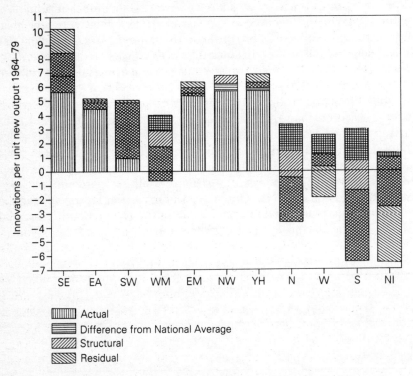

Figure 2.8 Standardized innovation rates by region.
Source: Harris, R. I. D. 1988. Technological change and regional development in the UK: evidence from the SPRU database on innovations. *Regional Studies* **22** 364.

Table 2.8 New firm formation rates, 1980–83

	Relative rate	Structural component	Fertility component
North	0.95	−507	−616
Yorkshire–Humberside	0.95	−689	−1,696
North West	0.97	−833	−1,036
East Midlands	0.98	−455	−501
West Midlands	1.00	−364	+532
East Anglia	0.98	−70	−457
South West	0.94	−371	−2,558
South East	1.08	+4,450	+11,622
Scotland	0.86	−348	−5,267
Wales	0.91	−607	−1,797
Northern Ireland	0.74	−305	−3,405
Unallocated	1.21	+102	+5,187

Source: Storey, D. & M. Johnson. 1987. Regional variations in entrepreneurship in the UK. Scottish Journal of Political Economy **34** 165.

indices for four small firm policy instruments: the Loan Guarantee Scheme, the Business Expansion Scheme, the Enterprise Allowance Scheme, and the Small Engineering Firms Investment Scheme. The index, Storey has found, also correlates strongly with rates of new firm formation as shown by registrations for value-added tax. He concludes that current policies, which aim at encouraging new firm formation, are actually likely to widen the gap between the prosperous and the less prosperous regions.[45]

Neo-Marxist explanations

An alternative explanation lies in the thesis that there has emerged, both internationally and inter-regionally, a new spatial division of labour that is no longer based on traditional specialization by the industrial sector but rather on what Massey calls spatial structures of production: the patterns of control and dependency within capitalist organization, from the smallest, autonomous single-region firm to the giant transnational corporations.[46] Goddard and Gillespie conclude that 'essentially, the existing geography of the information economy reveals the structure of regional dependency in Britain. Many of the information jobs located in London and the South East control economic activities in other parts of the world as well as in other parts of Britain ... The spatial division of information labour which exists in Britain suggests a structure of core–periphery interdependencies which are far from neutral or symetrical [sic].'[47] Thus the fact that the North has 80 per cent of its manufacturing controlled from outside provides the reason for the low proportion of informational occupations.[48] As Hudson puts it, the North East is now a global outpost in the scheme of international capitalism.[49]

The fullest statement of control theory comes from Marshall. He argues that during the inter-war and early post-war period, regional differences were dominated by the contrast between the traditional heavy industries in the North and the consumer goods industries of the Midlands and South. This was an inter-sectoral division between raw material and capital goods production, and areas producing for the emerging mass consumer market. However, since the mid-1960s there has begun to appear a new division based on intra-sectoral distinctions between territories specializing in different stages or processes of accumulation within the same industries. As a result, there is now a threefold regional distinction in the British space economy: first, mass-production and assembly-line production in areas of semi-skilled, low-waged and poorly organized labour, in old mining or heavy industrial regions; second, non-standard, non-assembly line, non-automated production in old northern industrial cities that are, however, disinvesting; and third, higher management, R&D and other control functions in financial and professional service centres.[50] This explains why industrial decline is no longer concentrated in the traditional heavy industrial sectors, as in the inter-war period, but has now spread into the consumer manufacturing industries that spearheaded the post-war recovery, especially the industrial heartland of the West Midlands. During the 1950s the West Midlands was last into each cyclical trough and first out, but after 1966 it began to synchronize with national trends. The decline of UK manufacturing has accentuated with each successive cyclical recession since 1966.[51]

Regional case studies

The South East

The South East is outstandingly the United Kingdom's best-performing region. Yet within it a glaring contrast exists between a depressed London and a booming ring of counties around it, which constitute the 'Rest of the South East' (ROSE) in official statistical terms (Table 2.9). Between 1971 and 1981 there was a loss of over 400,000 jobs in London, nearly 11 per cent of the 1971 total, and a gain of 322,000, just under 10 per cent, in the rest of the region. The gain in the Outer Metropolitan Area was some 6.3 per cent, that in the outer South East nearly double that. The net result was that the entire South East region shed 92,000 jobs, a slightly lower rate of loss than the country overall (1.3 per cent against 2.3 per cent).[52] Throughout the region the picture was consistent: the job loss was in

Table 2.9 Employment changes by sectors, South East region, 1971–81

	Manufacturing		Services		Total	
	Change	Percentage	Change	Percentage	Change	Percentage
Inner London	−186.7	−40.6	−105.0	−6.1	−318.3	−13.9
Outer London	−192.0	−32.6	+98.7	+10.2	−95.7	−5.8
Greater London	−378.8	−36.1	−6.4	−0.2	−414.0	−10.5
Outer Metropolitan Area	−125.4	−18.4	+237.2	+25.0	+111.6	+6.3
Outer South East	−15.3	−3.5	+231.1	+23.9	+210.5	+13.6
Rest of the South East	−140.7	−12.5	+468.3	+24.5	+322.1	+9.7
Western Crescent	−11.1	−3.0	+208.2	+32.1	+196.6	+17.6
Of which:						
Four growth areas	−2.0	−0.9	+133.6	+35.0	+134.0	+20.9
Rest of the South East remainder	−129.6	−17.2	+260.2	+20.5	+125.5	+5.7
South East	−519.5	−23.9	+462.0	+10.0	−91.9	−1.3
Great Britain	−1,963.2	−24.9	+1,703.0	+14.5	501.2	−2.3

Source: SERPLAN (1985) based on Census of Employment, 1971 and 1981.

manufacturing. The only exception was London, where there was also a marginal loss in service jobs, and particularly in inner London, where this loss was more serious. Outside London, the loss in manufacturing jobs was less and the growth in service jobs greater. There was a combination of two processes here: the general secular decline of manufacturing and the rise of service employment was accompanied by a relative decentralization of jobs from the region's core to its outer rings.

The latter is significant: London's job decline has occurred despite its apparently favourable industrial structure, showing that a differential 'London effect' is at work.[53] Inner London suffered massive manufacturing job losses, and compounded these by losses in service jobs. This is despite the fact that the area contains the information-processing, financial and media centre of the United Kingdom. But inner London was also a centre for goods-handling service jobs, and these in docks, in warehousing and in transport suffered particularly badly from the closure of the London docks and associated activities.[54] Figure 2.9, based on analysis from the ESRC's inner-city studies, demonstrates that the loss of manufac-

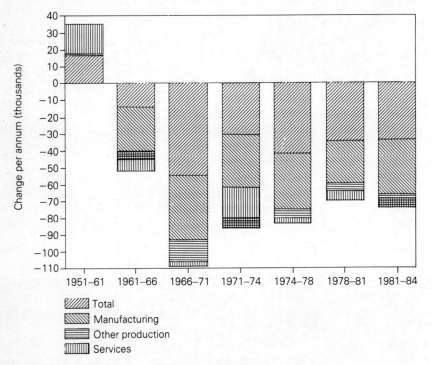

Figure 2.9 Changes in London employment, 1951–84.
Source: Buck, N., and I. Gordon, 1987. The Beneficiaries of Employment Growth: An analysis of the experience of disadvantaged groups in expanding labour markets. In Hansner, V. A. (ed.). Critical issues in Urban Economic Development. Vol. 2. Oxford: Clarendon Press.

turing jobs from London has been a long-term process, beginning early in the 1960s and continuing almost without pause and fairly consistently to the early 1980s. The countervailing gains in service employment, in contrast, have been very irregular, occurring in two short bursts in the early 1970s and the early 1980s; otherwise, as already seen, service employment has actually contributed to the general contraction.

The outer South East, in contrast, is generally regarded as the United Kingdom's outstanding growth area. It has been particularly attractive to two dynamic sectors: high-tech manufacturing and producer services. Indeed, in popular legend the M4 corridor west of London is seen as an internationally outstanding high-tech area, a kind of European Silicon Valley. The reality is somewhat different: the area, which is not a corridor but more of a wide crescent sweeping around London from Southampton–Portsmouth through Reading to Milton Keynes, Stevenage and Cambridge, has seen a modest increase in high-tech jobs, but this has done no more than compensate for a loss of such jobs in London. This helps reveal the underlying mechanism. Over a period of more than a century there has been a progressive decentralization of high-tech industries away from their origins as small-workshop bases in central and inner London as they required more and more space as the scale of industry grew. The process reached outer London in the 1920s and 1930s, and Berkshire and Hampshire in the 1950s and 1960s. Additionally, a very significant role has been played by the Defence Research Establishments, which are located in a crescent west of London for reasons that partly reflect rational location procedure and are partly contingent.[55] Overwhelmingly, high-tech industry in the crescent proves to be dependent on contracts from these establishments, often involving close and continuing contact. In some respects the entire area resembles an old-fashioned urban industrial quarter writ large, a point made by Scott about the much larger but similar complex in California's Orange County.[56]

The South West

The large and diverse South West region has shown one of the best regional performances in the 1970s and 1980s. At the region's north east corner, Bristol–Gloucester–Cheltenham represents an extension of the M4 corridor and is a diversified high-tech industrial region with a particular emphasis on aerospace. Bristol had a relatively strong economic structure derived from its previous industrial history: in the late nineteenth century traditional industries like tobacco and chocolate were transformed and new industries such as metal-working, engineering, printing and packaging and the production of boots and shoes expanded.[57] In the early twentieth century, new industries like electrical goods, vehicles, aircraft and precision instruments provided large pools of jobs. The city came to be dominated by large companies like Imperial Tobacco, Fry, Mardsons and Robinson (paper and packaging) and the Bristol Aeroplane Company; employment was heavily male-dominated. By 1961 manufacturing provided 44 per cent of total employment, with 64 per cent in three sectors: vehicles; food, drink, and tobacco; and paper, printing and publishing.[58]

During 1971–8, trends in Bristol were an exaggerated version of the national picture. Manufacturing employment fell 14.3 per cent; service employment grew by exactly the same percentage. Nearly 80 per cent of the loss in manufacturing came from the three big sectors. Not all the services grew. In utilities, transport and communications and distribution it declined or grew slowly. The really big increases were in miscellaneous services, insurance, banking, finance and professional and scientific services. The growth in financial services was no less than 61 per cent. More than 80 per cent of the growth in service employment was female.[59]

During 1978–81 total employment shrank, and now service jobs too declined, albeit by only 1.4 per cent, less than the national average. Two of the three main manufacturing sectors suffered big losses but vehicles (aircraft) gained. Within services, only banking, insurance, finance and miscellaneous services grew. Overall, between 1971 and 1981 Bristol lost 23,663 manufacturing jobs (22 per cent), and gained 21,799 service jobs (12.6 per cent). By 1981 manufacturing contributed only 28.1 per cent of employment, while services contributed 65 per cent. Male jobs had fallen by over 14,000 and female jobs had risen by nearly 13,000; females represented 41 per cent of the workforce.[60]

High-tech industry has remained an important element in manufacturing, and as in the Western Crescent at the other end of the M4 corridor it is heavily defence-based. Services have been dominated by financial services, much of them 'basic', reflecting the city's increasing role as a regional financial and business centre and the associated growth of headquarters; and by miscellaneous services, reflecting a big income multiplier. Thus growth in some areas has been associated with decline in others; and Bristol's apparent success reflects the fact that employment has declined here less than elsewhere.[61]

However, the benefits are not evenly distributed across the labour market, as is emphasized by the growth in youth and long-term unemployment among less-qualified school leavers and less-skilled workers shaken out of other jobs, who find few openings in the expanding high-tech and office sectors. The result is increasing polarization of the labour market.[62] Nevertheless, the fact remains that these high-risk groups have a better chance of a job in a growth area such as Bristol, or a new town, or one of the faster-growing towns around the metropolis.[63]

The growth of the South West economy is not restricted to the Bristol region, though; it extends widely, into remote rural areas like north Devon, 100 miles to the south west, where employment, especially in manufacturing, grew rapidly during the 1970s. The fastest growth was in female employment, with part-time female jobs growing by one-half. The search for cheap rural labour thus appears more important than 'voluntarist' explanations of population growth, e.g. businesses started by early retirees. However, both explanations may be relevant and they may intertwine.[64]

The West Midlands

The West Midlands is perhaps the most spectacular case of the collapse of the industrial heartland. Between 1966 and 1981, 369,500 manufacturing jobs were lost in the region, with especially heavy declines during 1971–6 and 1978–81. Until 1978 these were partly offset by gains in the service sector, which made up about half the manufacturing loss; from then to 1981 total employment declined by 12 per cent, nearly twice the national rate of decline, and unemployment rose.[65]

The root cause of the decline was not a fundamental shift away from traditional industries, but a failure of the local economy to diversify. The contraction of the key manufacturing sectors was faster than in the nation as a whole, and under-investment has led to poor and declining productivity across the whole of manufacturing. In addition, there was insufficient compensating service growth; the economy remained overly dependent on manufacturing. Shift–share analysis shows that structural and differential factors were about equally important in the 1971–81 period.[66] 'The present downturn in the West Midlands', Spencer and his colleagues conclude, 'is not therefore a response to the cyclical slump in the national and international economy but part of a long-term, structural decline in the economy.'[67]

The preconditions were evident well before the 1970s, if anyone had been willing to see. Over 50 years, the economy had become concentrated into certain sectors and also into a few companies that were not committed to expansion in the region and were increasingly independent of local subcontractors who became dangerously dependent on big companies tending to expand elsewhere.[68] With increasing internationalization of production and the growth of transnational corporations, the local economy was opened up to the global strategy of a small number of companies.[69] Diversification did not help local economies, as their most influential firms acquired companies elsewhere and based production there.[70] The region's top companies recorded disproportionate job losses, 38 per cent in the top ten companies, representing 18 per cent of the total manufacturing job loss.[71]

Added to this, there was local decentralization out of the region's industrial core. Advances in transport and communications technology made this feasible; obsolescence and congestion, coupled with capital–labour struggles, made it desirable.[72] Many observers have concluded that government dispersal policies aided the process, but Spencer and his colleagues disagree.[73] It would be wrong, however, to dismiss the role of policy entirely, but its effect on employment has often been overstated.[74]

Central Scotland

Glasgow's economy, like Bristol's, was based on eighteenth-century 'colonial industries' but was then transformed by the nineteenth-century industries of iron and steel, heavy engineering and shipbuilding.[75] Employment in these basic industries has fallen over a long period. In the whole of the west Central Scotland conurbation, employment fell from 844,000 in 1952 to 686,000 in 1981 and 640,000 in 1984. The rate of loss grew steeper, from a negligible 400 a year in the 1950s to 4,700 a year in the 1960s and early-mid 1970s, and 25,000 a year in the late 1970s and early 1980s. Manufacturing employment fell from 387,000 in 1961 to 187,500 in 1981, with faster losses in the late 1960s and late 1970s; service employment grew from 350,000 to 431,000 in 1978, and has remained stable since then. The manufacturing loss was most serious in Glasgow itself during the 1960s, but the crisis shifted to the outer towns during the 1970s.[76]

Service employment, in contrast, rose from 42 per cent of the total in 1961 to 63 per cent in 1981. Lever and Moore observe: 'It is not an exaggeration to say that in twenty years Glasgow – and the whole conurbation – has gone from being an industrial city with 60% of its labour in manufacturing, to a service centre with 60% of its labour in service occupations.'[77] Reality is now at variance with the myth:

> The external, and indeed the self-image of Clydesdale, is of an area populated by hard people doing hard manual jobs. This image still pervades the media and the literary world of books and plays. And yet the modern reality is quite different ... a large proportion of Clydeside's labour force does not have a job at all. Of those that do, relatively few are in traditional sectors. For example, in 1981 for every Glaswegian male employed in manufacturing there were two employed in the service sector.[78]

In the outer conurbation, manufacturing jobs declined dramatically as a proportion of the total, from 46 per cent in 1978 to 35 per cent in 1981; in Glasgow, the change was less dramatic, because both sectors declined. As a result, a massive increase in unemployment has occurred over the last 10 to 15 years. In the latter half of the 1960s there were on average about 35,000 unemployed, but by 1984 there was an average of 147,000. Unemployment has risen more sharply in the outer conurbation and among females.[79]

Shift–share analysis shows that during the 1970s the country caught up with the city's decline: much of the loss during the 1970s follows national trends. The city had a favourable structural component but its industries performed poorly in relation to the national average; the outer conurbation had a poor structure and it was no longer helped, as in the 1960s, by competitive growth.[80] The root cause of the lack of competitiveness was that labour productivity was low; the best available measure gives only 65.9 per cent of the national average for manufacturing in 1978, and this helps explain why employment declined so rapidly relative to national trends.[81] The sample of firms 'experience low value added per head because they overspend on labour despite lower wage costs in the conurbation'.[82] One important reason is that they are biased toward the labour-intensive, small-batch, specialist, often customer-specific end of the production spectrum.[83] The answer would be product innovation. Lever's study concludes that although product innovation is not synonymous with the creation of new firms, they are often related. The failure of Clydeside to generate an adequate rate of new firm (or new product) formation represents a severe problem.[84]

More recently, in the mid-1980s, the city itself has seen a spectacular revival in the form of service industry growth, associated both with the development of a regional office function and also with the city's position as a centre of education and culture – the latter symbolized by its having been designated European City of Culture in 1990. These factors, combined with the spectacular physical rehabilitation of the east side as part of the Glasgow Eastern Area Renewal (GEAR) scheme, and with a vigorous publicity campaign under the slogan 'Glasgow's Miles Better', have transformed the city's international image. However, in reality, they have only helped compensate to some degree for the deindustrialization of the 1970s. The GEAR scheme has had relatively little success in attracting new manufacturing employment to the depressed east side.

Postscript: the economic revival of the 1980s

During the economic recovery of 1983–5, employment at last grew again, but the growth benefited only female and self-employed people. There was a gain of 594,000 female workers and a further loss of 129,000 male workers.[85] More than 80 per cent of the additional female jobs were part-time (Fig. 2.10). There is thus an underlying secular trend toward the use of part-time female labour.[86]

At the same time, a huge regional gap became evident: the regions of the country that suffered the least from the contraction during the recession gained most from the expansion of employment in the recovery. The main impact of the recession was felt in the 'Peripheral North' and the 'Industrial Heartland', which recorded 40 per cent of employment in 1979, but 57 per cent of the subsequent decline (Fig. 2.11). The three southern regions, plus the East Midlands, had 52 per cent of employment in 1983 but accounted for 52 per cent of the subsequent increase. The gap is even more notable if London is separated, because employment recovery there has been minimal. The rest of the South East (outside London), plus East Anglia and the South West, stood out as the only really buoyant labour markets. They were the only ones with any growth of male employment and have seen most of the growth in female jobs and service jobs, which two phenomena are closely related.[87] The regional dualism is being reflected during the recovery, because of changes in the occupational composition of the manufacturing labour force;[88] the Heartland regions, as already noted, have very low new firm formation and weak representation of innovative, technologically advanced industry.[89] There is some

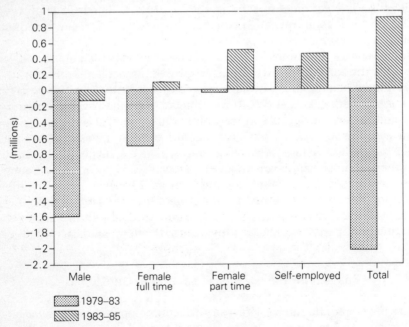

Figure 2.10 Employment changes, 1979–83, 1983–5.
Source: Martin, R. L. 1986b. In what sense a jobs boom? Employment recovery, government policy and the regions. *Regional Studies* **20** 465.

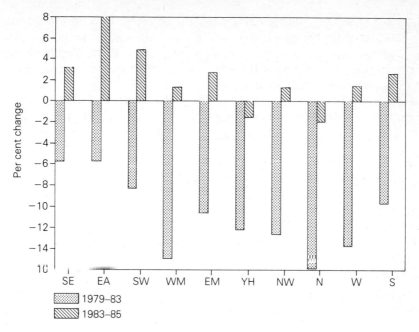

Figure 2.11 Regional employment changes, 1979–83, 1983–5.
Source: Martin, R. L. 1986b. In what sense a jobs boom? Employment recovery, government policy and the regions. *Regional Studies* **20** 466–7.

evidence of a revival of the West Midlands region since 1987 based on the development of Birmingham as a regional and national service centre, particularly through the expansion of the National Exhibition Centre and associated business tourism activities, which is promising, but it is too early to say whether this represents a long-term structural recovery.

A huge paradox remains. The regions that suffered the most massive deindustrialization in the 1970s have shown the least capacity to recover in the 1980s. They have the weakest entrepreneurial traditions, the lowest rates of new formation and the least evidence of technological innovation. They are locked into a dependency relationship at the periphery of the contemporary capitalist system of organization. In this sense, the Marxist and the Schumpeterian explanations coincide. However, the question remains, what policy prescriptions are appropriate to this dilemma? On that, there is little agreement and little inspiration.

Conclusion: the search for policy

The period of maximum deindustrialization from 1979 to 1983, and the period of recovery that followed it, both occurred under radical right-wing Conservative administrations led by Margaret Thatcher. It would clearly be impossible to ascribe any causal connection to the former, although government supporters would certainly claim it for the latter. More significantly, the first Thatcher government carried out the most radical restructuring of British regional policy that had occurred since its genesis in the Distribution of Industry Act, 1945. The map of regional aid was redrawn, with the Assisted Areas shrinking to a small fraction of the size they had achieved under the Labour administrations of the late 1970s. The main negative element of regional policy, the Industrial Development Certificate, which was a prerequisite for any significant industrial development, was simply abolished.

The end of traditional regional policy came soon after Thatcher took office, but it might have come anyway. Traditional policy had worked moderately well in the boom years of the 1960s, when there were plenty of new industrial jobs to be diverted from prosperous to less prosperous regions. Industrial firms were making major job-creating investments. In the trough of 1975–83 firms made very few investments of any kind; if they made them subsequently, they were likely to be job-saving rather than job-creating. The instruments of policy would have had to change because the old ones were no longer relevant. In the transition from an industrial to an informational economy, a different set of instruments, working at a different spatial scale, becomes appropriate.

The Thatcher policy involved two major innovations in spatially targeted industrial development policy. The Enterprise Zones (EZ) involved a package of measures, including virtual removal of planning restrictions and a ten-year holiday from local property taxes, the shortfall being made good by central government. The Urban Development Corporations (UDC) programme took a wide variety of planning and other powers away from local government altogether in the interests of promoting the fastest possible recycling and redevelopment of derelict industrial and other areas. There has been much argument about the success or otherwise of the EZ policy. The fairest conclusion seems to be that it did generate a modest number of new jobs, a substantial proportion of which were simply transfers attracted by the incentives, at a rather high cost per job. The verdict on the UDCs cannot yet be reached, because – with the exception of the London Docklands and Merseyside corporations, established in 1980 – most are relatively new. It does appear that, just like the EZs, their performance has been quite variable depending on the location and inherent attractiveness of sites.

Two incontestable points can be made about the EZ and UDC experiments. First, both involved fine spatial targeting, in that the designated zones were relatively small. (Most, though by no means all, were derelict inner-city areas.) Both experiments were thus *urban* rather than *regional* policy initiatives. Second, these policies were deliberately [to] encourage the development of land, on the assumption th[at if] development took place, economic growth would follow.

The first feature is significant but not novel. British spatial developme[nt] policy was clearly moving from a coarse-grained regional stance to a fine-grained urban one from 1977, with the publication under a Labour government of the consultants' studies of the problems of the inner cities and the enactment, a year later, of the Inner Urban Areas Act. Only the chosen instruments changed under the Thatcher government. This shift reflected a recognition of a changed reality: by the mid-1970s, the major differences in the British economic landscape had already become intra-regional rather than inter-regional.[90] In particular, both in the north and the south, the fate of the older industrial and port cities stood in sharp contrast to the relatively prosperous shire counties around them.

The second feature is more original but not entirely so. Physical regeneration of substantial areas was one of the objectives of the Partnership and Project schemes created under the 1978 Act, and of the similar Glasgow Eastern Area Renewal (GEAR) in Scotland, although there was a great deal of controversy as to whether their provisions could have been entirely effective in this regard. Ironically, the New Towns Act of 1946, which provided the model for the UDCs, was a creation of the radical Labour government of Clement Attlee. What was new, and what was certainly contentious about the Thatcher policy, was its emphasis on the most rapid possible commercial realization of the value of the site. Spokespersons for the UDCs might well claim, with some pride, that they had not neglected wider social policies such as provision of low-cost housing or training programmes for unemployed residents. They do not attempt to deny that these policies came second.

There is some relationship between these two features. The bundle of policies that is desirable or necessary to achieve the economic regeneration of small areas of dereliction is likely to be substantially different from those needed to turn around an entire decaying region, as GEAR was already demonstrating by the end of the 1970s. In this sense, by 1979 regional policy in Britain was more or less dead, apart from a residue made necessary to enjoy access to European Community regional funds. In its place had come a mixture of policies, some spatially specific, some not, to try to regenerate the worst-hit pockets in each region, as appropriate. This may be the most significant policy lesson that British experience has to offer to the rest of the industrial world.

3

Deindustrialization and regional development in the Federal Republic of Germany

FRANZ-JOSEF BADE & KLAUS R. KUNZMANN

Introduction

LIKE MOST industrialized countries of the western world, the Federal Republic of Germany is experiencing a growth of jobs in the service sector, while employment in the manufacturing industries is more or less declining. This growth and decline, however, is not equally distributed among the Federal Republic's regions and states. While in most of the south the decline in industrial employment has been rather modest and has been compensated for by the growth of the service sector, the north, primarly the Ruhr region, the former industrial heartland, is suffering from the structural repercussions of its gradually changing economic base. Since this has been accompanied by strong social effects such as high rates of unemployment, and a diminishing income and financial base, political and academic debates on north and south in the Federal Republic, on uneven development and on new regional disparities have arisen.[1]

In fact, a division between north and south does not give an adequate picture of present regional economic development trends. A more differentiated view is required to explain the spatial distribution of growth and decline, to reveal its underlying causes and to evaluate the regional strategies and policies initiated to cope with deindustrialization.

This chapter consists of five sections. Following the Introduction, the second section deals with general national trends in structural change. This is followed by a third section that offers a comprehensive quantitative description of the spatial differences in economic development. A fourth section then summarizes and assesses qualitatively the variety of 'hard' and 'soft' factors influencing the spatial particularities of the deindustrialization process, using primarily the states of Nordrhein-Westfalen and Baden-Württemberg as spatial reference. Finally, efforts undertaken to adjust continuously the regional economic base in the two states are summarized in the fifth section.

In order to facilitate the assessment of the West German deindustrialization process from an international point of view, the chapter begins with a comparison of structural change and economic development features of the Federal Republic and the United States.

The expansion of services between fear and hope

There is a curious discrepancy between US and West German debates on economic change. A large part of the US contribution is rather pessimistic about the future economic prospects of services.[2] A primary concern of many contributions relates to income effects. The deindustrialization process is regarded as a danger to the wealth and welfare of a nation. In these analyses, wealth and welfare are interpreted on the basis of productivity figures,[3] wages and quality of the new jobs created in the services sector,[4] or of the social effects of job losses in the manufacturing industries.

However, in the Federal Republic the increase in services is interpreted much more optimistically. Compared internationally, the Federal Republic has a relatively low proportion of service employment. Thus some authors believe that unemployment will be reduced as services expand.[5]

There is an easy psychological explanation of the differences between US and West German attitudes toward the expansion of services. One main reason lies in the different developments of the US and West German economies in the last decades. Compared to other countries, the service sector in the US economy has a high share of total employment. The total number of jobs has increased by 25 per cent in the last two decades. The West German economy has seen just the opposite effect. While the proportion of services is the lowest among most industrial countries, the Federal Republic has suffered a loss in total employment of 6 per cent in the same period.

West Germans, therefore, expect a positive relationship between share of services and economic development,[6] while Americans are concerned about the quality of jobs currently being created in the service sector.

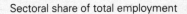
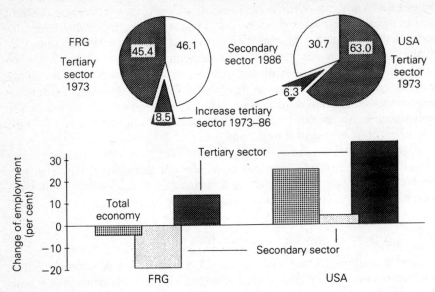

Figure 3.1 Sectoral changes in employment 1973–86, Federal Republic of Germany and United States.
Source: National Account Systems; authors' calculations.

They see the choice to be either deindustrialization with additional, but 'bad' jobs in the service sector, or the preservation of the manufacturing sector with the risk of total employment losses.

However, the reality is too complex for such simplifications. In absolute as well as in relative changes in employment, the deindustrialization process has been much more intensive in the Federal Republic than in the United States. While US industry has expanded its employment by 5 per cent, industry in the Federal Republic has lost 17 per cent of its jobs. Furthermore, the growth of services has been much lower in the Federal Republic than in the United States.

Conceptual aspects

Deindustrialization may be a response to several quite different situations. Moreover, the amount of deindustrialization evident in an economy depends very strongly on how it is measured. For example, the absolute decline of West German industry is apparent only in reference to employment. In contrast, measured by monetary figures, i.e. by value added, it shows nearly the same growth as US industry.[7]

In principle, there are two ways of considering structural changes in an economy.[8] The first is more traditional and concentrates on changes in *output*. In this respect, the transition from a manufacturing to a service economy is primarily the result of structural changes in the demand for goods and services. A classical output-oriented argument is the three-sector hypothesis postulating a greater income elasticity for services.[9] If income increases, the increase in demand for services is higher than for goods. With regard to the deindustrialization debate, the output interpretation means an absolute or relative decline in industrial production, stressing the effects on income and employment possibilities in manufacturing firms.

The second interpretation of structural changes relates to shifts in *input*. It is incontestable that the tertiarization process has been more intensive in the kind of labour input than measured by employment shifts from manufacturing firms to the services sector (Fig. 3.2). This tendency is particularly true for the Federal Republic, nearly two-thirds of all employees are performing some kind of service, which is an internationally comparable share. The service sector, on the other hand, employs only a little more than half of the total workforce.

The input-oriented delineation, which takes account of the kind of function performed by a person, is called 'functional structure', in contrast to the above mentioned sectoral structure based on the kind of firm (or organization) in which persons are employed. Functional changes in types of labour input have been analysed in more sociologically oriented studies,[10] for example, those contrasting impacts on 'white-collar' and 'blue-collar workers'. Presently, there has been considerable interest in 'information activities'[11] and the need for 'human capital' in order to preserve international competitiveness.

The tertiarization of input does not only refer to the kind of labour involved. Structural changes are also taking place in the overall combination of production factors. As mentioned below, the composition of total intermediate input for national production has been continuously shifting to services. Last, but not least, the character of final as well as intermediate goods is changing toward the more 'intelligent'. For example, a lot of hardware is increasingly enriched by software, for example as 'compacks' (complex packages) in the electronical industries. In a more hidden, but nevertheless important way, the service content of products is increasing in so far as the contribution of activities such as R&D or marketing seems to contribute more and more to the competitiveness of products.

The possibilities for investigating sectoral and functional changes vary widely. *Labour input* is usually studied through analyzing the occupational structure of employees. Data about shifts in the structure of *intermediate input* may be found in input–output tables. (However, at

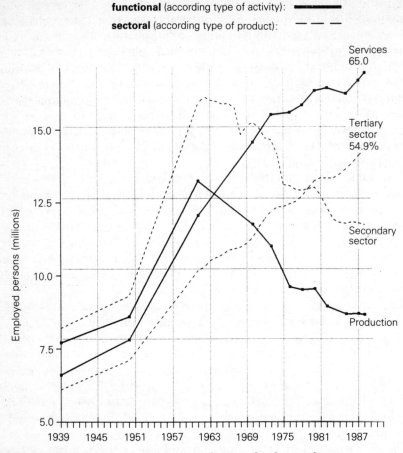

Figure 3.2 The tertiarization of employment in functional and sectoral terms. Source: Census, 1939, 1950, 1961, 1970; Mikrozensus, National Account System; authors' estimations.

least for the Federal Republic, tables for different years are hardly comparable.)

Unlike functional data, sectorally differentiated information is principally *output*-related. Interpretation of this data may differ, however, according to the kind of measurements used. Sectoral production values are obviously output figures, for, beside some other minor components, production values primarily consist of sales of firms. Consequently, sectoral changes in production values reflect changes in the demand for the products. *Value added* equals production value minus input value. Therefore, value added shows the (total) income gained by certain firms or branches (no matter who gets the money finally). Consequently, deindustrialization in value-added figures 'only' says that total income in service producing firms or organizations has grown faster than in industrial firms. As will be shown below, the change of income does not necessarily conform to developments in production and demand.

Finally, sectoral *employment* figures can only show in which firms or organizations employment has diminished or increased. Consequently, although sectorally differentiated, employment figures are a measure for input and do not immediately imply changes, whether in output, demand, or income. However, employment figures may be used as a proxy variable as there is a rather strong statistical relationship between change in value added and employment.

National trends in structural change

Production

Contrary to some expectations, even in 1987 more goods than services were produced in the Federal Republic. The share of services was less than half of total production value. Even in its dynamic aspects, the employment growth of services has been less strong than the increase of value added (for more details see Appendix Table 3A.1). With regard to *demand*, the economy of the Federal Republic is still a long way from becoming 'post-industrial'. If only *final* demand is considered, then the

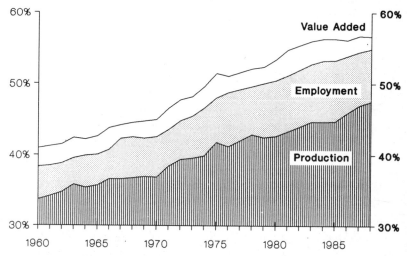

Figure 3.3 The expansion of the service sector in terms of production, income and labour input.
Source: National Account System; authors' estimations.

Table 3.1 The final demand for services, 1980

Economic sector	Consumption		Investment (gross)	Export	Final demand
	Private	Public			
			in millions of DM		
Agriculture	112.0	16.2	23.6	48.9	200.7
Goods production	476.3	78.7	441.1	524.8	1,520.9
Service production	575.5	263.8	99.6	164.2	1,103.1
Total production	1,163.8	358.7	564.3	737.9	2,824.7
			Sectoral share (percentage)		
Agriculture	9.6	4.5	4.2	6.6	7.1
Goods production	40.9	21.9	78.2	71.1	53.8
Service production	49.5	73.5	17.7	22.3	39.1
Total production	100.0	100.0	100.0	100.0	100.0
			Share of demand component (percentage)		
Agriculture	55.8	8.1	11.8	24.4	100.0
Goods production	31.3	5.2	29.0	34.5	100.0
Service production	52.2	23.9	9.0	14.9	100.0
Total production	41.2	12.7	20.0	26.1	100.0

Source: Author's calculations from Stille Frank et al. 1987. *Strukturverschiebungen zwischen sekundärem und tertiärem Sektor*. Berlin. Tab. 2.3/2.

share of services is even lower. These facts mean that a relatively high proportion of services is now included in intermediate consumption (Fig. 3.3).

Table 3.1 reveals one main reason for the permanent high share of goods in final demand. In 1980 (no more recent input–output table is available), more than one-quarter of total production was exported, and among the exports the share of services was 22 per cent. At present, the export share has risen to one-third of total production.

Unlike final demand, the intermediate consumption of services has grown to a much larger extent. Between 1974 and 1986, sales in firms producing business services increased by more than 200 per cent. In the same period, total consumer-oriented services increased by only 40 per cent. Furthermore – in contrast to the US experience – sales figures for restaurants, hotels and similar services did not rise much more than sales for the whole of manufacturing industry, i.e. by less than 100 per cent in 12 years.

Value added

The largest shift to services has happened in terms of gross value added (for more details see Appendix Table 3A.2). From 1960 to 1986, the share of the service sector in value added increased from 40 per cent to 57 per cent. In other words, services are much more important to the national income of the Federal Republic than production values alone reveal.

Internationally compared, this level is still rather low. In the United States, the share of the service sector in gross value added amounts to 65 per cent.

Although there is no direct information available for business services, there is some evidence that this group has made the largest contribution to the growth of national value added. As mentioned above, the production of business services has grown more rapidly than other segments of the sector. As business services are used mainly for intermediate consumption and as the difference between production and value added tends to be higher in products for final than for intermediate demand, the growth of value added in business services conforms to the evolution of production.

Sectoral changes in employment

Sectoral employment figures indicate where people work. Most of the new jobs in the services sector have been created in business and in consumer services. As for production and value added, the highest growth rates for services have been in business services. With regard to the total volume, however, consumer services have made the largest contribution of new jobs, above all in health and education (see Appendix Table 3A.3).

In spite of the clear shift of employment to service firms and institutions, 'German' expectations for the employment potential of the service sector have not been fulfilled. From 1960 to 1986, the total number of employees in the service sector rose by 4 million. In the same period, still more jobs (4.3 million) were lost in the manufacturing and agricultural sector.

The particular course of employment development in the service sector has also been discouraging. Growth in the service-producing and goods-producing sectors have proceeded more in a parallel than in a mutually compensating way (Fig. 3.4). In exactly those periods when the need for new jobs has been urgent (due to losses in manufacturing industry), employment opportunities in the service sector have decreased, too.

Employment trends in the Federal Republic have differed remarkably from those in the United States, given that economic variables mentioned up to now show a rather similar development. Total value added has increased in the Federal Republic more rapidly than in the United States (Table 3.2). However, the total number of jobs has decreased in the Federal Republic but has increased in the United States. Therefore, US firms have used much more labour input to achieve the same increase of net output than their West German counterparts. In other words, US complaints about diminishing productivity[12] and West German concerns

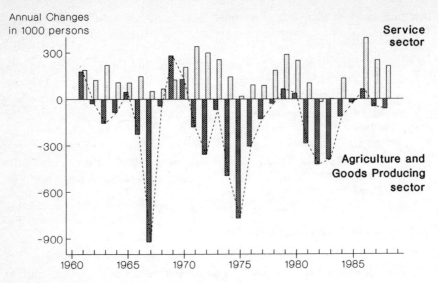

Figure 3.4 Sectoral gains and losses of employment.
Source: National Account System; authors' calculations.

Table 3.2 Employment vs. production changes: United States and Federal Republic of Germany

	Total economy		Goods production (including agriculture) Changes up to 1987		Service production	
	1961 =100	1973 =100	1961 =100	1973 =100	1961 =100	1973 =100
Change of value added[1]						
FRG	225.4	127.5	194.5	112.5	256.8	142.1
US	211.8	133.4	181.2	119.2	233.4	142.6
Change of employment						
FRG	98.6	95.7	73.7	80.9	138.8	113.6
US	159.7	125.1	114.8	104.1	193.3	137.4
Marginal productivity[2]						
FRG	2.29	1.33	2.63	1.39	1.85	1.25
US	1.33	1.06	1.59	1.15	1.20	1.04

Notes: 1) Gross value added in prices of 1980 (FRG) and of 1982 (US); 2) Ratio of change in gross value added to change in employment.

about high unemployment are, apart from other factors, two sides of the same coin. The lower propensity of West German firms to create new jobs and the lower labour productivity of US firms can be observed in both the goods and service sectors (Table 3.2).

Functional changes of employment

In the Federal Republic, the tertiarization of labour input is much more advanced than is indicated by sectoral figures. Since 1950 there have been more jobs in service functions than in manufacturing functions (Fig. 3.2). Also, as mentioned above, the share of services differs much less from those in countries like the United States if services are delineated functionally.

At first glance, changes in the functional structure of the economy appear to have paralleled developments in sectoral terms. On balance, since 1961 more than 4.5 million jobs were created for service functions, more than the increase of employment in the service sector (see Appendix Table 3A.4 for more details). Despite these global similarities, however, functional and sectoral shifts are not congruent, i.e. different individuals are affected. In only one group, personal and consumer services, are sectoral and functional classification close to identical.

In contrast, manufacturing functions are characterized by a rather heterogeneous sectoral structure. This is particularly true for producer services. As complementary functions they are (more or less) necessary for all kinds of production. Because of the variety of environments in which producer services are used, they are difficult to classify in a universally valid manner. However, given the limited usefulness of statistics, there are not many different ways in which to classify them. Consequently, in the literature, producer services frequently encompass

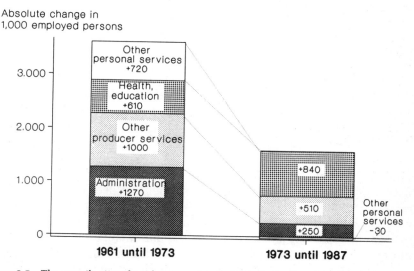

Figure 3.5 The contribution of producer services to the tertiarization of employment.
Source: National Account System; Census; authors' estimations.

technical and administrative services, and also include planning and strategic functions, such as marketing or consulting.

More than one-third of all producer services are performed *within* the goods-producing sector. In some sub-groups, such as technical activities, the share is twice as much. Likewise, two out of three persons performing R&D are employed in manufacturing industry.

By an increase of 2.8 million jobs, producer services have contributed the largest part to the tertiarization of labour. However, seen dynamically, the importance of producer services has decreased substantially (Fig. 3.5). After 1980, only one-third of new jobs could be considered producer services.

Spatial differences in the shift to services

The national overview may be enhanced by considering regional differences in deindustrialization. In contrast to national analyses, regional investigations must be restricted to employment. Monetary figures are available only at the level of the 11 states of the Federal Republic. Because of strong inter-regional relations between firms, data about sales and value added are difficult to estimate, and exist only on a very aggregated level.

Among the 11 states (see Fig. 3.6), Nordrhein-Westfalen is the largest, with one-quarter of total population and employment (Table 3.3); next in order of population and employment are Bayern, Baden-Württemberg and Niedersachsen. As the area of Nordrhein-Westfalen is smaller than that of Baden-Württemberg and only half that of Bayern, it is also the most densely populated state, apart from the city states of Hamburg, Bremen and Berlin.

Although in Nordrhein-Westfalen the increase in production and value added have been below the average growth rate, even today 27 per cent of GDP is produced there. A worker in Nordrhein-Westfalen produces more value added than the federal average. Baden-Württemberg and Bayern present the opposite case; GDP per employed person is still below average, although the growth in GDP has been clearly stronger.

The differences indicated in Table 3.3 between the city states of Hamburg, Bremen and Berlin and the remaining areas reveal that economic performance and development are strongly influenced by settlement structure. In this respect, the various states are extremely different. Nordrhein-Westfalen, for example, includes the large Rhein–Ruhr-metropolitan area, which consists of several more or less equivalent agglomeration centres (see Fig. 3.7); in addition, the remaining parts encompass other middle agglomerations such as Bielefeld or Aachen. Baden-Württemberg, by contrast, is strongly dominated by the Stuttgart agglomeration. Karlsruhe is another agglomeration in the state, but it is much smaller.

The differing spatial structures of the states obscure any common tendencies in structural change. Recent studies have suggested that contrary to popular opinion regional development occurs in rather

Figure 3.6 The states (Länder) of the Federal Republic of Germany.

Table 3.3 The states of the Federal Republic of Germany

	Area (thousands of sq. km.) 1986	Inhabitants (millions) 1986	Employed persons (millions) 1986	Gross domestic product (billions of DM) 1986	Percentage rate of change		
					Inhabitants 1960–86	Employed persons 1960–86	Gross domestic product 1970–86
Federal Republic	248.7	61.0	26.6	1,831.0	10.1	1.6	163.2
				Regional shares of Federal Republic (percentages)			
Schleswig-Holstein	6.3	4.3	4.3	3.5	13.9	15.6	178.3
Hamburg	0.3	2.6	2.6	4.6	−13.0	−23.0	147.1
Niedersachsen	19.1	11.8	11.1	9.8	9.4	0.0	172.7
Bremen	0.2	1.1	1.0	1.4	−4.6	−20.6	136.4
Nordrhein-Westfalen	13.7	27.3	25.4	26.9	6.3	−3.6	154.9
Hessen	8.5	9.1	9.2	9.9	17.0	9.2	191.9
Rheinland-Pfalz	8.0	5.9	5.8	5.2	7.0	−2.2	166.7
Baden-Württemberg	14.4	15.2	16.1	15.7	21.9	8.7	173.3
Bayern	28.4	18.0	19.7	17.6	16.8	10.0	201.9
Saarland	1.0	1.7	1.5	1.5	−0.3	−1.5	180.0
Berlin (West)	0.2	3.0	3.2	3.8	−15.7	−17.9	159.3

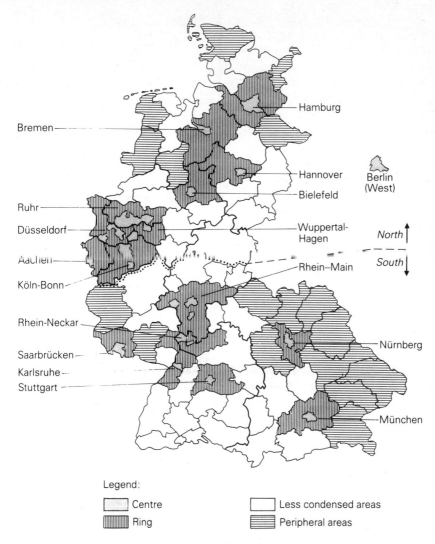

Figure 3.7 The spatial structure of the Federal Republic of Germany.

autonomous patterns that may be explained only in small part by sectoral development within the regions.[13] Regional particularities of growth can be (more or less) recognized in most of their sectoral components in so far as the economic sectors within each region develop along lines particular to that region.

An adequate evaluation of the spatial impacts of deindustrialization needs therefore to take into account the general tendencies in spatial

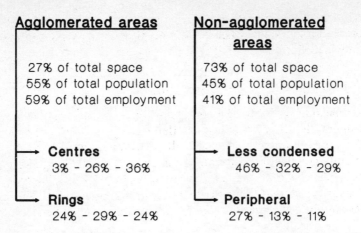

Figure 3.8 The hierarchical delineation of the spatial structure of the Federal Republic of Germany.

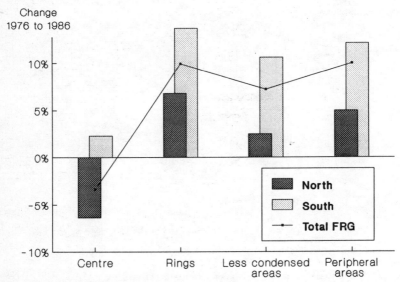

Figure 3.9 Spatial trends in the change of employment.
Source: Social Security Data.

structure. As structural changes take place on different levels of spatial hierarchy, the Federal Republic may be divided into 89 regions by two steps. First, the largest agglomerations have been identified and, second, their borders have been defined by delineating core and ring areas of each agglomeration (see Fig. 3.8) An analogous subdivision was made for the regions outside the agglomerations. The regions that are characterized by an obviously unfavourable location have been classified as peripheral. The remainder is a mixture of regions the position of which in the spatial hierarchy is difficult to evaluate.

General trends in the spatial structure

The empirical basis for the 89 regions is given by the social security data, beginning with 1976. The last available year is 1986. The data base includes more than 80 per cent of all employed persons; the remaining parts mainly consist of self-employed persons and civil engineers.

From 1976 to 1986 the number of total employees (contributing to social security) increased by 4 per cent. Employment growth was distinctly higher outside the agglomerations; the share of agglomerations in total

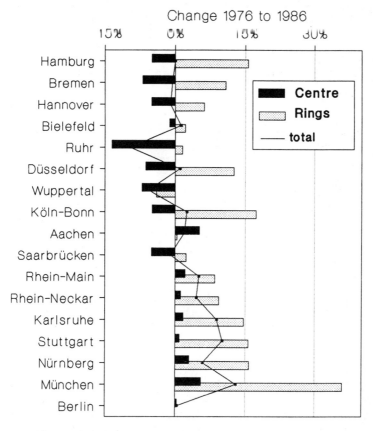

Figure 3.10 Change of employment in the agglomerations of the Federal Republic of Germany.
Source: Social Security Data.

employment fell from 61 per cent to 59 per cent. This process of 'deglomeration', however, is not homogeneous. First, the main losers were the centres of the agglomerations; in most of these employment declined. The winners have been the rings of agglomerations, which had the highest growth rates among all space categories (see Fig. 3.9).

Second, and somewhat surprisingly, this suburbanization process seemed to lose momentum at the borders of agglomerations. Employment in the areas lying between agglomerations and periphery has increased less than in the peripheral areas. Third, a south–north decline can be observed which is reflected in the suburbanization process as well as in the good performance of peripheral areas. In all categories of spatial hierarchy, from the centres of agglomerations to the peripheral areas, the southern regions have performed relatively better. Thus, the unfavourable development of total agglomerations is, above all, most significant in the northern areas. Most of the southern agglomerations maintained their employment levels even in their centres (see Fig. 3.10).

Sectoral differences in regional development

Regional development of total service sector

Between 1976 and 1986 national employment in the goods-producing sector decreased by 5 per cent while in the service sector it rose by 14 per

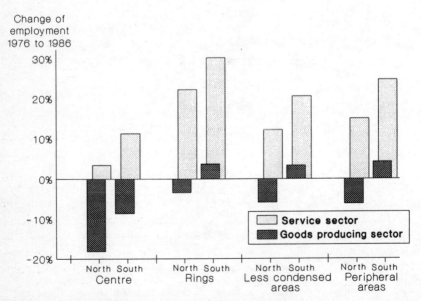

Figure 3.11 Spatial trends in the changes of sectoral employment.
Source: Social Security Data.

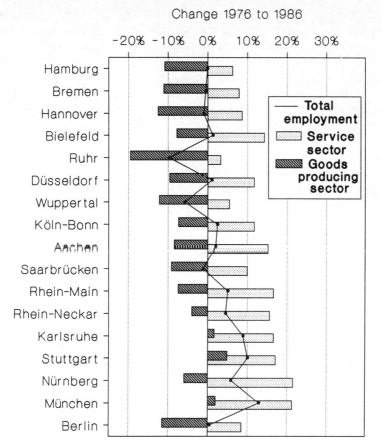

Figure 3.12 Sectoral changes of employment in the agglomerations.
Source: Social Security Data.

cent. However, large differences exist between regions in the growth of goods-producing and services-producing sectors. These differences clearly conform to general trends in regional employment (see Fig. 3.11), including regional deviations from federal average growth. The centres of agglomeration have had the lowest growth, both in the manufacturing and the service sector. On the other side, rings as well as peripheral areas have attained the largest gains not only in the goods-producing sector, but also in the services-producing sector. Furthermore, the south–north decline is evident in both sectors. Figure 3.12 shows parallel trends in sectoral growth for the four types of areas.

Statistically, the parallel quality of regional development trends may be expressed by a very strong correlation between the sectoral growth rates across all 89 regions ($r^2 = 0.83$). More than 80 per cent of regional variations in the growth rates of the service sector conform to regional differences in employment change in the manufacturing sector.

Regional differences in the growth of consumer and business services
In spite of the strong correlation, it cannot be excluded that parallel growth rates are restricted to the large sectoral aggregates. For example, it has been assumed that business services have a preference for central locations and, consequently, show the largest deviations from general changes in spatial structure.

In fact, however, their supposed preference for central location is not very strong (Fig. 3.13). Once again, the rings of agglomeration and the peripheral areas have attained the largest gains (up 65 per cent) while the centres of agglomeration have gained jobs at a below average rate.

In addition to the sectorally independent pattern of regional change, consumer and business services differ in so far as the growth of the former has spread much more uniformly over the regions. In consumer services the regional deviations from the national average growth rate are considerably smaller than in the business services industry. Between the most successful regions (rings, peripheral areas) and the relative losers (centres) the difference in growth rates amounts to 16 percentage points in favour of consumer services, and nearly double that for business services (30 percentage points).

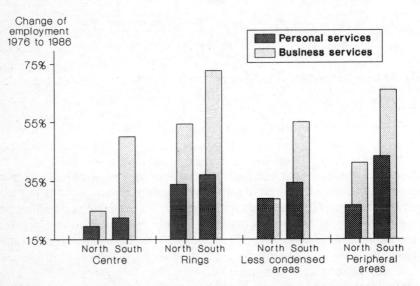

Figure 3.13 Spatial trends in the employment of selected service industries.
Source: Social Security Data.

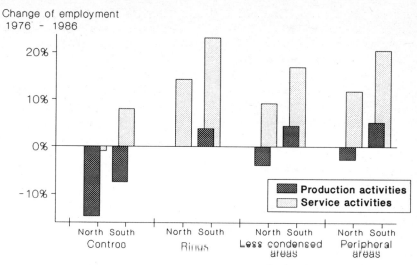

Figure 3.14 Spatial trends in the changes of functional employment.
Source: Social Security Data.

For both industries, there is a strong correlation with the growth rates for all services. However, while the (linear) relationship is proportional for consumer services, it is disproportionate for business services. With regard to the respective increase of total services, the winning regions have attained relatively high growth rates (some regions have increased by more than 100 per cent). At the other end, regions with a poor performance in total services have seen only very low increases in business services. For example, in the centre of the Ruhr agglomeration consumer services grew by 20 per cent, while business services increased by only 6 per cent. Above all, regions in the south of the Federal Republic, particularly the rings of the southern agglomerations, are positively affected by the disproportionate relationship.

Functional differences in regional development

As explained above, functional and sectoral delineations are not identical because they consist (partly) of different individuals. However, if services are functionally defined the picture of regional variations in growth rates is very similar to the sectoral ones (see Fig. 3.14). That is why it is unnecessary to discuss the regional growth patterns of production and service functions in more detail.

One slightly disproportionate relationship can be observed between the regional growth rates of total employment and of service functions. As in the case of sectorally defined business services, the expanding regions

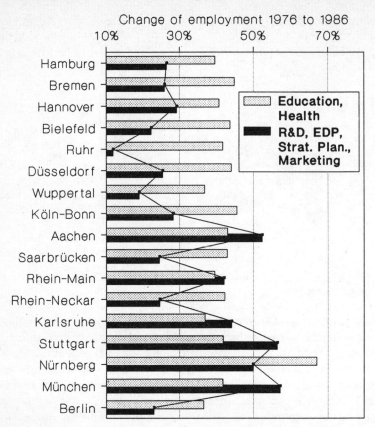

Figure 3.15 Development of selected service functions in the agglomerations.
Source: Social Security Data.

have put relatively more services into their production processes than have the declining areas, above all, those in the north of the Federal Republic.

If only producer services are considered, the disproportionate input of services is even more evident. This is particularly true for those producer services that are deemed essential to business competitiveness, i.e. R&D, strategic planning, marketing and electronical data processing (EPD). In particular, the southern agglomerations had a strong increase of these producer services (Fig. 3.15). In the rings of the southern agglomerations, for example, the employment of these producer services has increased by more than 65 per cent. In the northern rings the growth rate was only 38 per cent.

This particular nature of the southern agglomerations is depicted in Figure 3.15, where the growth of the selected producer services is contrasted with the expansion of education and health services. The regional increase in personal services is rather uniform; apart from Nürnberg all agglomerations have expanded education and health services by (more or less) 40 per cent. By contrast, regional growth rates of producer services vary to a much larger extent, from not more than 11 per cent (Ruhr) up to 57 per cent (Stuttgart, München).

Trends in the functional specialization of regions

The higher growth rates in the southern agglomerations cannot be interpreted as a backlog of demand. In fact, since 1976, the beginning of

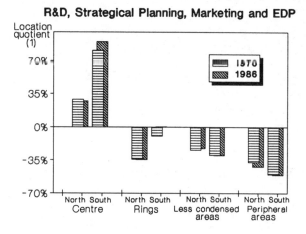

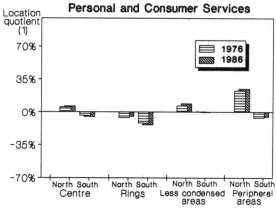

(1) Regional share in selected functions related to regional share of total employment

Figure 3.16 Spatial trends in functional specialization.
Source: Social Security Data.

the period of analysis, the share of these services has always been clearly higher than in the northern regions (Fig. 3.16). Furthermore, the southern centres have intensified their specialization in producer services more than all other regions.

The increase in the share of producer services is caused by two factors: on the one hand, growth in the southern centres has been slightly higher than the federal average. On the other hand, employment in their remaining functions has more or less stagnated (apart from personal services, education and health).

In the northern centres, the share of the selected producer services in total regional employment has risen only below federal average. In addition, it is mainly the second factor that has caused the increase of share: due to large losses in the remaining parts of employment, even below average growth has been sufficient to produce a rather strong shift of employment toward the selected producer services.

Outside the agglomeration, the relations between both factors are reversed. On the one hand, selected producer services in these areas have expanded beyond the federal average. On the other hand, however, the remaining functions have also expanded. Even those functions that have had the largest average federal losses, production (down 3 per cent) and stock-keeping (down 7 per cent), have only stagnated outside the agglomerations. Consequently, in spite of their above-average growth, the share of producer services in total employment did not rise in these regions.

In order to sum up briefly, two general observations may be made concerning the regional expansion of services. First, employment in the FRG is spatially deconcentrating. In particular, the northern centres have lost jobs, while the areas outside the large agglomerations have gained in employment. The expansion of services does not make any exception of these general trends in spatial structure. Most of the various service sectors or functions have had a higher growth rate outside the agglomeration centres than inside them.

Second, a process of regional specialization in certain services is taking place. The specialization is most marked in producer services such as R&D, strategic planning, marketing and EDP. In particular, the southern agglomeration centres such as München, Frankfurt, and Stuttgart have strongly increased their share of these services in total regional employment. However, the specialization in producer services is only partly caused by additional jobs in these services. The other more important reason is the decline of jobs in the remaining sectors and functions.

Some key factors explaining regional differences in deindustrialization in the Federal Republic

There is a multitude of 'hard' and 'soft' factors explaining regional differences and particularities of regional deindustrialization in the Federal Republic, factors that have had and still have an explicit or implicit influence on regional economic development in the different states. Using the states of Baden-Württemberg and Nordrhein-Westfalen to represent the 'South' and the 'North' respectively, these differences are briefly described below:

Baden-Württemberg

The state of Baden-Württemberg in the South (35,000 square kilometers 9.4 million inhabitants in 1988) owes its existence to a popular vote in 1951; it covers the territories of the two former dukedoms of Württemberg and Baden. Both territories and their regional economies survived the Third Reich's centralization relatively untouched. This fact and the area's liberal tradition made it a relatively independent state with its own profile and a powerful conservative political voice in the country. Its present economic strength is based on its former peasant tradition which provided skilled labour (for the early textile industry) based on the Protestant work ethic, on the related entrepreneurial spirit and on the early commitment of its former ducal rulers to general education and training, to universities and polytechnics and to support for science and research. On such foundations a wide variety of innovative and productive small and medium-scale industries has evolved over the last two centuries, turning modern Baden-Württemberg into the most industrialized state of the Federal Republic. Baden-Württemberg, particularly the Stuttgart agglomeration, is the location of some of the most important and powerful enterprises of the whole Republic such as Mercedes-Benz, Bosch, SEL and IBM. Its sound economic base, its image of success, the fact that the natural and physical environment of the state is still very attractive and the common prejudice that the quality of life in that part of the Federal Republic is supposed to be superior to that of the North, have turned Baden-Württemberg into the most prosperous region of the country.

There are several key characteristics to the economic structure in Baden-Württemberg. In 1986 manufacturing in the state contributed 39 per cent to GNP. The federal average is 34 per cent. Investment goods are 57 per cent of total sales (as compared to 41 per cent in the Federal Republic as a whole). Of all world exports, 1.5 per cent originate in Baden-Württemberg. The three most important economic sectors in the state are mechanical engineering, automobile industries and electronical engineering. These employ 49 per cent of the labour force and account for 46 per cent of the country's total sales. The service sector, particularly the production-oriented service sector, is relatively weak and internationally less competitive than the modern industrial sector.

Nordrhein-Westfalen

The state of Nordrhein-Westfalen (34,000 square kilometers, 16.8 million inhabitants in 1988) has a historical record similar to Baden-Württemberg. It was established in 1948 by merging three former Prussian provinces: Rhineland, Westphalia and Lippe-Detmold. Based on rich deposits of natural resources (coal and iron), heavy industry has evolved since the mid-nineteenth century between the Ruhr and Rhine rivers. This area has been Europe's industrial heartland for decades. Coal, energy and steel production dominate, but there are also considerable amounts of chemical industries and mechanical engineering, textile industries and a variety of consumer goods industries producing for the regional market.

In 1961 the industrial sector still accounted for 62.7 per cent of Nordrhein-Westfalen's labour force while only 35.4 per cent were employed in the service sector. Since the early 1970s the industrial sector has lost its extreme dominance and in 1987 only 49.4 per cent of the labour force was still employed in this sector, compared with 49.7 per cent in services. This radical change reflects a loss of 1.5 million jobs in the traditional sector, while only 420,000 new jobs in the service sector had been created. Job losses in the mining industry are the most spectacular. In 1957, 560,000 employees produced 133 million tons of coal per year, in 1989 the 98,000 employees left in the industry produced 65 million tons. Structural crises have in fact affected Nordrhein-Westfalen's heavy industry (except the chemical industry) since the mid-1930s. It has been argued that without the extensive war preparation activities and the enormous post-war reconstruction efforts, the need for structural adjustment would have been felt two decades earlier.[14]

Some of the key factors explaining regional differences in deindustrialization in the Federal Republic are described below.

The dominance of traditional heavy industries

The dependency of a region, its population and its labour force on one industrial sector, and one dominant single source of income, obviously makes it particularly vulnerable. If this sector is a traditional resource-based industrial sector, as it is in the case of the iron, steel and coal industries in Nordrhein-Westfalen, this dependency has considerable impact on regional economic development. As a rule, the vested interests of the dominant industries control the public sector infrastructure, including utilities and facilities, and their needs have shaped forward and backward linkages and private sector services.

The actors involved in the concerted public-private support for the dominating industries, such as politicians, trade unionists, bankers, public sector administrators and their dependent clientele, do not act independently. Their attitudes toward regional innovations usually reflect possibly negative impacts on their own vested interests. In this situation the indigenous regional tradition may become a heavy burden rather than an asset for the future if technological changes occur or international market competition becomes tight. The large coal, steel, and energy complexes (Thyssen, Krupp, Veba, RWE, VEW, Hoesch, Mannesmann) in the Ruhrgebiet dominated industrial development in Nordrhein-Westfalen for decades. Heavily subsidized by the public sector (in 1987 the coal industry alone received subsidies of more than DM19 billion), the large corporations had to be forced to innovate and diversify. In such an environment, modern, future-oriented services could not grow on their own. The local influence of large industrial corporations has usually created a socio-political environment that has been less favourable for small and medium-sized firms outside the traditional industrial sector.

Baden-Württemberg is known as a modern industrial region where innovative, flexible small and medium-sized industries have contributed to the economic wealth of the state. However, the dominance of one powerful industry, the Daimler-Benz Corporation, is continuously growing. Whether it will lead to similar regional dependencies, as has been the case in the Ruhrgebiet, will depend on the state's regional and economic policies.

Recently, joint efforts by the federal and state governments were succesfully undertaken to merge Daimler-Benz with MBB, the powerful München-based aircraft and military equipment enterprise, to create a diversified giant company capable of competing worldwide. One aspect of this 'elephant-marriage' was the intention of engaging Mercedes-Benz in the production and financing of the heavily subsidized European Airbus. Another motive may have been the wish to diversify further the enterprise, which some critics consider to be too vulnerable because of its high dependency on automobile production. Such critics welcomed the results of a recent study that had shocked the regional public with its finding that future rationalization efforts in the monostructured automobile sector in the Stuttgart agglomeration may cause a loss of 20,000 industrial jobs in the near future.[15]

The availability of a qualified labour force

The skill of the regionally available labour force is one essential element of regional economic development. Manpower qualified for future-oriented industries (e.g. innovative electronic engineers and technicians or specialized creative labour for producer-oriented services) is a key to the successful continuous adjustment of industries and services.

Considerable regional differences in the locally available labour force have contributed to the regionally divergent development of industries and services. In traditional industrial regions, such as the Ruhr region or the Saarland, the labour force was traditionally trained to meet the requirements of the dominant coal, iron and metal industries, and related firms and services. Over time, and given the dualistic vocational training system, this has caused the neglect of other, more future-oriented education and training.[16] Tradition, lack of information and mental and physical immobility have been the major reasons why young trainees favour a secure, almost tenured job in the dominating industrial sector rather than enter new, unknown and insecure professions.

A second fact is worth mentioning: until 1965 the Ruhr region did not have a single university. Apart from a teachers' college and a miners' college no further or advanced education or training institutions were located within the region. The only technical university in Nordrhein-Westfalen was that in Aachen. The new industrial regions in the South, however, (München, Stuttgart, Karlsruhe) could draw their local labour force from well-established and respected technical universities. The formal and informal linkages between local industries and the local institutions of higher education guaranteed the practice-oriented exchange of knowledge and technologies. Baden-Württemberg in addition profited from the far-sighted initiatives of F. von Steinbeiss, who in the mid-nineteenth century heavily promoted the development of non-agricultural jobs in the country.

The intensity of producer services

Strongly linked to the available labour force, as both consequence and cause, is the degree of producer services required; this factor is likely to be decisive in regional development. According to the empirical results presented above, there is an obvious correlation between the high level of producer services in the southern regions and their economic performance. In fact, there are good arguments that producer services such as strategic planning, R&D, or electronic data processing are essential preconditions for the innovativeness and competitiveness of firms, in particular in the manufacturing industries. Although it is nearly impossible to give a clear definition of such innovation-relevant activities, their primary feature is

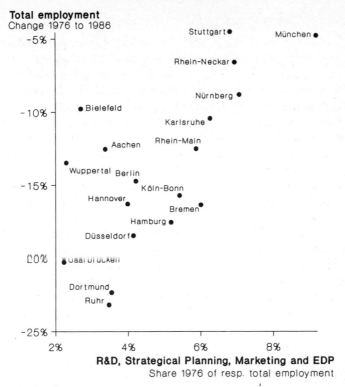

Figure 3.17 Producer services and employment change in the manufacturing industry of agglomeration centres.
Source: Social Security Data.

'looking ahead,' characterized by an 'offensive' search for relevant information. Obviously such search and planning can be done in various ways, depending on the particular kind of market environment a firm may meet. However, all efforts are likely to include research and planning activities in a more or less intensive manner.

In Figure 3.17 the regional intensity of producer services is correlated with the change of employment in the manufacturing industry for each of the 17 agglomeration cores. The southern agglomeration cores, especially Stuttgart and München, have both an above average level of producer services and an above average change of employment. Dortmund, however, shows both a very low level of producer services and a low change of employment. Nevertheless, regional differences in the level of producer services are hardly influenced by the sectoral structure of the regions. Even in industries characterized by a relatively low need for producer services (e.g. clothing or leather), firms are usually more innovative than in the Ruhr or Hannover agglomerations.

The impact of defence spending

The rise of traditional industrial regions in Germany was based on railways and shipbuilding, two major steel-consuming sectors. It was also linked to the defence industries, which have always been technologically innovative. Hardly any other sector has such a high R&D component. The labour force required by the R&D-intensive defence industries is highly skilled and motivated, well-paid and conservative. This sector is totally state dependent and thus heavily subsidized, with no control other than the general auditor's. Its technological spin-offs and R&D by-products are essential innovations for the regional economy. The defence industries and their regional forward and backward linkages profit from defence spending by the federal government. This amounted to more than DM51 billion in 1988. The spatial implications of defence industries have repeatedly been investigated in the United States and the United Kingdom. A study in the Federal Republic has shown that military production has continuously moved south during the past four decades. While Krupp, once one of the most powerful producers of weapons before the Second World War, was located in the Ruhr region, the new modern R&D and production centres of the defence industries are located in the München region and in Baden-Württemberg.[17]

Wage structure and job security

One of the major complaints and excuses of industries in traditional industrial areas and of their economic advisors is high income levels.[18] It is often said that wages in these regions are higher than elsewhere, particularly higher than those in the south. Recent studies, however, have shown that such arguments do not reflect reality.[19] On the contrary, due to the tight labour market in München, for example, the average wage for an engineer in the München region is slightly above that of an engineer in the Ruhr region. However, given the fact that the cost of living in München and Stuttgart is more than 10 per cent higher than in the Ruhr region or in Hamburg, the respective household has a higher income and more room for consumption.

Land blockage and land use control

Powerful large corporations in traditional industrial regions have also had a detrimental impact on the land market by blocking their own or adjacent land for other development.[20] Once fearing that new industries would attract scarce industrial workers, they successfully hindered the allocation of new companies in the 1960s. Contaminated, difficult and costly to recycle, the land reserves have not been available for new industrial parks or urban renewal. The automobile plant installed in Bochum by General Motors in the 1960s is one of the very few examples of a new successful siting of a new enterprise.

Given the overall population density of Nordrhein-Westfalen (492 per square kilometer) and Baden-Württemberg (262 per square kilometer) land use control procedures have become an important factor in controlling local economic development. Comparing these states, there seems to be more motivation to avoid urban sprawl and to limit urban expansion in Nordrhein-Westfalen than in Baden-Württemberg.

The socio-political climate of regional development

The socio-political and economic climate of regional economic development (the 'milieu') is another decisive location factor for both local and foreign investors. Southern Germany, for example, is a traditionally conservative environment, where, for a long period, labour unions played a minor role, and where social experiments and achievements roused less interest than in those areas where traditional class conflicts had evolved decades earlier.

Potential investors and their corporate lobby groups in traditional industrial regions such as the Ruhrgebiet continuously complain about protracted decision-making processes, inflexible bureaucracies and about costly and time-consuming environmental control mechanisms and procedures, although such complaints are often based on prejudice rather than on real day-to-day experience. It is true, however, that the complexity of problems in traditional industrial regions and the sensitivities of the socio-political environment in such regions render decisions much more difficult than elsewhere.

The regional, national and international image of regions and cities

In recent years the general, and particular the cultural image of regions and of cities has become an increasingly important factor in regional economic development and for attracting qualified labour and branch plants from outside the Federal Republic.[21]

The recent 'success-stories' of the restructuring of Pittsburgh and Dortmund have demonstrated that the image of success, too, is very influential in creating a positive and innovative business environment.[22] The media are very influential in creating the international image of a city. The 1972 Olympics in München and other cultural events have given the city an international reputation. Frankfurt's tradition is that of a world financial centre and as a centre of international fairs (e.g. International Bookfair, Achema). It is no coincidence that both cities house the headquarters of an increasing number of large companies.

In contrast, traditional industrial regions have little to offer in terms of social and cultural life. Although their bad image is hardly justified and does not necessarily reflect the preferences of the people in such areas, it is difficult to overcome. Efforts to improve the local image through city and regional marketing and by media campaigns have multiplied in the past decade. The Ruhr region is particularly anxious to build up a new image and a new regional identity.

Regional policy responses to deindustrialization

The complex institutional structure of a federal state like the Federal Republic renders it difficult to describe briefly regional policy responses to deindustrialization. Regional policy is predominantly the proper responsibility of the states or city states. The federal government, however, apart from its sector policy and in accordance with the states, sets some guidelines for regional policies. The 'Gemeinschaftsaufgabe Verbesserung der regionalen Wirtschaftsstruktur' is an important regional policy instrument of the federal and state governments. It supports investment projects for industry and industry-related infrastructure in regions lagging behind the national trend.

Due to the relative independence of the politically strong local governments in the Federal Republic, local economic development depends very much on both the indigenous economic potential, the qualifications of the local labour force and the variety of local conditions, including image, cultural identity, appropriateness for corporate headquarters and other factors.

Although most efforts to cope with the consequences of deindustrialization are linked to or depend on federal and state initiatives, and, increasingly, European policies and strategies, they differ from state to state. To demonstrate the similarities and the differences of state policies, two examples are briefly described: Baden-Württemberg as a state that profits from the positive economic effects of its modern and innovative industrial base, and Nordrhein-Westfalen, as a state suffering from considerable industrial decline.

Baden-Württemberg

Despite its liberal tradition, regional economic development has been given substantial support by the state government of Baden-Württemberg during the last two decades. The financial support that enterprises, firms and potential investors are receiving from the state is based on clear political guidelines of the state government. They are designed to avoid favouring particular economic sectors, to avoid granting continuous subsidies that merely keep declining industries alive and to avoid supporting projects and measures that may have detrimental environmental impacts and negative effects on water resources and urban amenities. They are to promote only such projects that are to be implemented within the state. These policies effectively favour small and medium-sized firms.[23]

State support for small and medium-sized firms has included efforts to enhance vocational training, to encourage counselling, training and co-operation, to promote the establishment of new firms, to open international markets to small and medium-sized firms and to improve their access to information and research. In particular, technological innovations have been promoted, as have measures to reduce environmental pollution. Local priorities for support were synchronized with the 'Gemeinschaftsaufgabe Verbesserung der regionalen Wirtschaftsstruktur' and favoured in comprehensive physical regional planning.

The spirit, aims and dimensions of the state's future-oriented structural and technological policies may be further illustrated by the following three examples.

The city of Ulm, a medium-sized city with more than 150,000 inhabitants, halfway between Stuttgart and München, is being developed by the present state government into the country's technopolis. Based on its strong defence-related electronics industry (AEG) and on a tradition in medical research and industry, the city is to become a modern centre of industrial R&D. The state activities are based on the fashionable concept of public–private partnership in which state government policy has assumed a centralistic top–down approach. The private sector, in turn, is supposed to assist in the implementation of public efforts and government strategies. Investments of DM750 million are being planned that will establish 2,500 jobs in a science park with numerous future-oriented institutes.[24]

Two other cities of the state are deliberately being developed into regional centres of R&D with an international profile. The city of Karlsruhe, the former capital of Baden, with more than 250,000 inhabitants in 1988, is one of the Federal Republic's foremost centres of R&D. No other West German city has a higher percentage of R&D-related jobs. A strong technical (state) university and various public research institutes profit from considerable contracts from both private and public sectors, including defence-related research of major significance. Heidelberg and its university are also systematically being turned into an international centre of biogenetic research and development, in which public and private sectors are joining forces to achieve international stature.

Being aware of the importance of favourable cultural environments for a qualified labour force, and of the role played by the arts in regional development, the state government is supporting the development and expansion of the region's cultural image. One of the flagship projects is the projected 'Zentrum für Kultur und Medientechnologien' in Karlsruhe, a high-tech centre for the arts. Increasing public investments in public cultural infrastructure and high culture events is an element in the state's regional policy, which combines the interests of both economic and cultural sectors. The political philosophy underlying regional economic development in Baden-Württemberg has been widely disseminated by its prime minister, and this has contributed to the good image of the state's economic policy.[25]

Nordrhein-Westfalen

It was the famous Marshall Plan that supported the economic reconstruction and renaissance of regional industry in Nordrhein-Westfalen in the first decade of its existence. In the following decades, the state government intervened in the process of structural adjustment by means of comprehensive structural plans and investment programmes. This government was reacting to economic crises and to political pressure from below, particularly from local governments and powerful steel workers' and miners' labour unions. Such investment programmes were the 'Entwicklungsprogramm Ruhr' in 1968, the 'Nordrhein-Westfalen-Programm' in 1975, and the 'Aktionsprogramm Ruhr' in 1979, paralleled by legally based regional development plans (Landesentwicklungspläne I bis VI).

As a rule, the investment programmes combined measures designed to renovate the monostructured and extremely immobile industrial complex with measures cushioning the most serious social and environmental repercussions of the structural adjustment process. It has been argued that these programmes have hindered or slowed the necessary structural adjustment process in the region. Others, however, say that they have been quite successful in introducing innovative elements into traditional regional policy. The most obviously rewarding long-term regional investment, initiated by the first state programme, was the establishment of a set of new universities in the state, doubling the number of existing institutions of higher education and tripling the number of students between 1960 and 1980.

A more recent initiative of the state government was the introduction of the so called 'Zukunftsinitiative Montanregionen' (ZIM, 'Future initiative coal and steel regions'), followed by the 'Zukunftsinitiative Nordrhein-Westfalen' (ZIN). In contrast to earlier programmes these initiatives did not formulate projects 'from above', but aimed at mobilizing local potential, creativity, and commitment. The only guideline for projects designed at the local level, to be submitted 'from below' for financial support 'from above', were five 'action areas' of structural importance: measures to enhance innovation and new technologies; measures to improve the qualifications of the labour force; measures to create new jobs and protect existing ones; the modernization of the infrastructure; and measures to improve environmental conditions and to conserve energy.

One prerequisite criterion for submitting projects turned out to be extremely useful: the requirement to gain broad consensus among local interest groups. This criterion forced local groups to seek contact with others, communicate their ideas, and compromise in joint proposals. It also contributed considerably to the accelerated implementation of innovative projects.

Other activities of the state government aimed at renovating the economic and infrastructural base of the country include the usual multiple repertoire of measures, such as supporting technology transfer between universities and regional industries, promoting small and medium-sized firms trying to enter the international market and the establishment of new research institutions and the extension of existing ones. They are paralleled by supportive sectoral policies (for culture, transport, urbanization and the environment) to improve the quality of life.

A last point is worth mentioning: Nordrhein-Westfalen, particularly the Ruhr region, suffers nationally and internationally from a negative regional image. Public relations campaigns to alter this image have been run. Recently the chief executive officers of large corporations in the Federal Republic undertook an initiative, the 'Initiativkreis Ruhrgebiet', to promote the economic renaissance of the industrial heartland of the country. In addition, new efforts are being undertaken to change its environmental and physical appearance. Under the label 'Internationale Bauausstellung Emscherzone', a new approach toward urban transformation has been programmed for the next ten years.

Outlook: the European dimension

The deindustrialization processes described in this chapter, and the different regional policy responses to the decline in industrial employment in the Federal Republic must also respond to the European dimension of regional economic development. The first studies assessing the regional implications for the country of the European Community's single European market have already been undertaken. They all come to the same basic conclusions: the transition from a national to a European market will test the regions' capabilities for structural adjustment and international competitiveness. Regions that have been competitive in the past, which are export-oriented and which are less protected by regulations and state subsidies should profit from open boundaries (e.g. Baden-Württemberg). Regions such as the Ruhr will have to undergo serious additional structural adjustment before the positive effects of the single European market can be felt. It is possible that south–north disparities in the Federal Republic will widen. Whether more pronounced regional dualism will increase internal political tensions will depend on the overall effects of the single European market. Its long-term positive employment effects may offset potential political conflicts.

The possible regional and economic impact of the considerable political changes in Eastern Europe are more difficult to predict. There is some reason to believe that this new situation may slow down the process of restructuring in traditional industrial regions, while the overall economic effects of a larger, more interconnected Europe are undoubtedly positive for the German economy as a whole.

4
Spatial impacts of deindustrialization in France

JEAN-PAUL de GAUDEMAR & RÉMY PRUD'HOMME

Introduction

OVER THE PAST two decades, France, like most other developed countries, has undergone rapid structural changes that have had a major impact on the spatial distribution of people and activities. Three background conditions are significant. First, the population, which was 50.7 million in 1970, reached about 57 million in 1989, a total population slightly smaller than those of the Federal Republic of Germany, Italy and the United Kingdom, and a growth rate significantly higher than those registered in these countries. Second, because France is nearly twice as large as each of these countries, the population density is much lower, and indeed much lower than in most other European countries. Third, over the 1970–87 period, French GDP increased at a rate of 2.7 per cent per year, or a 50 per cent increase of GDP for the entire period.

One of the major, or at least most apparent features of this rapid structural change has been a shift from secondary to tertiary activities. This chapter will investigate this shift and its regional components and implications.

A few words about the meaning of 'regions' in France are appropriate. France is a unitary country that for the last two centuries has been divided into about 100 *départements* (the equivalent of counties). Over the last 20 years, regions, defined as groupings of *départements*, were instituted, first as administrative and planning units, then as political units. They are political units in the sense that they are managed by elected councils that raise taxes and make expenditures. There are 22 such regions, as shown on Figure 4.1. They are often referred to as European Community level II regions. They differ substantially in area and population ranging from 240,000 inhabitants in Corse and 740,000 in Limousin to 5 million in Rhône-Alpes and 10 million in Ile-de-France, the Paris region. For the purpose of analysis, it will be useful to consider 'large regions', by aggregations of European Community level II regions.

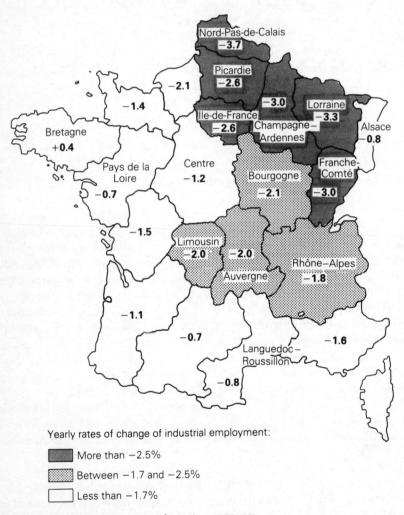

Figure 4.1 Deindustrialization rates by region, 1976–86.

The chapter begins with an overview of the scope and the causes of deindustrialization for France as a whole, continues with an analysis of the regional dimensions of the phenomenon, then examines what happened in three different regions, or groupings of regions: the Paris region, the Mediterranean and the Northeast. It will also study the case of the automobile industry.

Deindustrialization in France

There are at least four ways to assess a possible deindustrialization trend. One is to look at industrial production, measured in value added at constant prices. Another is to look at industrial employment. A third is to look at relative production, i.e. at the share of industrial value added in GDP. The last is to look at relative employment, i.e. the share of industrial employment in total employment. They tell rather different stories in the case of France.

Four measures of deindustrialization

Absolute industrial production did not decrease in France over the last 20 years. As a matter of fact, it increased very rapidly until 1979, and stagnated thereafter. The quantity of industrial goods produced in 1987 was more than double that in 1967 or 1970. In that sense, one could say that there has not been any deindustrialization in France.

The pattern is different for absolute industrial employment. Industrial employment, which had been increasing for decades, and was still increasing rapidly in the late 1960s and early 1970s, reached a peak in 1974, and declined thereafter. The decline was particularly rapid in the early 1980s. In total, between 1974 and 1987, industry recorded a net loss of 1.4 million jobs, or 26 per cent of the 1974 number.

Relative to the entire economy, the share of industry was about constant until 1974, then declined, both in terms of production and in terms of employment. In 1970, industry accounted for about 24 per cent of employment and also of value added; in 1987, it accounted for 18 per cent

Table 4.1 Four concepts of industrialization, 1967–87

	Value added	Employment
Absolute values	Increase until 1979, then stagnation	Increase until 1974, then decrease
Relative values	Stagnation until 1974, then decrease	Stagnation until 1974, then decrease

of employment and 19 per cent of value added. These trends are summarized in Table 4.1. These patterns can be explained by three main kinds of changes: changes in demand, changes in production processes and changes in prices.

Changes in demand

Changes in the structure and volume of demand for goods and services can explain part of the industrial production profile. Industrial goods are used either by households for final consumption, by enterprises for intermediate consumption, or by enterprises (and to a lesser extent households) as investments. In the final demand of households, an income elasticity of demand for industrial goods lower than one combined with low income growth, particularly after the second oil price shock, account for stagnation in the demand for industrial goods. This trend was exacerbated by the decline in investments, particularly in the 1980s. Not enough is known about trends in intermediate consumption by enterprises. However, available evidence suggests that the demand for raw materials or primary inputs per unit of finished products declined in many sectors. This is well established in the case of steel, for instance: less and less steel is needed to produce a given product, such as a car. The same is true for glass: less and less glass is needed to produce a bottle. In our societies, what is known as 'materials productivity', i.e. the ratio of finished products to materials utilized has increased constantly, and increased particularly rapidly after the oil price shocks. It is one of the main forms taken by technological progress. It clearly decreases the demand for industrial goods.

Did increased international trade also contribute to decreasing the demand for industrial goods, as has been argued for the United States? During the past two decades, the share of exports, and of imports, in GDP, did increase significantly, from 16 per cent to 22 per cent. The latter figure, by the way, is nearly twice as high as the comparable figure for Japan, and three times as high as that for the United States. More detailed figures are needed to provide an answer.

Table 4.2 details the changes that occurred, over the 1970–86 period, in domestic demand, foreign trade, and output, for both industrial goods and services. (The case of agricultural goods, which are relatively unimportant, is not reported here.)

Table 4.2 shows why and how demand for services increased faster than demand for industrial goods (76 per cent as against 43 per cent) over that period. First, household consumption increased at a higher rate for services than for goods. Then, public consumption of industrial goods (which in national accounting is defined as a consumption of services;

Table 4.2 Changes in final demand, foreign trade and output: industrial goods and services, 1970–86

	Industrial goods		Services	
	Absolute (billions of 1980 francs)	Relative (percentage)	Absolute (billions of 1980 francs)	Relative (percentage)
Domestic demand, changes in:				
Household consumption	+413	+60	+336	+85
Public consumption	—	—	+241	+65
Investment	+133	+27	+3	+26
Inventory	−6	−21	—	—
Total, final demand	+540	+45	+580	+73
Foreign trade, changes in:				
Exports	+297	+131	+68	+106
Imports	+352	+116	+39	+90
Net foreign trade	(+55)	(+71)	+29	+138
Output, changes in	+484	+43	+610	+76

Source: Calculated from INSEE. 1988. *Rapport sur les comptes de la nation de l'année 1987*. Paris: INSEE. pp. 30, 94, 99, 103, 112.
Note: (+55) means that the net foreign trade for industrial goods was negative in 1970 and in 1986, and that this deficit increased by F55 billion.

goods consumed by the public sector are considered as intermediate consumption) also increased at a rate higher than private consumption, although not much higher, and not higher than GDP. More importantly, investment, which consists primarily of industrial goods and accounts for an important share (about 40 per cent) of the final demand for industrial goods, increased at a low rate over the period.

Net foreign trade played a negative role in the demand for industrial goods. Even though exports of industrial goods increased somewhat more rapidly than imports of them, because the net balance was negative at the beginning of the period, and because the growth of both imports and exports was very rapid, the magnitude of the negative net balance increased and therefore reduced the demand for domestically produced industrial goods. This contrasts with the role played by foreign trade in the case of services; the net balance was positive in 1970, exports increased faster than imports, and, as a consequence the net balance increased and foreign trade increased the demand for services in France.

Changes in prices

It has been suggested that changes in relative prices could explain a decline in industrial production in volume, i.e. in value added measured in constant industrial prices. This would be the case if average prices were increasing faster than industrial prices. Industrial output in current prices corrected with an implicit GDP deflator – an equally legitimate indicator of industrial production – would increase faster than (or would not decline as fast as) industrial output in current prices corrected with an industrial price index. This did not happen in France over the period considered. Surprising as it might seem, industrial prices increased at about the same rate as average prices, and, as a consequence, industrial value added in constant average prices or in constant industrial prices followed the same pattern.

Changes in production processes

More important are the changes that took place in industrial production processes. One is the vertical deconcentration movement. More and more, industrial enterprises tend to contract out a number of functions – such as maintenance, packaging, or accounting – that used to be performed in-house. In addition, industrial enterprises increasingly prefer to utilize temporary workers provided by specialized enterprises rather than hire permanent staff. In all these cases, tasks that were performed by an industrial enterprise, and considered as industrial production done by industrial workers, will now be considered as tertiary production done by tertiary workers. This leads to a shift of both value added and jobs from manufacturing to services that is largely a statistical artifact or illusion. Measured deindustrialization is greater than effective deindustrialization.

There are, unfortunately, no estimates of the magnitude of this statistical effect. A recent study for Saint-Gobain, a large manufacturing company, indicates that between 1980 and 1987 the share of purchases of services in total production costs increased from 27 per cent to 32 per cent, an indication of the increased importance of contracting out.

We have already referred to progress in 'materials productivity', that has decreased the demand for industrial goods, and therefore industrial output and employment. More significant and more obvious, but quite different, have been advances in labour productivity. They have been very important. In the course of the past 20 years, productivity doubled. It increased in every year (except for the year 1975, when industrial output first declined), even in the later part of the period, when industrial production no longer increased. Increasing labour productivity is the main single factor explaining deindustrialization in terms of employment. It can be argued that this increase was caused by the growing involvement of France in international trade, in the sense that productivity improvement was a condition of survival in an open economy. This is only true in part, because cutting jobs is only one way of cutting costs, which in turn is only one way of competing on international markets.

A given increase in value added (v) can be associated with different sets of increases in labour productivity (p) and in employment (l):

$$v = l + p$$

A 4 per cent increase in value added can either mean a 3 per cent increase in productivity plus a 1 per cent increase in employment or a 5 per cent increase in productivity combined with a 1 per cent decrease in employ-

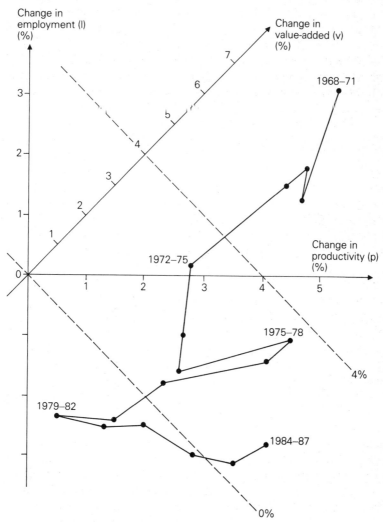

Figure 4.2 Changes in value added, employment and productivity in French industry, 1968–87 (three-year periods).

ment. As we shall see, for a given period, different combinations may be found in different regions. And, for France as a whole, different combinations can be found for different periods. Figure 4.2 indicates the values of v, p, and l for each of a series of three-year periods, beginning with 1968–71 and concluding with 1984–7. Until 1979–82, industrial value-added growth rates declined constantly, from about 8 per cent per year to −1 per cent. At the same time, productivity and employment growth rates also declined, with productivity rates being always two or three points higher than employment growth rates; as a result, productivity rates declined from about 5 per cent to 0 per cent and employment rates from 3 per cent to −2 per cent. After 1979–82, in the 1980s, a rather different pattern occurred. A modest increase in value added, from about −1 per cent to +1 per cent, was accompanied by no increase in the employment growth rate, which continued to deteriorate – employment continued to fall at a rate of −2 per cent or −3 per cent – and a significant increase in productivity growth rates, from a low 1 per cent to more than 4 per cent.

Deindustrialization in the European context

It is interesting to contrast deindustrialization in France with that in Europe as a whole, considering that trade barriers between European Community countries have virtually been eliminated for industrial products. The relative position of France changed over time. In the early 1970s, industrial value added increased faster in France than in Europe as a whole. The reverse was true for the 1980s. In terms of employment,

Table 4.3 Value-added and employment growth[1] in industry: France and Europe, selected periods

	1970–75	1975–80	1980–86
Value added			
Europe	60.2	45.0	30.2
France	88.0	45.0	24.6
Structural component[2]	59.6	45.3	30.0
National component[3]	28.4	1.1	−5.2
Employment			
Europe	−5.8	−2.4	−14.4
France	2.9	−6.7	−14.1
Structural component[2]	−5.7	−2.0	−13.8
National component[3]	8.6	−4.7	0.3

Source: Calculated from EUROSTAT. National accounts. 1983 and 1988.
Notes: 1) Growth percentages relate to entire five- or six-year periods. 2) The structural component is the rate at which industry would have grown if each sector had grown in France at the rate at which it grew for Europe as a whole. 3) The national component is the difference between the effective growth rate and its structural component.

France did better than Europe in the 1970s, worse than Europe in the 1975–80 period, and as poorly as Europe in the 1980s.

Can this performance be attributed to the sectoral distribution of industry in France at the beginning of each period? The results of a shift–share analysis that attempts to answer this question appear in Table 4.3. This suggests that in the early 1970s industry in France was doing better than industry in the rest of Europe, for both value added and employment, but that the picture changed after 1975. For value added, the national component became negligible in 1975–80, and negative in 1980–6. For employment, it was strongly negative in the late 1970s, and became about neutral in the following period.

Regional deindustrialization patterns

Where did deindustrialization occur in France, and how can differential deindustrialization patterns be explained?[1] Among the many indicators that can be considered for many periods, let us choose the most sensitive, industrial employment, for the 1976–86 period. During this period, industrial employment declined at an average yearly rate of 2 per cent for France as a whole. The yearly growth rates for each region are given in Figure 4.1.

Three kinds of (nearly contiguous) regions appear. The Northeast, with the exception of Alsace (a small region close to Germany and Switzerland), but including the Paris region (Ile-de-France), exhibits a higher than average industrial decline. The eastern central parts of the country (Rhône-Alpes, Bourgogne, Auvergne and Limousin) have lost industrial employment at rates close to the national average. Finally, in the South and in the West deindustrialization proceeded at rates significantly lower than in the rest of France; in one region, Bretagne (the most western region), industrial employment even increased slightly over the period. These differential growth, or rather decline, rates can be related to a number of variables or explanatory factors.

The role of industrial structures

Deindustrialization has been particularly important in the more industrialized regions. This is not only true in absolute terms (which is tautological) but also in relative terms. There is a reasonably good correlation between the degree of industrialization (ratio of industrial employment to total employment) at the beginning of the period, and the rate of deindustrialization (rate of decline of industrial employment) during the period. Since the less industrialized regions were also the regions in which industrial growth had been strongest in the 1960s and the early 1970s, this suggests that newly established industrial enterprises resisted the deindustrialization trend better than older enterprises.

Can this phenomenon be explained by differences in the sectoral structure of the different regions? Shift–share analysis does not support this hypothesis. It is true that some of the rapidly deindustrializing regions, such as Lorraine (in the Northeast) were strong in mining, steel, textiles and other rapidly declining sectors, and that they exhibit a much below average 'structural effect'. However, regions such as Alsace or Languedoc–Roussillon did relatively well (in terms of industrial employment) in spite of an unfavourable initial industrial structure. Conversely in other regions, such as Ile-de-France, industrial employment declined rapidly in spite of better than average structural effects. The industrial mix appears to be a rather poor predictor of deindustrialization.

It follows that regional 'specificities' account for the bulk of the relative performance of regions in terms of industrial employment. The map of the 'regional (or residual) effects' of the shift–share analysis resembles the map of deindustrialization discussed above. All the rapidly deindustrializing regions of the Northeast, as well as the Paris region, have negative regional effects. Most of the slowly deindustrializing regions of the South and the West have positive regional effects.

We have no simple explanation of this regional 'dynamism' to offer. It is not related to the distribution of enterprises by size. Some of the regions in which the share of big enterprises is largest displayed dynamism, while others did not. The most concentrated regions, in terms of the share of the 15 largest enterprises in regional value added (Franche-Comté and Auvergne) did just as poorly as the least concentrated regions (Ile-de-France, Picardie, Champagne–Ardennes). Dynamism is equally unrelated to the degree of specialization. Some regions that are highly specialized did well – because they were specialized in growing industries, but others did very poorly – because they were specialized in declining industries.

One indicator is positively correlated with the regional effect: the degree of dependence of the region, defined as the share of industrial employment in enterprises controlled from outside the region. The more 'dependent' regions turn out to be the most dynamic. This is a surprising finding, at variance with what is often said about the role and importance of endogenous development.

The role of services

In France, as elsewhere, tertiarization is the other side of the deindustrialization coin. During the 1977–86 period, 1.6 million jobs were created in the service sector, while 1.4 million disappeared in industry (and

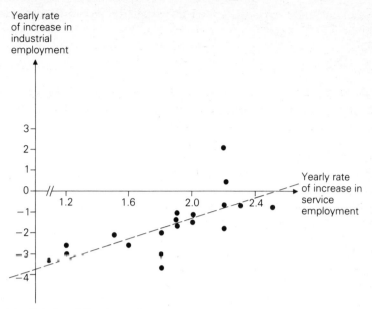

Figure 4.3 Deindustrialization and tertiarization, 22 regions, 1976–86.

nearly 0.5 million in agriculture, thus leaving a total job deficit of 0.3 million). This raises the question of the relationships between service employment (or output) and industrial employment (or output). Are service activities a substitute or a complement to industrial activities? If they are a complement, are service activities led by industrial activities (as is implied in the classical base theory), or is it the other way around? The analysis of regional data can help throw some light on these important issues.

Deindustrialization has not been associated with tertiarization. As Figure 4.3 shows, there is a rather good correlation between changes in employment in industry and changes in employment in services. Industry did better in the regions where services did best. This brings support to the complement thesis, as opposed to the substitution thesis.

However, it does not tell us anything about the nature of the complementarity. For certain services, such as commerce, transportation and telecommunications, the base theory must hold; a decline in industrial employment leads to a decline (or a lower relative growth) in these services. For other services, such as insurance, finance, or services to industry, it can be assumed that the causal relationship is different, and that it is the availability (and the efficiency) of these services that contributes to industrial development. A third alternative, probably the most likely, is that the elusive 'regional dynamism' factor demonstrated in the case of industry explains the relative performance of both industry and services.

The role of national policies

In the entire post-war period, the role of governmental policies has been important in France. An important part of the economy (in banking, insurance, power, coal, rail and air transportation and automobile production) was state-owned, and a large number of industrial enterprises was added to the list in 1982. A national economic plan, providing a common framework for day-to-day business and administrative decisions, has been prepared every five years, a result of a co-operative effort between the various parts of French society. Regional and spatial concerns were part of this policy environment. In the early 1960s, a specific administration, DATAR, was created to help define and carry out regional policy. The policy objectives were rather simple and the policy instruments fairly classic.

The main policy objectives, as they were defined in the 1960s, included fostering development, which meant the industrialization of the South and the West, and slowing down the growth of the Paris region. As deindustrialization quickly appeared to be a problem in some parts of the Northeast, leading to painful unemployment levels in some cities, a third objective was added: to contribute to the restructuring of deindustrializing areas.

The policy instruments utilized were the following. First, some contraints were placed on industrial and service development in the Paris region. A special tax on office floor construction was created, and special authorization was required for large-scale investments, whether by new enterprises or extensions of existing ones. Second, efforts were made to foster the development of infrastructure (in transportation, telecommunications and education) in the targeted regions. This is difficult to achieve, because investment decisions are basically taken by technical ministries or agencies whose primary concerns are not regional policies. DATAR, through a skillful use of political influence and financial contributions (it controlled a fund for regional policies), managed to influence the location of infrastructure investments. Third, there were subsidies for enterprises locating or expanding in designated areas. The exact delimitation of these areas changed over time, but it always included the entire western part of the country and most of the South, as well as limited areas in the Northeast, as represented in Figure 4.4.

It is difficult to assess the impact of these regional policies upon deindustrialization patterns. A comparison of Figures 4.1 and 4.4 is a first step in that direction. It suggests that, to a certain extent, stated regional policy objectives have been achieved. Two of the main goals materialized, namely, the slowing down of the Paris region and the industrialization of the West and the South. With a few exceptions (Alsace and Centre), the regions that did best in terms of industrial employment, i.e. lost fewer industrial jobs than others, are regions that were targeted for development, particularly industrial development, by policy. However, the third goal of regional policy, facilitating industrial restructuring in the Northeast, was not equally successful. Deindustrialization was severe in Lorraine, a region fully designated for subsidies. In Nord–Pas-de-Calais, a region partially designated for subsidies, the situation was just as severe as in regions that were not aided at all.

This partial correlation between policy and achievements cannot be considered causal. Did the aided regions that did relatively better than others succeed as a result of aid? Several studies have attempted to

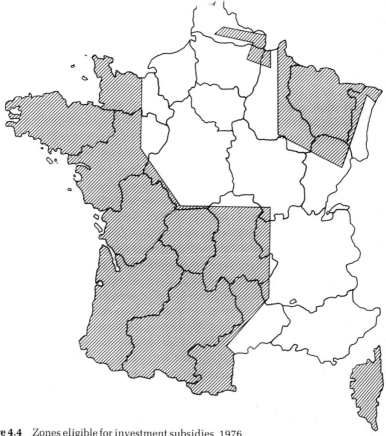

Figure 4.4 Zones eligible for investment subsidies, 1976.

answer that question, particularly in relation to investment subsidies.[2] They rarely reach the same conclusions. They agree that respondents to surveys on 'the major reasons for locating here' never cite subsidies as an important consideration. One econometric study[3] conducted on the 21 regions, using changes in industrial employment over the 1975–82 period as the dependent variable, and many factors, including the amount of regional development subsidies awarded, as explanatory variables, fails to show any impact of the subsidy. Another econometric study,[4] on 18 *départements*, reaches the opposite conclusion, and shows that regional development subsidies are a solid explanatory variable of the residual component of a shift–share analysis.

The truth is probably in between these contrasted findings. The number of jobs aided (jobs promised to be created in enterprises benefiting from such subsidies) was around 40,000 per year. This is about one-third of the number of industrial jobs that were disappearing yearly. This is not enough to be a major countervailing factor, but it is enough to be influential. Without investment subsidies, deindustrialization patterns would have been different, with an even more pronounced decline in Nord–Pas-de-Calais and in Lorraine, and perhaps with a less satisfactory performance in the West and in the South.

It is even more difficult to assess the impact of the other policy instruments. The constraints imposed upon development of Ile-de-France are probably associated with the poor performance of this region. There are no good indices of change in infrastructure availability by region, and *a fortiori* no study of the impact of such changes on deindustrialization. However, it can be speculated that DATAR did influence public investment patterns in favour of targeted regions, and there is every reason to believe that better infrastructure favoured industrial investments and development.

In addition, there is also a sort of involuntary instrument or policy that probably helped shape deindustrialization patterns: the tax and expenditure policy of the central government. The tax system, and to a large extent the expenditure system, are not designed with a good understanding of their spatial consequences. Nevertheless, they affect regions differently. A study conducted on the regional impacts of the French budget[5] shows that regions gaining at the budget game are precisely those that are eligible for investment subsidies, as shown on Figure 4.4, and therefore also the regions that did best in terms of deindustrialization.

The role of regional strategies

Regions as political and financial entities are rather weak in France (they did not even formally exist as such until 1982) and cannot be said to

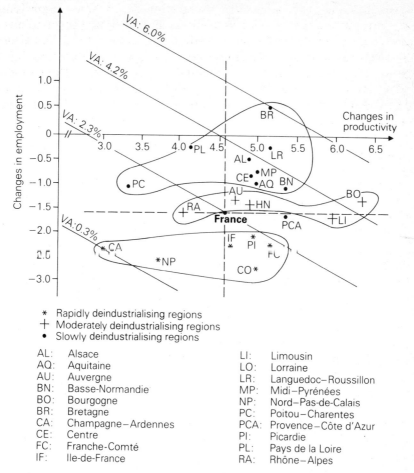

Figure 4.5 Changes in industrial value added, employment and productivity, 1975–84.

conduct policies or to devise strategies. Nevertheless, regions as geographic units exhibit different behaviours, which for our purposes here we shall consider 'strategies', particularly relative to the allocation of value-added changes to employment and productivity changes. Figure 4.5 shows, for industry in each region, changes in these three related indicators for the 1975–84 period. The three types of regions identified above on the basis of Figure 4.1 are represented differently. Not only do regions differ in terms of value added, which increased for all regions, but they differ in their 'usage' of value-added gains, so to speak.

In every region (except for one) productivity gains were higher than value-added gains, thus leading to employment losses. However, there is no simple relationship between value-added gains and productivity achievements. In some regions, such as Franche-Comté, Lorraine, or Picardie, that did relatively poorly in terms of value added, productivity gains were nevertheless high, around 5 per cent per year, as high or higher than the average for France. The dispersion of productivity gains is much smaller than the dispersion of value-added gains. This would tend to suggest that competitive pressures, whether national or international,

Figure 4.6 Large regions selected for further analysis.

were very strong, and that enterprises had to improve productivity, even at the cost of laying off workers. Employment, or rather unemployment, was an adjustment variable. In many cases, of course, this adjustment process took the form of the disappearance of less-productive enterprises.

Yet there are also regions that exhibit, for a given value-added growth rate, rather different trade-offs between productivity gains and employment losses. Consider for instance Limousin and Pays de la Loire, two regions in which industrial value added increased at a relatively high rate of 3.5 per cent per year. In Limousin, this increase was accompanied by a high productivity rate of about 6 per cent, and consequently by a sharp employment decline. In Pays de la Loire, this increase was accompanied by a productivity growth of the same magnitude, leading to a very low employment decline.

A tale of three regions

To explore some of the issues discussed above, three regions, or rather groupings of regions, were selected for further analysis: the Northeast, the South and Ile-de-France.[6] They are shown on Figure 4.6. This grouping of contiguous regions is not arbitrary, but puts together regions that have (as has been seen) much in common. The three groups have about the same importance in terms of population, about 10 million people each. However, they have many contrasting features, some of which are indicated in Table 4.4.

Table 4.4 Selected economic indicators: Northeast, South, Ile-de-France, 1976–86

	Northeast	South	Ile-de-France
Population 1975 (millions)	9.3	10.3	9.8
Population 1986 (millions)	10.5	11.4	10.2
Population change (percentage)	+12	+11	+4
GDP per capita 1986 (EC average = 100)	96	98	163
Industrial employment 1976 (thousands)	1,430	682	1,227
Industrial employment 1986 (thousands)	1,024	610	946
Industrial employment change (thousands)	−416	−72	−281
Industrial employment change (percentage)	−29	−10	−23
Service employment 1976 (thousands)	1,580	1,881	2,778
Service employment 1986 (thousands)	1,834	2,299	3,115
Service employment change (thousands)	+254	+418	+337
Service employment change (percentage)	+16	+22	+12
Service employment: industrial employment 1976	1.1	2.8	2.3
Service employment: industrial employment 1986	1.8	3.8	3.3

Sources: EUROSTAT; INSEE. *Statistiques et indicateurs des régions françaises 1988*.
Note: Employment figures are for wage earners only, excluding self-employed.

The Northeast: uncompensated deindustrialization

The group of five regions defined collectively here as the Northeast, with a population of about 10.5 million people and 16 per cent of French GDP, offers a classic case of deindustrialization. In the first half of the century, the Northeast was one of the most industrialized and prosperous parts of France.

In the early 1970s, the Northeast could be characterized as follows. First, the economy was dominated by traditional industries, such as textiles, mining (coal mining in Nord–Pas-de-Calais and iron ore mining in Lorraine), steel, machinery and automobiles, and basic chemicals. Second, much of the employment and activity was concentrated in large, oligopolistic firms, often controlled from outside the region. Third, levels of skill and education were low, much lower than for France as a whole. In 1982, for instance, 7.8 per cent of people aged 15 or more in France had a graduate or post-graduate degree, the figure for the Northeast was 5.9 per cent, about 25 per cent lower.[7] Fourth, there were no major urban centres in the Northeast. The region appears to be highly urbanized, because urban areas are defined as agglomerations of more than 2,000 inhabitants. However, this is a misleading statistical illusion; in reality, many cities in the Northeast were company towns, rather than large labour markets and service centres.

The shock of deindustrialization was particularly brutal for the Northeast. As indicated in Table 4.4, more than 400,000 industrial jobs, representing nearly 30 per cent of the initial industrial employment, disappeared beteen 1976 and 1986. All types of industries, all sorts of enterprises and all regions of the Northeast suffered. This was true of intermediate industries (steel, building materials, smelting) and of equipment industries (automobiles) and more so of consumption goods industries (textiles, garments, wood, furniture). Large steel, textile, or automobile enterprises had to lay off people in great numbers. So did small enterprises, in textiles, mechanical equipment and furniture, either because their fate was linked to that of the large firms (although subcontracting was not very widespread) or more generally because their markets, and their shares in these markets, were shrinking.

This industrial collapse – the phrase is not too strong – was not offset by a boom in the service sector. Employment in the service sector did increase in the Northeast at a rate close to that of the national average (16 per cent as opposed to 18 per cent). However, because the initial number of jobs in this sector was small, the total absolute number of jobs created in the service sector remained modest: about 250,000 between 1976 and 1986. Half the increase took place in what is called 'non-commercial services' (government, education, health), that is in services that are financed out of taxes, through local and, more importantly, national budgets. Employment in trade, both for food and other goods, increased more slowly in the Northeast than it did for France as a whole; for non-food trade it even declined (−1 per cent, as opposed to +6 per cent for France).

This seems to suggest that in the Northeast services played an ancillary role, following the industrial decline. They did not lead an industrial revival. The economy and the society of the region were industry-based. When industry failed, or more precisely when the existing industries failed, there were not enough resources – intellectual, entrepreneurial, financial, institutional – to develop successfully many new activities, whether industrial or tertiary. Deindustrialization was not offset by tertiarization.

As a consequence, total employment declined sharply. There was net outmigration for all the regions of the Northeast (except for one, Picardie).[8] However, there was also a higher than average rate of natural increase. Far from declining, total population even increased, nearly as fast as the national average. Fewer jobs for more people has meant lower activity rates and increased unemployment. Unemployment levels in the Northeast are therefore higher than for the country as a whole, and they are particularly high for long-duration unemployment.

National and local policy makers were of course fully aware of these developments. What has been done and what is envisaged in terms of policy? A convenient way to find out is to look at the so called 'Plan contracts' that were negotiated and signed between each region and the central government (and prepared by DATAR). In these contracts, each region outlines its economic and social priorities, and lists the investments and actions it intends to undertake over the next five years. Both the regional and central government are committed to the financing, or co-financing, of some of these investments. Four lines of action have been devised.

The first has to do with the large enterprises that have been at the same time the strength (because they created activity, employment, value added and income) and the weakness (because they prevented the development of small and medium enterprises and suffered from the deindustrialization process) of the Northeast. These large enterprises have a role to play. The trying period of retrenchment and restructuring seems to be over. They benefited from various national policies, for finance (the steel industry was nationalized in the process) and for trade (automobile or textile industries were, to some extent, protected from Asian competition). They improved productivity, developed new products and are now making profits. They may not in the future significantly increase their contribution to the development of the Northeast, because most of them operate in relatively low-growth sectors, but they should no longer be a problem for the region.

Second, efforts were made to help small and medium-sized enterprises (SMEs). As mentioned earlier, many but not all parts of the Northeast were eligible for investment subsidies, and large amounts of subsidies were granted. For 1985 and 1986, the Northeast received more than 30 per cent of the total amount spent in investment subsidies. At that time, specific schemes were developed, in the framework of the 'Plan contracts', to make SMEs more efficient. For instance, special funds, co-financed by the region and the central government, have been set up to help SMEs hire the services of consultants. In the same vein, the ministry of research and technology in combination with the ministry of industry has set up public consultancies that offer SMEs technological advice.

A third line of action has been to try to develop research and training. This is particularly needed in the Northeast, where expenditures on research and development relative to GDP are lower than they are in the rest of France. The central government is partly responsible for this (because research funds are allocated to researchers, who are less numerous in the Northeast) and it has tried to redress this misallocation, particularly in favour of Lorraine. In the meantime, the regions of the Northeast are generously funding research out of their own limited budgets. The same is true of education and training. At all levels, be it technical or university, the Northeast is less developed than the rest of the country. The need to improve this situation is presented as a first priority in many of the Plan contracts, and regional resources are allocated to that purpose.

Fourth, improving the infrastructure endowment of the area is seen as another important line of action. The emphasis is put on transportation. The Northeast is potentially rather well located in Europe, close to three of the most important economic world centres (London, the Randstadt and the Ruhr, not to mention Paris). Many important highways have been created. The Channel Tunnel currently under construction will provide an easy connection to the United Kingdom. High-speed rail links, also under construction, will put the region closer to Paris, London and the Federal Republic of Germany.

So far, the policies undertaken have not been enough to prevent large-scale deindustrialization in the Northeast, nor to mitigate its consequences. This area, once one of the most prosperous in France, is now one of the most impoverished.

The South: compensated non-industrialization

The South, as defined here, consists of five regions and has about the same population and GDP as the Northeast. Yet the two areas had very different structural characteristics, and went through very different deindustrialization experiences.

To describe the South in the early 1970s, one can take the characteristics of the Northeast, and turn them around. First, the economy of the South was not dominated by industry; there was some industry, mostly consumption goods industries (textiles, garments, shoes, wood, paper, food), but its share in value added or in total employment was much lower than the national average; the regional policy efforts of the 1960s to promote industrialization in the West and the South were more successful in the West than in the South. The regional economy was based on agriculture (wine, fruit, vegetables), on services and on central government expenditures. Second, most of the relatively few industrial enterprises were SMEs; the share of employment in large firms was much smaller than for France as a whole. Third, levels of education, if not of skills, were high; for instance, the share of people aged 15 and more holding a graduate degree was in 1982 about 7.5 per cent, compared with 5.9 per cent for the Northeast and 6.5 per cent for France excluding the Paris region. The reason is that public education at all levels was provided free of charge and financed by the central government, and the local demand for education was greater than elsewhere, partly because there was no industry to provide employment. Fourth, there was in the South a network of old and well-functioning cities, such as Bordeaux, Toulouse, Montpellier, Marseille and Nice; in fact, one of the problems of the South – more acute than ever – was the contrast between these relatively prosperous central places and their decaying hinterlands.

This low level of industrialization did not, by itself, protect the South from deindustrialization. On the contrary, the small-scale, underfinanced, consumption goods-oriented enterprises of the South could have been expected to be particularly fragile. Yet, as indicated in Table 4.4, industrial employment did not suffer too much, not only in absolute terms (−70,000 jobs between 1976 and 1986), but also in relative terms (−10 per cent, to be compared with −19 per cent for France as a whole, and −29 per cent for the Northeast).

Similarly, one might have expected that the high level of tertiarization of the South would have slowed or deterred employment creation in the service sector. However, this is not what happened, and the service sector developed rapidly in the South in the last decade. Employment in that sector increased by 22 per cent in the 1976–86 period, faster than in France during the same period (18 per cent).

Several related explanations may account for these patterns. They are summarized in Figure 4.7. A closer look at industrial developments in the South shows that great changes occurred. Although hard data on the decomposition of net changes are not available, there are reasons to

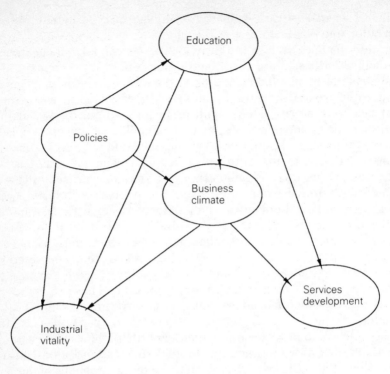

Figure 4.7 Factors in the development of the South.

believe that the relatively small net losses in terms of industrial employment are the result of large losses and of significant gains. As could have been expected, traditional enterprises suffered from deindustrialization in the South. Employment in the consumption goods industries declined by 21 per cent in the 1976–86 period, slightly more rapidly than for France as a whole. However, employment in the equipment goods industries hardly declined (by 5 per cent, in contrast with a 20 per cent decline for France as a whole). This relatively good performance is due to aircraft industries (Airbus are partly manufactured and assembled in Toulouse, Midi–Pyrénées), electronics, computers and other technology-intensive industrial products.

The relatively good performance of the industrial sector can be explained by several factors. Policies must have played a role. The South benefited from French and from European Commission subsidies. Provence–Côte-d'Azur, which was not eligible for French budget investment subsidies (as indicated on Figure 4.4) benefited from European Community regional funding, through a procedure called 'Integrated Mediterranean Programmes'. More generally, as indicated earlier, the

South has been heavily subsidized by the French budget, to which it contributed much less than it gained. In addition, the entry of Spain and Portugal to the European Community, while a threat to the agriculture of the French South, opened up new opportunities.

The higher level of education in the region also played a positive role. It provided enterprises or potential enterprises with a skilled or easy-to-train labour force. Largely as a result of policies, research was relatively well developed in the South. The ratio of researchers to total labour force, 2.6 per cent, was much higher than in the rest of France (excluding Ile-de-France), 1.5 per cent.[9]

A third factor was what can be called, for lack of a better word, a good business climate. The regional specificity (i.e. the residual of a shift–share analysis conducted on industrial value added) is positive for three of the four regions of the South.[10] This can in turn be explained by education, policies, perhaps industrial structures (small-scale enterprises are better than large enterprises for entrepreneurship) and certainly the physical climate and natural amenities of the area. What is certain is that enterprises were created locally, and were attracted from outside, including from abroad. Examples of such enterprises include Alcatel, Ford, IBM, Matra, Motorola and Thomson.

Finally, services contributed to the industrial vitality of the South. The growth of tertiary employment associated with industry was impressive: +45 per cent for financial services, +31 per cent for personnel and industrial services,[11] or +16 per cent for telecommunication. In all these types of services, employment was significantly higher in the South than in France as a whole. In other more consumer-oriented services, such as food and non-food commerce, employment was also increasing faster in the South.

It seems that the service sector was not being pushed by industry, but that it was leading the entire economy, including industry.

Ile-de-France: compensated deindustrialization

The case of the Paris region resembles neither the Northeast nor the South. The Paris region, officially called Ile-de-France, is a single region, but its population, about 10 million, is about the same as that of each of the other two groupings of regions. It differs from these other two areas, and from the rest of France, in a number of ways. First, as is obvious, it is highly urbanized; the entire region is centred on the Paris agglomeration and functions as one large and reasonably efficient labour market. Second, Ile-de-France is by far the wealthiest French region; the Paris region's GDP is 71 per cent higher than that of the North, and 65 per cent higher than that of the South (both in 1986). Third, Ile-de-France has traditionally been an important industrial centre; in 1986, for instance, the number of industrial jobs in the Paris region was nearly twice as high as in the South, and as much as 86 per cent of the number of industrial jobs in the Northeast. This industry was very diversified and included many SMEs. Fourth, the Paris region was a major tertiary centre, as could be expected for the capital city of a highly centralized country; it dominated government, finance, education, research and information to a rare degree. To give but one, significant, example, in 1983, 57 per cent of the people involved in research in France were concentrated in the Paris region.

The evolution of Ile-de-France over the past decades offers an interesting paradox. The Paris region did not do as well as the rest of France in either industry or services (and also of course for agriculture); yet it did better than France as a whole.

The three elements of this paradox are easy to document. First, Ile-de-France suffered greatly from the deindustrialization process. It was already losing industrial employment in the 1960s and the early 1970s, at a time when industrial employment was growing in the rest of France. When industrial employment began to decline, it declined faster in the Paris region than in the rest of France.[12] This is partly the result of national policies that explicitly aimed at diverting to the West and the South some of the industrial activities located in Ile-de-France, and partly the result of changing comparative advantage. It is reflected in the negative value of the regional effect in the shift–share analysis conducted on changes in industrial value added for the 1979–84 period. The growth rate of industrial value added for 1982–7 is also lower than the same rate for France as a whole.

Second, and perhaps more surprisingly, the Paris region did not do very well in terms of services, at least in appearance. Tertiary employment increased more slowly than in the rest of the country (12 per cent and 18 per cent respectively, for the 1976–86 period). This is also true of value added in the tertiary sector. This is in part explained by the slower population growth of the Paris region, by the delocalization of some tertiary activities, and by the budget-induced transfers of income out of the Paris region.

Yet overall, the Paris region did rather well. Unemployment, for instance, has been consistently lower than in the rest of the country. In 1988 it was 7.6 per cent, as opposed to 10 per cent for the rest of France. And this unemployment was relatively less structural and relatively more frictional. Unemployment of long duration was less important in the Paris region than in the rest of the country. The most significant overall indicator, GDP, increased faster in the Paris region than in France as a whole, in spite of lower-than-average population growth, which means that, on a per capita basis, the gap between the Paris region and the rest of the country increased. Hence, the paradox, which can be summarized in Table 4.5.

One can think of two explanations. The first is classical, and fairly obvious. Ile-de-France did better than France because the share of services in the Ile-de-France economy was larger, and because services growth was higher. The structural effect was higher than average, even though the regional effect was lower than average (i.e. negative).

The second explanation is that qualitative as well as quantitative considerations have to be taken into account. Jobs in the Paris region require higher qualifications, command higher salaries and involve greater responsibilities. They are qualitatively different, and, most importantly, are becoming more so. The Paris region may not have done well in terms of tertiarization, but it did very well in terms of 'quaternarization'. Evidence to support this proposition can be obtained by considering office construction. Quaternary functions, be they managment, finance, consulting, research, information, etc., take place in offices, as opposed to many tertiary functions, performed, for example, in shops, hospitals, or buses. Table 4.6 documents office construction in France over the past decade.

For the 1976–84 period, figures did not change much from year to year (which is why they are averaged in Table 4.6). In this period of rapid deindustrialization and slow tertiarization, the 26 per cent of the Paris region in office construction was significantly higher than its share in total

Table 4.5 Growth of value added by sector, 1982–86: Paris and France

	Paris region	France
	(percentage per annum, constant prices)	
Agriculture	−4.8	−3.7
Industry	−0.8	−0.2
Services	2.1	2.3
Total, or GDP	1.7	1.3

Source: INSEE, Premiers Résultats **165**. Décembre 1988. p. 2.

Table 4.6 Office construction: France and Paris region, 1976–88

	France (in millions of m^2)	Ile-de-France (in millions of m^2)	Share of Ile-de-France (percentage)
1976–84 (average)	2.5	0.7	26
1985	3.3	1.3	39
1986	4.1	1.8	43
1987	5.3	2.2	42
1988	6.4	2.7	42

Source: Comité de Décentralisation. Rapport 1988. p. 227.

population. However, the real change occurred in 1985, when this share increased to about 40 per cent. This is a powerful indicator of how attractive the Paris region is for quaternary activities, an observation reinforced by data on office floor-space prices, which are about three times higher in this region than in the rest of France.

The case of the automobile industry

The automobile industry exhibits some of the trends that have been examined so far. France has always been a major producer (presently the fourth) and a major exporter (presently the third) of motor vehicles, and the automobile industry has always played an important role. In 1987, it accounted for about 10 per cent of both value added and employment in manufacturing and about 16 per cent of industrial exports.[13] In addition, it is a 'leading sector', one that buys a lot from other sectors: the ratio of intermediate consumption to value added, which is 1.4 for manufacturing industries as a whole, is 2.1 for the automobile industry.

Over the 1970–87 period, production and employment followed a pattern similar to that of industry as a whole. Employment increased rapidly in the 1960s and in the early 1970s, until 1973, when it reached a figure of about 530,000. It stayed at this high level for about six years, then began a sharp and continuous decline. In 1987, it was down to 380,000, a 28 per cent decline in eight years.

Output, measured in terms of production (sales) or of number of motor vehicles, increased until 1979, then fluctuated around a horizontal trend. It is interesting to note that output measured in terms of value added fared less well. It peaked earlier, in 1977, and never returned to this high level. This difference reflects increased outsourcing and 'externalization' in the automobile industry. Some of the parts and of the support services that automobile assemblers used to provide themselves are now purchased from outside contractors, and appear in the accounts as intermediate consumption instead of as value added.

This tends to exaggerate the decline in terms of value added and also, of course, in terms of employment. We can assess the magnitude of this 'exaggeration'. If the ratio of value added to production that existed in 1970 had prevailed throughout the period, value added and employment in 1987 would have been about 10 per cent higher than they actually were. The corresponding activity, and jobs, did not disappear. Some of the jobs are reclassified as service. Some are counted in other sectors (such as electrical equipment). Other jobs have been transferred to countries with lower wages, where parts and components are now manufactured and from where they are imported.

Figure 4.8 The automobile industry.

Traditionally, the motor vehicle industry was concentrated in two regions: the Paris region, and, to a lesser extent, in Franche-Comté, in the eastern part of the country (Figure 4.8). Three of the four main firms that dominated production in the 1960s, Renault, Citroën, and SIMCA (a subsidiary of Chrysler), were located in Paris or its immediate suburbs. The fourth, Peugeot, was based in Sochaux, in Franche-Comté. In addition, Berliet, the leading French truck and bus manufacturer, was located in Lyon, in the Rhône–Alpes region. Three important developments took place: horizontal concentration, industrial decentralization and increased professionalization.

First, a reduction in the number of manufacturers occurred. Peugeot absorbed SIMCA, which disappeared altogether, and Citroën, which remains as a trademark. Renault absorbed Berliet, which became Renault

Véhicules Industriels, leaving only two major motor-vehicle manufacturers, Renault and Peugeot. The spatial impact of this horizontal concentration was to centralize decision-making in Paris. Peugeot's headquarters had already been in Paris for many years, but the absorption of SIMCA and Citroën made it even more necessary to concentrate marketing, finance and research activities in Paris, thus turning the Sochaux plants into mere production units. Similarly, the takeover of Berliet, which was headquartered in Lyon, by Renault led to further concentration of decision-making in Paris.

Second, and probably more importantly, in the 1960s and in the 1970s a number of manufacturing units were created in the West and in the North. This was partly the result of the national regional policy and of the wishes of automobile manufacturers. The French government in general, and DATAR in particular, wanted industrial jobs to be delocalized or created in these lagging regions. The automobile industry, in which employment was increasing and which had many links with the government (Renault was and still is state-owned), was an obvious target. Automobile manufacturers were approached and induced or pressured to expand into these areas, where they would benefit from regional policy investment subsidies. It was made clear to them that they would not be allowed to grow in the Paris region.

Manufacturers hardly needed either the carrot or the stick. They were looking for unskilled labour in large quantities, which could be found in the West as agricultural employment declined rapidly, and in the North because of the rapid decline in textile and mining employment.

French manufacturers built large plants in Rennes, Bretagne (14,000 jobs in 1985); in Caen, Normandie (3,000 jobs); in Metz, Lorraine (5,000 jobs); in Cléon, Normandie (9,000 jobs); in Sandouville, Normandie (12,000 jobs); in Douai, Nord–Pas-de-Calais (8,000 jobs); in La Rochelle, Poitou–Charentes (2,000 jobs). In addition, several foreign firms were persuaded to create plants in these same regions, as part of their world-wide manufacturing networks. They included Ford, in Bordeaux, Aquitaine, and Opel (GM), in Strasbourg, Alsace.

Third, the share of unskilled labour in total employment decreased, from 47 per cent in 1975 to 39 per cent in 1983. This was induced largely by automation, the substitution of robots for unskilled labour and also by increased subcontracting.

As a result of these three movements, the geography of the automobile industry in France changed substantially, as is shown in Table 4.7. The share of the original centres of automobile production, namely Ile-de-France and Franche-Comté, was sharply cut, both in absolute and in relative terms. In absolute terms, employment in these strongholds of the automobile industry was nearly halved (−45 per cent). Their share in total

Table 4.7 Employment in the automobile industry, 1976–86, by groups of regions

	1976	1986 (thousands)	1976–86	1976–86 (percentage)
Strongholds[1]	226	125	−101	−45
West and North[2]	136	127	−11	−8
Rest of France	167	119	−48	−29
Total France	529	371	−158	−30

Source: INSEE. 1988. Statistiques et indicateurs des régions françaises. p. 228.
Notes: 1) Ile-de-France and Franche-Comté. 2) Basse-Normandie, Haute-Normandie, Bretagne, Pays de la Loire, Aquitaine, Nord–Pas-de-Calais.

automobile employment declined from 43 per cent in 1976 to 34 per cent in 1986. The newly created plants located in the West and in the Northeast fared much better. In these areas, employment did not decrease much, and, as a result, their share increased substantially.

The decline of the strongholds in terms of employment is impressive. However, when one looks at the various types of employment in these areas, and particularly in the Paris region, one gets a rather different picture. Over the 1975–83 period, the number of engineers, technicians and professionals in the Paris region actually increased in absolute terms. In spite of its sharp decline in total employment, the role of Paris in the French automobile industry significantly increased in the last decade.

The jobs that were lost in the Paris region (and to a lesser extent in Franche-Comté) in the automobile industry were therefore mostly unskilled jobs, often occupied by guest (foreign) workers. Some were substituted by capital or technology; others were transferred to the West and the Northeast regions; still others were 'exported' abroad; and some, finally, were contracted out to other sectors, including the service sector. In the meantime, sophisticated and well-paid jobs in the sector were being created in the region. This explains how the Paris region could lose about half its automobile employment without really suffering much from it, and even benefit from the changes.

Conclusion

Structural change in France over the last decades has been marked by a major shift from industry to services. This shift, however, has not been homogeneous over space and the analysis of its regional dimensions yields some interesting insights on the very nature of this process.

First, industry and services are more complementary than substitutable or antagonistic. Industry did not do relatively well in those regions where services did (relatively) poorly. The opposite is true. In the South, both services and industry did better than the average, while in the Northeast and in Ile-de-France, both sectors did worse than average. Given this evidence, the concept of a 'shift' from industry to services is misleading. Structural change is not a zero–sum game and industry does not decline because services increase. A regional economy, or a national economy for that matter, depends on the expansion or contraction of both types of activity. It walks on these two legs but does so more or less rapidly.

Second, structural change took place within sectors as often as between sectors. This is shown by the fact that net deindustrialization was much more pronounced in the Northeast and in Ile-de-France, in the more heavily industrialized regions than in the South, the less industrialized part of the country. Restructuring industry was more difficult where and when industry was 'old', heavy, oligopolistic and dominant. A comparison of the Northeast with Ile-de-France shows the importance of intra-sectoral changes. In both cases, there was a sharp decline in industry and a moderate increase in services, at rates that were not very different. One could be tempted to conclude that both regions experienced a similar structural change. Such a conclusion would be completely wrong. In reality, as is shown by unemployment or income data, Ile-de-France did much better than the Northeast, because within industry and even more so within services, Ile-de-France managed to specialize in tertiary and quaternary activities.

Third, when industrial output or value added declines or grows at a relatively low rate, a choice has to be made between labour productivity and employment. Some regions tried to protect jobs at the expense of productivity. Other regions emphasized productivity growth at the expense of employment. For France as a whole, the first strategy prevailed until 1982, the second was preferred later. Data seem to suggest, although not very strongly, that the first strategy is more desirable, because it leads ultimately to a greater increase in value added, which is likely to lead to an increase in employment.

Fourth, human and cultural factors are essential ingredients in the process of adaptation to structural change. Why did the South and Ile-de-France do better than the Northeast? Not because they were favoured by their location in Europe or by their infrastructure endowment or by regional policy: they were not. The most convincing answers have to do with a better 'business climate', a more effective 'system of cities', and a more highly qualified labour force.

Fifth, the impact of structural change upon inter-regional disparities is mixed. On the one hand, structural change decreased disparities and inequalities between regions. At the beginning of the period, income was higher in the Northeast than in the South; at the end of the period, it was roughly the same. On the other hand, structural change probably increased differences between Ile-de-France and the rest of the country. High-level functions in management, finance, research and culture are more than ever concentrated in the capital.

This chapter has been deliberately focused on the past. Can we infer from this analysis something about the coming years? More specifically, can we forecast a continuation of the shift from industry towards services? Or should we rather predict for France what Lester Thurow is now predicting for the United States,[14] namely, a return of industry?

Many, if not most of the points made by Thurow are valid for France. The rapid increase in service employment over the 1978–87 period was concentrated in five sectors: non-market services, services to household, services to enterprises, restaurant and hotel services, and financial services. For most of these five sectors, the prospects are not bright at all. For employment in restaurants and in finance or banking, the rapid increase of the last decade was, to a large extent, a one-time-only phenomenon for the reasons indicated by Thurow. This is equally obvious for the so called 'non-market services', mostly government services: it is widely agreed that government is already too big and that retrenchment, not expansion, is the order of the day. Prospects are better for household services – from travel agencies to hairdressers – and for services to enterprises. However, it should be clear that these services cannot grow faster than the output of enterprises ultimately dependent on the output of industry. If they do, as noted above, it is because industrial enterprises contract out a growing part of their activities and the implied shift from industry to services is then largely a statistical artifact or illusion. It is therefore hard to see in what sectors employment in the services could continue to increase at a high rate.

Furthermore, the future of industrial employment is not as dark as its past. The major, job-killing structural adjustments in steel, mining, automobiles, shipbuilding, or textiles have been made. International competition, particularly within the European Community has forced French industrial enterprises to increase productivity in order to become competitive. Many are by now. Their output is likely to rise, spurred by investments at home. Industrial output can be expected to increase as fast, or faster, than productivity and therefore to create net growth in industrial employment. As a matter of fact, this is exactly what happened in 1988, for the first time since 1974.

Deindustrialization will perhaps have been only a temporary phenomenon. What is likely to remain permanently, however, are two processes, associated with deindustrialization. One is the tertiarization (or shall we say the deindustrialization?) of industry: industrial processes and even products will be more and more information-intensive. The distinction between the secondary and the tertiary sectors will be less and less meaningful. The other is structural change, fed by changes in demand and in technology, of which deindustrialization is but a major example.

5

Regional deindustrialization and revitalization processes in Italy

ROBERTO P. CAMAGNI

The deindustrialization condition

Definition

ALTHOUGH THE TERM 'deindustrialization' is widely used, a close inspection of the existing literature shows that an accepted definition does not exist (Chapter 8). At the macroeconomic, global, or national level, deindustrialization may be regarded as the statistical counterpart of the process of tertiarization of the economy, characterized by an increasing share of service employment. This process is determined by a higher growth rate of industrial productivity compared with the other sectors, the increasing importance of intangible and 'soft' production factors such as information, and the shift of consumer demand toward superior goods and services.

At the regional level, and therefore assuming a relative perspective with reference to a wider economic context, deindustrialization may mean a loss of competitiveness and the disappearance of entire industrial sectors. At the metropolitan or urban level, the same concept often refers to a process of delocalization and movement of industrial plants towards the outskirts of the city, the 'rings', as service activities come to occupy the 'core'.

This chapter discusses deindustrialization from a regional perspective rather than from macroeconomic or metropolitan viewpoints. However, the term must still be defined properly, in order to avoid naive and superficial interpretations.

In our opinion, deindustrialization is not the simple absolute or relative decrease of industrial employment, a fact that may well be explained by physiological processes, and that may emerge well before an absolute decrease of industrial output shows up. If the research objective is to come close to an *economic* interpretation of this complex process rather than a simple statistical description, we cannot rely upon a single indicator, such as employment or output growth. For example, a large reduction in industrial employment may well be associated either with a condition of reduced international competitiveness of the local productive sectors, or with a condition of wide restructuring and relaunching of local competitiveness, two situations with completely different economic meaning, as is easily understandable.

As I have previously suggested, *two* indicators have to be inspected at the same time to detect a deindustrialization condition: employment growth (and in particular, *relative* employment growth) and productivity growth, the former indicating a social problem and the latter explaining its causes and the future prospects of the local economy.[1] In a historical period characterized by rising industrial productivity (albeit at declining rates) and growing service employment, regional deindustrialization should be considered and defined as: a *relative* phenomenon, with respect to the *average* behaviour of the entire economic system encompassing the region, taking place when job losses *and loss of competitiveness* occur simultaneously, driving the local economy into a vicious cycle. Thus construed, deindustrialization becomes a rather general process by which a region is unable to carry out successfully a process of industrial transformation.

In analytical terms, it is possible to chart the economic performance of each region on a cartesian graph where on the X axis we put the relative employment growth (REG) of region r, namely:

$$\text{REG}_r = (N^1/N^0)_r / (N^1/N^0)_n$$

and on the Y axis we put the relative productivity growth (RPG) of the same region (value added per employee), namely:

$$\text{RPG}_r = (P^1/P^0)_r / (P^1/P^0)_n$$

where n denotes the nation (or the European Community), N and P are employment and productivity and 0 and 1 the initial and final year.

An interesting feature of this methodology is that in the relevant region of the graph, a 45° negatively sloped line passing through the origin approximates a condition of regional product growth equal to the national average. In fact, a region may develop at the same GNP rate either if both employment and productivity grow at the respective national rates, if productivity increases at a lower rate but employment at a proportionally higher than average rate, and vice versa.

On the same graph we may therefore control three indicators for each region, and chart their relative performance in different sub-periods. In Figure 5.1 the six possible patterns of regional growth that emerge may be designated as:

1 'virtuous cycle', when higher than average productivity growth generates good performance in both employment and output;
2 'restructuring', when a higher productivity growth is reached through severe employment cuts, leading nevertheless to good output performance;
3 'dropping-out', when productivity growth is reached by closing down inefficient production units, generating lower than average output growth;
4 'deindustrialization', defined as a vicious cycle in which employment cuts are unable to restore competitiveness, a condition that perpetuates job losses and low output growth;

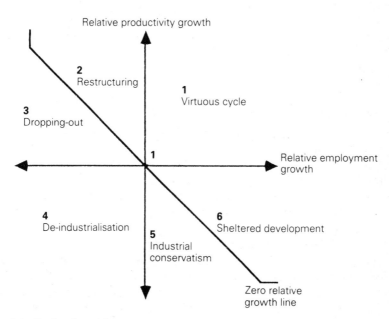

Figure 5.1 Regional growth patterns.

5 'industrial conservatism', where poor productivity performance is accompanied (and partly explained) by a better than average employment growth, thanks to public assistance and industrial rescues;
6 'sheltered development', where similar explicit or implicit assistance policies spur the initial development of (mainly backward) areas, notwithstanding low productivity performance.

Deindustrialization is depicted in condition 4, but conditions 3 and 5 are also worth inspecting. Both cases reflect slow output growth, the good employment condition comes from artificial intervention policies ('industrial conservatism') and the apparently good productivity condition is the result of the simple suppression of inefficient units with no positive counterpart ('dropping out').

A previous inquiry on the performance of European regions in the 1970s has shown that many areas were suffering from deindustrialization. In the years 1970–4, the West Midlands, North West and South East in the United Kingdom, Hamburg and Hessen in the Federal Republic of Germany; in the years 1974–8, Bremen and Nordrhein-Westfalen in the Federal Republic, Wallonia in Belgium, Nord–Pas-de-Calais in France, West and Zuid Nederland in Holland. Many old industrial areas in Italy and France were benefiting from some direct or indirect support from national sectoral or macroeconomic policies in the second sub-period, and fell into an apparently less distressed situation in terms of job losses. These included Liguria, Piemonte and Lombardia in Italy, and Lorraine in France. The condition of 'industrial conservatism' seemed to be the Mediterranean counterpart to the Northern European 'deindustrialization' condition.[2]

A taxonomy of deindustrialization conditions

As we can see from the types of regions mentioned above, different structural conditions may lead to industrial distress. Accurate identification of prevailing conditions is essential to determine the strategies and the chances for a 're-industrialization' or 'revitalization' policy. It seems useful therefore to define a possible taxonomy of the deindustrialization condition according to the main causes that may lead to it.

A useful taxonomy may be derived by a two-entry table (Fig. 5.2), charting the endogenous or exogenous origin of the forces driving an industrial crisis vs their macroeconomic or microeconomic nature. Into the first, macroeconomic-exogenous case, fall, e.g. distress arising from sudden exchange rate appreciations (originated by capital movements or trade changes in non-manufacturing sectors); the so called 'Dutch disease' of the late 1970s, generated by the sudden reversal of that country's need

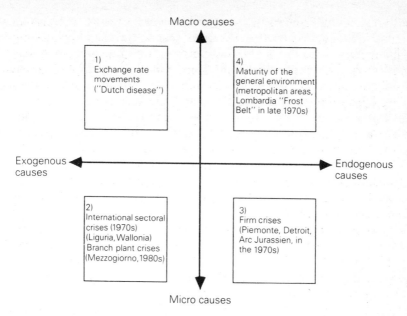

Figure 5.2 Taxonomy of deindustrialization processes.

for imported energy; and perhaps a similar process in the United Kingdom, underscored by a policy of exchange rate 'grandeur' of the Conservative government in the early 1980s.

The second case, the exogenous-microeconomic case, encompasses instances of industrial crisis linked to the performance of specific sectors on a worldwide scale. Regions specialized in shipbuilding, coal mining and iron and steel were severely hit by the irreversible market crises in their respective sectors in the 1970s: Wallonie (Belgium), Liguria (Italy), Lorraine and Pas-de-Calais (France), Yorkshire–Humberside (United Kingdom) are but a few examples of this situation.

The third case, the endogenous-microeconomic one, stems from the difficulties of specific regional firms or sectors in facing external competition. Although this case is also likely to happen especially in periods of slow growing world demand, the sectoral constraint is less tight than in the previous one, as the presence of successful competitors elsewhere reveals the existence of technological or managerial alternatives to the models employed in the region. The problems of the automobile industry in Detroit, Birmingham, or Torino in the 1970s, those of the Arc Jurassien in Switzerland (the watch manufacturing area), or of the northern Italian regions in the electrical appliances industry in the early 1980s fall into this class, as in all cases external success areas were showing up at the same time in the same industries (Japanese areas in the first and second case and West German areas in the third, respectively).

The fourth case, the macroeconomic-endogenous one, refers to the condition of many old-established, highly urbanized, diversified industrial areas in which the economic and environmental structures have matured. These areas are characterized by physical congestion, high land prices and labour costs, union aggressiveness and social resistance to change and bureaucratization of industrial and managerial practices. Under these circumstances, a region loses its locational advantage relative to other newly industrialized areas, and a selective decentralization of sectors and phases of the production cycle takes place. Usually, tertiary activities show high rates of growth in these areas, but they cannot match the employment losses in manufacturing activities, especially when new small-scale industrial firms choose a non-metropolitan location. This was typical of many capital or chief-city regions in Europe and of the so-called 'rust-belt' in the United States in the late 1970s; in Italy, the Lombardia region has fallen into this category in some recent periods.

The four situations sketched above are highly abstract, and some regions often share the characteristics of two categories. Nevertheless, a prevailing feature can always be discerned in each regional situation, and this may indicate the best possible strategy and the most suitable policy instruments to cope with a region's difficulties. Most importantly, this taxonomy may help offset the common view that deindustrialization comes only from a coupling of weak external demand with local specialization in declining sectors. While this may be an important part of the story, it encompasses only one possible condition, the one described in case 2 (exogenous-microeconomic case). In the other three cases, and in particular in the endogenous ones, the ultimate source of deindustrialization lies in the lack of innovative capability and adequate 'context' characteristics, including the poor condition of the general environment and overhead infrastructure.

In recent years, the interpretation of spatial development has turned convincingly towards the 'indigenous', 'bottom-up', 'from below' nature of the processes of mobilizing local resources and potential for innovation in the local 'milieu'. By the same token, accurate interpretation of the crisis processes must be based on identifying the local, endogenous roots of industrial 'maturity' and senescence. Lack of 'response' capability to the turbulence of the external environment more than the turbulence itself must be seen as the source of a crisis, and the difficulty of raising productivity and external competitiveness more than the slowdown of export demand must be investigated as its 'prime cause'.

Possible revitalization strategies

If the sources of deindustrialization stem from several different processes and phenomena, different policy strategies must be developed. In general terms, our suggestion for an appropriate revitalization strategy may be defined as follows (Fig. 5.3):

(a) In case 1, characterized by lowering of external competitiveness in all industrial sectors, due to exchange rate revaluation, the best policy may be to focus on the most advanced products, the highest value-added phases within the production cycle (headquarters, finance, marketing, research, design functions), and the best performing firms, while accepting a selective international decentralization of the declining sectors and of direct manufacturing phases. An important goal may be to strengthen the high-tech *filière*, encompassing advanced production services, information technologies and high-tech production.

(b) In case 2, characterized by specialization in declining sectors, the natural medium-term strategy should be one of reconversion towards new production sectors.

(c) In case 3, characterized by firm crises, a viable strategy is the restructuring of existing production through an integrated process of product-and-process innovation and the reinterpretation of the tradi-

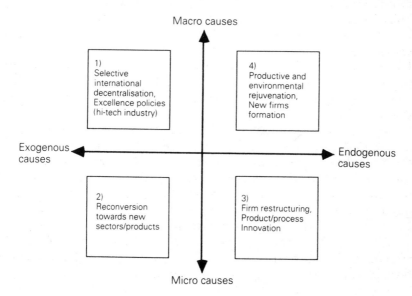

Figure 5.3 Revitalization strategies in different deindustrialization conditions.

tional 'vocation' of the local area in modern, market-oriented terms (successful examples of revitalization of traditional products through new marketing ideas include the Swiss Swatch; automobiles designed to meet customers' needs and comfort; textiles and clothing based on fashion creation; computers and chips customized for specific uses and applications).

(d) In case 4, characterized by senescence of the general urban or regional environment, the best strategy could be the creation of new social awareness, solidarity and a sense of regional 'patriotism'; rejuvenation of the built environment; construction of new, advanced information and communication infrastructures; and modernization of existing production through the adoption of advanced technologies.

New 'information' technologies may be of strategic importance in cases 3 and 4. Case 2 is generally more complicated, as existing activities may be of a good technical quality and as the creation of new activities is not just a technological affair. Nevertheless, as far as the sectors involved are not characterized by continuous processes (such as chemicals, or iron and steel) but by discontinuous processes (such as mechanical engineering, machine tools, textiles and other 'light' industries), the new production technologies may play an important role in raising local competitiveness and helping the regional industry widen its share of even a declining market.

In microeconomic terms, this last process means the possibility of generating an upswing of the S-shaped curve describing the product-life cycle after the maturity phase. This upswing is determined by the simultaneous adoption of a new technology and a new marketing philosophy, which pushes a tremendous improvement in the effort : performance ratio of the firm. But in a more macro-spatial setting, how to identify the most appropriate 'context' conditions for the firms to choose the right new technology at the right moment remains an open question.

Progress towards industrial maturity in a region is a natural process, tied to the ageing of infrastructure, bureaucratization of social and economic relationships, increasing labour and floorspace costs, unionism and congestion. Regional economies move from stage 1 to stage 3 or possibly 4 of Fig. 5.1 through a sort of 'regional life cycle'.[3] The revitalization process, however, is by no means natural or automatic, but needs a 'wilful and intentional' socioeconomic environment.[4] After Tornqvist and Andersson, the conditions for 'creative' regional development may be found in structural instability, competence (knowledge and information), and synergism.[5] These conditions are sufficiently self-explanatory, and they will be considered in reference to specific regional case studies later in this chapter.

Regional performance in Italy, 1970–84

The spatial structure of the Italian economy may be rapidly sketched in terms of the evolution of the four macro-regions: the Northwest, Northeast, Centre and South (see Fig. 5.4). Traditionally, development has taken place in the regions of the 'Industrial Triangle', the regions of Piemonte, Lombardia and Liguria in the Northwest. However, starting in the mid-1960s, a new phenomenon has emerged, namely the development of new intermediate and peripheral areas, belonging to the Northeast and Centre, which have followed a new industrialization pattern characterized by dispersed, non-metropolitan industrialization. Emilia-Romagna, Veneto and Trentino first, and then Toscana and the regions along the Adriatic coast (which has been called the 'Third Italy') until the

Figure 5.4 Regions of Italy.

end of the 1970s showed astonishing performance in industry and also in some, though not very advanced, market services. In the same years, the Northwest was experiencing clear signs of deindustrialization, due to adverse locational factors: rising costs of labour and floorspace, union aggressiveness, congestion and low innovation pace.

The Mezzogiorno regions of the South were historically left aside in the process of national development. In fact, after the unification of the country in 1861, the pre-existing industrial structure disappeared and a vicious cycle of poverty and emigration began. After the Second World War national policy toward the Mezzogiorno was characterized by three successive strategies: the provision of basic infrastructure and overhead capital formation until the end of the 1950s; the intervention through publicly owned enterprises and the attraction of big branch plants of private corporations, through a 'growth poles' strategy (until the end of the 1960s); and a strategy of 'diffused' industrialization based on small firms until the end of the 1970s. No new real strategy was launched in the 1980s, when public intervention became random, eclectic and decentralized.

In recent years, the Mezzogiorno problem reached dramatic proportions. The big development poles of the 1960s (iron and steel plants in Naples and petrochemicals production in Sicilia and Sardegna) are in crisis. The traditional stagnation of most parts of the territory continues (apart from some development episodes in Abruzzi and in some provinces in Campania, Puglia and Sicilia) and the new economic and social difficulties of big agglomerations like Napoli and Palermo are becoming explosive. The recent upswing of the national economic cycle after 1983, pushed by a new accumulation effort in anticipation of 1992 (investments in equipment grew at an average yearly compound rate of 7.2 per cent from 1983 to 1988), has relaunched the Northwest and the main metropolitan areas of the country, but has excluded the Mezzogiorno. The new 'technological paradigm' calls for higher education levels in the labour force, an organized and structured urban environment and the presence of advanced service activities, all elements that are likely to create further imbalance between the north and the south.

Employment and value added

The growth rates of employment and value added are given in Appendix Tables 5A.1 and 5A.2 for all regions and two strategic sectors, manufacturing and market services, from 1970 to 1984. Unfortunately, regional accounts are now under revision and stop at 1984, omitting the most recent years in which, as already mentioned, many 'revitalization' and 're-industrialization' processes have shown up in core areas.

As far as manufacturing is concerned, we can see that national employment was rising during the early 1970s, (5.3 per cent overall in the period 1970–6), stagnating in the second part of the decade, and declining sharply in the early 1980s (a 10.8 per cent overall fall in 1980–4). The Northwest was anticipating the general trend, with a clear stagnation in the first sub-period (especially Lombardia and Piemonte), some already apparent decline in the second sub-period (especially Lombardia and Liguria) and a strong decline in the third (with overall rates of decline of 18 per cent in Piemonte, 15 per cent in Liguria, and 11 per cent in Lombardia between 1980 and 1984).

Looking at the figures for total employment, some interesting elements emerge in regard to the last sub-period. Among old industrial regions, Piemonte performed badly, and Lombardia and Liguria experienced a condition similar to the national trend (stagnation). The different meaning of the apparently similar performance of Lombardia and Liguria will become clear in the next paragraph. Among new development areas, Lazio and Abruzzi performed well, although the Marche and Umbria regions saw some initial problems.

The analysis of value-added figures gives us important information on the depth of the national crisis in the recent sub-period 1980–4. In fact, in spite of the good performance of the economy in the terminal year, manufacturing value added decreased by an overall rate of 2.7 per cent. Also, we may better understand the nature of the relatively good employment performance of many regions of the South and Centre in the same sub-period. Stagnation, or even an increase in employment occurred despite large real decreases in output (see the case of Abruzzi, Molise, Puglia, Sicilia and Sardegna; Umbria and Marche were showing a decrease in both employment and value added) and therefore in a condition of decreasing efficiency or assisted development.

Many scholars have interpreted the performance of many regions of the South in the last decade as a sign of a reversal, equally distorted, of the development model of the 1960s. While the earlier model could have been labeled 'development without employment' (due to the capital-intensive nature of most investments), later on it became a model of 'employment without development', characteristic of many backward and assisted areas.

As far as the old industrial areas are concerned, Liguria and Piemonte show a strong absolute decrease in industrial production in the third sub-period, while Lombardia, which performed relatively poorly in the 1970s, showed a relatively good performance in the 1980s, despite major reduction in employment. What we may call an 'absolute deindustrialization' condition may also be seen in regions like Friuli and Trentino in the Northeast. In contrast, Lazio performed well in all periods.

Productivity

Productivity growth is perhaps the most important variable worth considering (Appendix Table 5A.3). From a statistical point of view, the productivity growth rate is approximated by calculating the difference between value-added and employment growth rates, but from an economic point of view it may be considered to be the most relevant single factor in the development of a region and of a possible deindustrialization condition.

According to a contrary, widely shared view, productivity growth is nothing but an effect of product growth: in fact, in the short run, employment is relatively fixed and each production expansion generates an increase in productivity; in the long run the expansion of demand determines the adoption of new and better equipment and allows learning processes. This view, which refers theoretically to Verdoorn's law and the Kaldor–Thirwall growth model,[6] may be accepted in some cases of externally driven development or crisis (in our debate, it may explain the crisis of some specialized areas, hit by the sudden collapse of world demand), but as a general explanation of regional growth it looks rather simplistic, suspicious and far from established.

This view presupposes that an exogenous and predefined demand for the products of a specific region could be identified. Unfortunately, in a world of differentiated products, continuously innovated production and inter-firm competition, the existence of this sort of demand is far from being assured. Like firms, local economies compete in order to create new markets and attract increasing shares of existing demand through efficiency and image elements. In other words, regional development is a supply-side phenomenon, closer to the microeconomic competition process than to the aggregate trend of the world economy.

If world sectoral demand were the first exogenous cause of regional development, we could never have seen the take-off of an agricultural area or of a region specializing in light and traditional industrial products, as by definition the demand for these products grows slowly. The phenomenon of the Third Italy should have never shown up![7]

Appendix Table 5A.3 confirms the relevance of autonomous technical progress in the performance of the national regional economies. In manufacturing activities at the national level, we see similar productivity growth rates in the first and second sub-periods (17–18 per cent) in spite of completely different growth rates of value added (25 per cent and 17 per cent respectively). Moreover, the divergence between the two series is even clearer if we shift to the third sub-period, when on the basis of the negative growth rates of GNP we would have expected a negative or a very low increase in productivity, and on the contrary we got a 9 per cent increase. This element shows the emergence of an autonomous process of technological change, innovation adoption and restructuring in the Italian industry starting in the second half of the 1970s, coinciding with the signing of important union agreements and with the turn round of the general climate of industrial relationships.

These autonomous tendencies are evident also at the regional level and may have had strong effects on the relative performance of the individual regions. In Lombardia, the acceleration of productivity growth in manufacturing activities from the first to the second half of the 1970s is much more evident than acceleration of product growth and this trend continued in the 1980s when an extraordinary result of a 14 per cent rise in productivity (in four years) was reached despite stagnation in manufacturing production (1.5 per cent growth). The same thing may be said about the Piemonte manufacturing industry, about three or four years after. What could have been judged as a continuing and deepening deindustrialization process on the basis of employment data for 1976–80 and 1980–4 (0.8 per cent and 18.7 per cent drops, respectively) and of the production slump of the early 1980s (a 9.4 per cent drop) proves on the contrary to be a dynamic process of crisis management through industrial restructuring and technical change (9 per cent and 11.4 per cent of productivity increase in the two sub-periods).

In Liguria the restructuring process of more recent years proceeded much more slowly than the national average, in spite of severe employment cuts. This case demonstrates quite a different response capability in the face of the crisis condition of the mid-1970s.

In contrast to the powerful restructuring process taking place in the main old industrial areas, Lombardia and Piemonte, is the lower than average productivity performance of *all* other regions in recent years. All the Third Italy regions have shown a low productivity change in the period 1980–4, compared to the astonishing performance of the previous decade. This poor performance has accompanied both a higher than average production growth (as in the case of Veneto), production growth rates similar to the national ones (as in Emilia and Toscana) and lower than average rates (as in Trentino, Umbria, Marche, Abruzzi and Molise).

In the Mezzogiorno regions, the 'employment without growth' model shows up once again in the low productivity change figures of the 1980s: all regions with the exception of Calabria have performed at lower than average rates.

As far as services are concerned, it is interesting to note their anticyclical, unemployment-absorbing role. In the last sub-period, in the presence of a strong rationalization process in industry, they worked as an employment stabilizer, diminishing average per capita productivity in almost all regions.

A more straightforward way of appreciating changes in inter-regional efficiency and competitiveness is to look at the ranking of regions on the basis of productivity levels and productivity growth in different sub-periods. This is done in Appendix Table 5A.4 with reference to all sectors.

In spite of the fact that the core regions of the Northwest have shown very poor performances and the lowest scores in all sub-periods in terms of productivity *growth*, it may be easily seen that the regional ranking in terms of productivity *levels* has remained quite stable in the 1970–84 period. Notwithstanding the deep changes that have taken place in these years, with the emergence of a new, intermediate belt of fast-developing areas the relative positions of the individual regions have undergone only minor shifts.

In fact, three regions in the Northwest – Lombardia, Liguria and Valle d'Aosta – have stayed in the first three ranks in terms of productivity levels, and the stability of the relative positions of all regions is confirmed by the very high values of Spearman's rank-correlation index. The rank-correlation index is much lower if computed in terms of productivity changes, especially in the last sub-period.

The effects of reallocation processes of regional resources

Important insights about productivity growth differentials may be derived from a sectoral analysis of the shift–share type. In fact, as in the growth rates of value added, regional productivity growth rate differentials may be the effect of different components, and in particular of a 'mix' effect, stemming from a favourable sectoral structure, and of a differential or competitiveness effect, stemming from the ability of the local sectors to boost productivity at higher than average rates.

In the case of productivity growth, the total aggregate shift with respect to the national average is determined not only by the two above-mentioned effects. In addition a third component has to be taken into account, which may be called a dynamic mix component or, better, a 'reallocation' effect.[8] If, during the period under consideration, the structure of the regional economy has *changed*, and a reallocation of local resources, capital and labour has been directed toward higher productivity sectors, the final productivity level may differ from the simple sum of a differential effect (each sector on the average performing better than the national counterpart) and of a static mix effect (a sectoral composition in the initial year favouring fast productivity-raising sectors).[9] Appendix Table 5A.5 gives us the results of this exercise for the entire period 1970–84. They are in my opinion extremely interesting:

(a) in the three deindustrializating regions – Lombardia, Piemonte and Liguria – the components of a poor aggregate performance are completely different. They include (i) a negative reallocation effect with positive, or neutral, differential and mix effects in Lombardia, the negative reallocation effect being explained perhaps by the shift towards the public administration sector; (ii) a strongly negative differential effect in Piemonte, stemming from the poor performance of all sectors on average; (iii) a strongly negative mix effect in Liguria, linked to the presence of crisis sectors like iron and steel or shipbuilding, coupled with a strongly negative differential effect;
(b) the good performance of all Third Italy regions, explained for the most part by a sectoral differential effect (Veneto, Emilia, Umbria, Marche, Abruzzi and Molise);
(c) a negative differential effect for all regions of the South not belonging to the Third Italy phenomenon (Campania, Puglia, Basilicata, Calabria, Sicilia, Sardegna), reflecting a weak productive structure, often a negative mix effect and always a positive and relevant reallocation effect, coming from the structural shifts towards industry in the last decade;
(d) other positive reallocation effects visible in Marche, Abruzzi and Molise, and other negative mix effects, besides Liguria, visible in Lazio and Valle d'Aosta, linked to the high share of non-market services, where productivity is traditionally stagnating.

The deindustrialization and revitalization process

Up to now, we have analysed the evolution of single indicators of regional performance (employment, value added, productivity) in the entire territorial system of Italy, and we have focused on absolute trends. In this perspective we were able to understand the following.

(a) A phenomenon of 'absolute deindustrialization' has taken place in the early 1980s in the country as a whole, in conjunction with the slump of the world economy, revealed by a decrease of real industrial ouput.
(b) The effects of this negative cyclical trend were emphasized on the employment ground by an 'autonomous' process of industrial restructuring, which began in the late 1970s in most advanced areas.
(c) Industrial productivity growth is an important exogenous determinant of regional performance and may be accompanied by completely different trends in employment and production.
(d) Service employment, and in particular non-market services employ-

ment, has been used in most advanced areas as social shelter in periods of strong job losses in industry.
(e) The mix component of product or productivity growth, and in particular an unfavourable industrial structure based on mature production, may be an important cause of the poor performance of a regional economy, but it may also be accompanied (or, on the other hand, it may be counterbalanced) by an insufficient (favourable) autonomous innovation or reconversion capability of the local industrial fabric.

We may now return to the methodology proposed at the beginning of the chapter, where a regional approach to the deindustrialization problem was built on the basis of two main reflections: first, deindustrialization has to be analysed on the basis of three *combined* indicators, productivity growth, employment growth and regional product growth; and, second, it has to be considered as a *relative* phenomenon, a differential occurrence with respect to the aggregate trends whose strong effects on regional accounts must be taken out of the picture.

Patterns of regional growth and decline

The proposed methodology, centred on the idea of a sequence of 'patterns' of regional growth, is applied to the recent industrial performance of Italian regions. In order to have a wider time perspective, especially in recent years, a different series of data is employed, namely, the regional accounts of manufacturing firms with more than 20 employees. This source has the advantage of reaching 1986 (therefore allowing an inspection of recent regional revitalization trends) and of presenting comparable data on value added and employment (in fact the national accounts data on employment come from different inquiries with respect to value added). Of course, the figures on past sub-periods are not comparable with those employed hitherto.

Three groups of regions are analysed in four sub-periods between 1973 and 1986:

(a) old industrial regions: Lombardia, Piemonte and Liguria (Fig. 5.5);
(b) new industrial regions: Veneto, Emilia Romagna and Toscana (Fig. 5.6); and
(c) regions of the South, both rapidly developing (Abruzzi) and characterized by the crisis of big branch plants, inheritance of the industrialization strategy of the 1960s: Campania (iron and steel, shipbuilding), Puglia (iron and steel, chemicals), Sicilia (petrochemicals) (Fig. 5.7). The basic figures on relative employment growth and relative

Deindustrialization in Italy 153

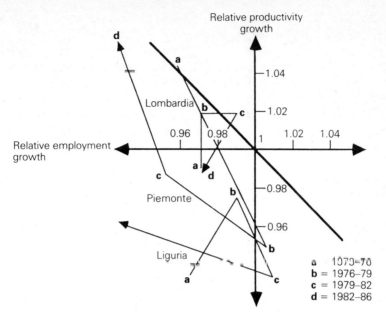

Figure 5.5 Patterns of regional growth: old industrial regions, 1973–6–9–82–6.

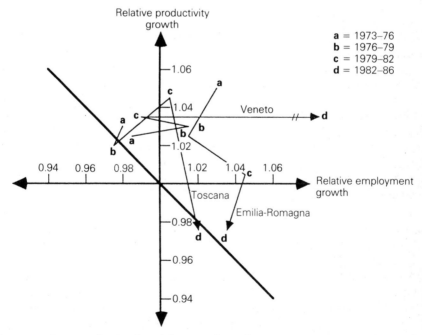

Figure 5.6 Patterns of regional growth: new industrial regions, 1973–6–9–82–6.

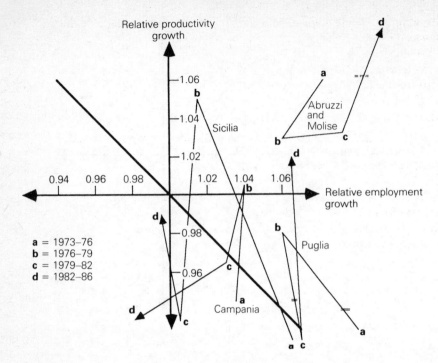

Figure 5.7 Patterns of regional growth: southern regions, 1973–6–9–82–6.

productivity growth (as defined analytically on pp. 138–9, Chapter 5), are presented in Appendix Table 5A.6.

The three old industrial regions of the Northwest were the driving areas in the boom period after the Second World War (1948–62) and experienced in the following decade all signs of approaching maturity. Investment rates fell dramatically, especially during the 1964–5 crisis and the 'hot autumn' of 1969, characterized by huge social contrasts, and a process of spatial decentralization of industry towards intermediate and peripheral areas.

From a structural point of view, the conditions of these regions fall into three categories of our taxonomy of the deindustrialization process (Fig. 5.2). Liguria suffered mainly a sectoral crisis (case 2) in the iron and steel and shipbuilding sectors, together with a poor urban environment. Piemonte saw mainly a managerial and market crisis in the car and computer sectors (Fiat and Olivetti), as a pay-off of its traditionally strong know-how in mechanical technologies and a lack of advanced services (case 3). Lombardia exhibited a congested and conflicting, but otherwise diversified industrial 'milieu', with a good urban services endowment (case 4).

According to our scheme (Fig. 5.5), these regions have in fact experienced a condition of deindustrialization in the recent past, but with different characteristics, pace and depth. Lombardia and Liguria first entered this condition in the early 1970s, but the response capability of the two regions was completely different.

The former in fact succeeded in a strategy of revitalization, first through a rationalization process (1976–9: 'dropping-out') and then through a restructuring of the local manufacturing industry (1979–82). In more recent years (1982–6) a deindustrialization condition was once again apparent, a sign that such a condition represents a structural danger for this kind of region. However, other symptoms show that this situation was overcome in the years 1984–8, when the region performed better than Italy as a whole in both employment and product growth.[10]

The interesting element in the revitalization process of the last years is that the manufacturing industry of the core area of the region, the Milano province, performed better than the region as a whole for the first time in 30 years. At the regional level, most of the recent revitalization process took the form of service employment expansion in peripheral provinces.

The case of Liguria is almost at the opposite extreme. A response capability to the crisis condition never appeared, and the regional economy has remained in a deindustrializing condition over the entire period, sometimes leaning toward an assisted development strategy (1979–82: 'industrial conservatism'), sometimes imposing deep employment cuts (1982–6). The rationalization processes were never sufficient to relaunch the competitiveness of the area.

The picture presented by our figures is very credible. The socio-economic life of the region was characterized in the last decade by a decline in entrepreneurial spirit, tight industrial relations and conservative attitudes on the parts of the unions and the labour force and general inability of the top management of big state-owned companies to find an industrial alternative to the hopeless local specializations (iron and steel, nuclear engineering, shipbuilding). The greatest attention was captured by the long struggle of the docks unions, that up to now have resulted in a complete paralysis of all the activities linked to the port. As we have seen before, the development of private service activities in this region was also lagging behind.

The experience of the Piemonte region is another story. There the crisis of the automobile industry after the 1974–5 oil crisis struck hard, and the major firm, Fiat, was not able to find the proper strategy to cope with it. After a period of declining productivity (1976–9: 'industrial conservatism') and of initial restructuring (1979–82: 'deindustrialization') competitiveness was revived in recent years, at the expense of severe employment cuts.

In conclusion, three apparently similar cases of deindustrialization of old industrial areas in fact have quite different resolutions and timing, due to the differing nature of the sources of industrial troubles and the different reaction capability of the local 'milieu'.

The case of the new industrial regions is of course very different (Fig. 5.6). The growth of productivity and employment have always been higher than the national average, with the exception of more recent years in the case of Toscana and Emilia Romagna. In these latter cases, in the 1982–6 period some competitiveness problems began to appear, linked to some initial problems in continuing the innovative processes of the past.

In some light industries such as textiles and apparel, a new organizational model of production has emerged in recent years, focusing on wide utilization of information and communication technologies (CAD–CAM, remote production control, telecommunication networks linking points of sale to the production management, etc.) and on tight relationships with fashion creation and international marketing, relaunching the role of the big, vertically integrated firm typical of other regions such as Piemonte and Lombardia.[11] In contrast, the dispersed pattern of industrialization that is typical of these two regions has not shown a similar innovative capability, especially in the case of Toscana (the financial crisis of Prato is telling in this respect).

In the Mezzogiorno regions (Fig. 5.7), apart from the case of Abruzzi, a region that has recently attracted many branch plants of high-tech firms (telecommunications equipment in particular) or of firms widely utilizing advanced technologies (like the automobile industry), the picture is a differentiated one. Up to 1982 the most common condition was industrial conservatism or sheltered development. As said before, a condition of 'employment expansion without development' and of strong protection of the existing jobs in big plants was typical of the entire macro-region, which was severely hit by the failure of 30 years of industrialization policies.

But if one overlooks the elements of competitiveness and productivity that are typical of this strategy, it is impossible to perceive any long-term alternative to the assisted development model. The only exception seems to be Puglia, the most effective and socially equilibrated of southern Italian regional structures, where higher productivity levels were reached in the last sub-periods and, as we have seen before, a relevant fabric of private services is developing rapidly.

In the other cases, deindustrialization may prove to be the only long-term effect of the absence of a solid and far-looking reconversion strategy. In some cases, plant closures are imposed by supra-national authorities like the European Community (the iron and steel plants in Campania); in other cases, closures are the long-term effects of private decisions and of the market mechanism. It is useful to remember, nevertheless, that the figures employed here refer only to firms with more than 20 employees, and that what remains outside may be of great significance for this kind of region.

As far as the core regions are concerned, the feed-back effect may easily be seen in the evolution of labour conflict. The Milano area in Lombardia, which accounted for 58 per cent of the working hours lost to strikes in the Northwest in the 'hot' period 1968–72, and for 13.2 per cent of hours lost in the country as a whole, has shown a substantial reduction in these percentages in recent years (1981–2) to 46 per cent and 10.1 per cent, respectively. The Lombardia and Piemonte share fell from 25 per cent and 12.6 per cent in the 1968–73 period to 22.9 per cent and 9.8 per cent, respectively, in the 1974–82 period.

Computing a relative index of industrial conflict for the Italian macro-regions, it appears that the general locational advantage of the Northeast and Centre shrank substantially over the last decade. In fact their index grew from 92 in 1969 (Italy = 100) to 99 in 1972 and 112 in 1982; in the same years the index of 'core' regions in the Northwest fell from 140 in 1969 to 117 in 1972 to the same 112 level in 1982.

Inter-regional interdependence and feed-back effects

The previous analysis suggests in a clear-cut way that it is necessary to adopt a dynamic approach to the problem of regional development, and to abandon the more traditional attitude linked to a 'structuralist' approach. There are of course important structural differences between regions that strongly influence the path and characteristics of their economic behaviour, but in the medium run two main elements appear that are not implicit in the initial structure:

(a) the birth of innovative and creative productive structures that, as said before, may spring from a condition of crisis and instability; and
(b) the existence of feedback effects in time that stem from the presence of some spatially fixed resources (physical space, environmental qualities, but increasingly also labour) and that force both success and crisis situations to destroy the spatial preconditions that brought them into existence. For example, structural change brought about by innovative development models (as in the Third Italy) may in the short run strengthen development itself through synergy effects and collective learning processes, but in the medium and long run it is likely to destroy previous preconditions of success, through increas-

ing labour and floorspace costs, congestion, and changes in social relationships and generational attitudes toward work and local society.

By the same token, a crisis situation may also help in overcoming the spatial conditions that generated it, as it pushes toward lower industrial conflict, decreasing labour costs and physical congestion of the built environment, and lower urban land rents.[12]

The idea of something like a regional like cycle emerges from the previous reflections. The subjective and innovative elements of local development processes may merge with a more readily foreseeable succession of 'objective' stages. This concept should not be taken as inevitable or mechanical, as it is subject to innovation and revitalization processes, but it tries to include them in a wider context of inter-regional interdependence.

In the early 1970s, the internal difficulties of core areas determined some 'objective' favourable conditions for the take-off of many intermediate regions, for example Third Italy regions, through the weakening of the exchange rate that they have brought about. Now the increasing labour costs in these latter regions, together with an upswing of competitiveness in core areas, stands in the way of unproblematic continuation of the previous development model, and requires a new wave of organizational, technological, and spatial innovations that are not yet apparent.

Some case studies of revitalization strategies

Let us now turn to a brief analysis of some successful (or unsuccessful) revitalization strategies. The first case regards the utilization of new flexible automation technologies in manufacturing for the relaunching of competitiveness, a process that proved effective in the Northwest in recent years. The second case regards the revitalization processes of big metropolitan areas like Milano (Lombardia) and Torino (Piemonte) that have reversed the previous trend towards deurbanization. The third case regards the management of the iron and steel crisis.

Flexible automation technologies in the Northwest

Flexible automation technologies employed in discontinuous processes look particularly appropriate for the revitalization of mature production. In fact, their characteristics are particularly suitable for the general socioeconomic and cultural context of old industrial metropolitan regions. They:

(a) are labour-saving technologies, allowing a substantial reduction of labour costs;
(b) are floorspace-saving technologies, especially when coupled with advanced organizational models of production management (like just-in-time practices);
(c) are flexible technologies, in three main senses: (i) they are versatile, in that they may be pre-programmed to operate on different products of the same family, in random sequence; (ii) they allow easy 'restyling' of products, even if manufactured with mass production equipment; (iii) they are convertible to other uses or production at low cost, estimated in 20 per cent of total investment;
(d) allow a rising standard in both labour productivity and quality control, reliability of the production process and managerial control;
(e) allow, through all the preceding characteristics, an easy introduction of product innovation, of new marketing strategies (for example, they allow 'customization' of the product at low marginal costs), and of new performing forms of procurement organization;
(f) are risky and complex technologies, and require continuous learning. This explains why these technologies diffuse slowly, at least in their 'integrated' forms of FMS and CIM (flexible manufacturing systems and computer-integrated manufacturing), but also give old industrial and metropolitan areas a lead in the adoption process itself, due to the presence of a long-standing 'industrial culture' and of a rich endowment in human capital.

Due to all these characteristics, the diffusion of robotic technologies in Italy has followed a 'conservative' spatial pattern. The location of production facilities is concentrated in two regions, Piemonte and Lombardia (with some new developments in Emilia), and these two regions alone accounted for more than 70 per cent of the installed robot equipment in 1984 (56.7 per cent in Piemonte and 17 per cent in Lombardia).[13] It is important to note that the rest of the robot park is installed in branch plants of big northern car producers (like Fiat and Alfa Romeo) located in the Mezzogiorno.

Through the adoption of these technologies a tremendous improvement in productivity was reached, even leading in many cases to more jobs. The most successful strategy in fact seems to be not to try to reduce costs, but to enlarge revenues through enhanced competitiveness, product differentiation and increasing product quality and reliability, all elements that may mean the possibility of raising prices.[14]

An important study of product and process innovation in the Milano metropolitan area (the province) was conducted in 1986.[15] This showed the importance of advanced technologies in the process of industrial revitalization and their strict relationship with the introduction of product innovations. In that year the Milano province was performing much better than the region (Lombardia) in terms of innovative behaviour, despite the traditional tertiary specialization of the area. This brings us to the second case study.

The case for metropolitan revitalization

The process of regional revitalization undertaken recently by Lombardia and Piemonte was based largely upon a revitalization process of their main metropolitan areas, Milano and Torino, respectively.

In fact, if the preconditions for economic and spatial creativity lie in the elements of competence and synergy, as stated above (p. 140), it may be maintained that generally speaking these elements are particularly apparent in large metropolitan areas, and that their importance is highly emphasized nowadays by the emergence of, in Freeman's words, the new 'technological paradigm' of information and communication technologies.[16]

As far as 'competence' is concerned, the large metropolis has always been marked by its ability to attract research facilities and advanced industrial and service production. The metropolis has traditionally been the right place for 'incubation' of small and medium-sized high-tech enterprises and for the location of vocational training facilities and technical and managerial schools. All these elements were found particularly in the Milano area.[17]

A new awareness of the need for a broader knowledge of the past and prospective role of the city has recently emerged among politicians and local governors. This has taken the form of two ambitious research projects sponsored by the two municipalities and the regional governments: Progetto Milano and Progetto Torino (Fig. 5.8).

Two other preconditions for revitalization may be found in the efficiency of the urban environment and in the general flexibility of the urban organizations, both public and private. The efficiency of urban structures may stem from the reduction of transaction costs allowed by the integration of different activities on the same territory, from the diffusion of communication networks linking the city within a global urban system, from the presence of large private and public decision-making centres and, last but not least, in the integration of the producers of advanced services with both industrial high-tech firms and firms undertaking profound modernization processes.

Flexibility in its turn may be considered to be one of the most interesting new conditions for revitalization. At the microeconomic level, this characteristic is linked to the development of new organizational and managerial styles in big metropolitan firms seeking to compete with the system of 'flexible' small firms. New techniques include the adoption of 'flexible' manufacturing systems, just-in-time logistic models, remote control systems through telecommunication networks and new flexibilities in the use of the labour force.[18] All these elements may be found in the recent revitalization processes of large industrial firms in the Northwest.

In the context of macroeconomic and social elements, a pronounced downward flexibility in relative factor prices has already been mentioned as an important process occurring during the crisis period of the 1970s. Another success factor in these two metropolitan areas was the ability of their public authorities to conceive and implement urban policies and infrastructure projects supporting the revitalization strategy. Examples of this kind are the advanced telecommunications projects launched at the regional level ('Lombardia cablata') or at the local level (in many 'systems areas' in the two regions), the plans for a rational use of derelict urban

Factors \ Relations	COMPETENCE	EFFICIENCY	FLEXIBILITY	SYNERGY
Macro-economic	Concentration of R&D "Incubator" effect of advanced industries	Presence of private decision centres Reduction in labour and floorspace intensity of production processes	Flexible automation technologies New managing practices in industrial relations and labour organisation Presence of small firms	New functional integrations (marketing–engineering–R&D–strategy) Integration between product-process innovation
Meso-economic	Presence of qualified factors in industry and services	Inter-industry linkages Sectoral diversification	Production services as adjustment cost-reducing factors	Interfirm integration (co-operation agreements, joint ventures)
Micro-economic	Vocational training and managerial schools	Inter-metropolitan communication networks	Downward flexibility of factors prices in a crisis condition	New industrial relations practices, no-strike agreements
Social	High educational levels	Increasing awareness of being part of an urban community	Consumer services as personal cost-reducing factors	New social solidarities Class integration
Political	Ambitious cultural projects	Presence of public decision-making centres	Urban revitalisation policies and projects	Reduced political conflict; private/public partnerships

Figure 5.8 Taxonomy of local success factors in metropolitan areas.

land and the plans for the development of new technology centres in Torino (Technocity) and Milano (Progetto Bicocca).

The synergy element is also a crucial factor in the revitalization process. At the microeconomic level, it may take the form of new strategies for integrating the main functions of the firm (marketing, engineering, research, strategy) for the purpose of accelerating the innovation process. Strategies of these kinds are visible in the 'baricentric' and therefore 'central' location (close to the Milano and Torino 'cores') of new integrated 'mission units' committed to the development of specific innovation projects by high-tech firms or by big firms engaged in advanced technological projects.

At the macroeconomic and social level, we find new industrial relations practices being adopted, together with new social solidarities and class integration. From the political point of view, we find reduced political conflict and the launching of a wide spectrum of partnerships between private and public institutions to run large projects ranging from advanced vocational training to infrastructure provision. Most of these undertakings were not even conceivable in a conflict-driven, short-sighted and politically static situation such as the one prevailing in the Liguria region.[19]

The difficulties of managing industrial decline: the case of iron and steel

The management of the iron and steel crisis in Italy is a telling example of the difficult trade-offs that confront politicians and managers when declining industries in declining areas threaten to trigger a vicious cycle of deindustrialization.

The first response of the national system to the crisis of the 1970s was typical of an assisted development attitude. While in all European countries employment in the iron and steel industry fell between 1974 and 1980 (from 232,000 to 197,000 in the Federal Republic of Germany, from 157,000 to 104,000 in France, from 194,000 to 112,000 in the United Kingdom), in Italy it grew. The state-owned financial holding in the sector, Finsider, did not present a balanced budget in any of these years, in spite of the short-term upswing of the years 1979–80.[20]

In these years, a sort of 'overcompensatory response' strategy was followed. This consisted of increased doses of old prescriptions to cope with the new demand problems (investing in capacity modernization and expansion), instead of trying new remedies (capacity reduction and rationalization).[21] The same strategy was followed in other contexts, for example in Belgium and France, where the response to the iron and steel crisis was often to invest even more in that sector.

In the following years, a series of restructuring plans, partly financed by the European Community, reduced jobs, but was not sufficient to restore competitiveness and financial equilibrium in big state-owned enterprises. Total national productive capacity fell by one-fifth from 1980 to 1986; the Finsider group lost 50,000 jobs in the same years but in 1987 still showed a loss of 1,600 billion lire (US$1.15 billion).

Managers and unions, the major players in the game, have found themselves in a critical position. Unions have tried to defend jobs, but in the longer run they have been forced to accept closures, with the effect of having to choose between the interests of labour in different regions. Liguria, hit by deindustrialization processes, was pitted against Campania, hit by never-ending and structural unemployment problems.

At the same time, the managerial structure of the sector, especially in state-owned enterprises, was weakening as the sector became increasingly unattractive in terms of the technical and managerial positions it could offer. Many strategic skills were lost, with the effect of decreasing response capability and decision-making ability. Conservative attitudes and high 'barriers to exit' (social, political and psychological) were the result of this condition. The introduction of 'exit prizes' assured by the European Community for plant closures is a typical example of a policy instrument created in order to overcome these resistances.

The general strategy followed in recent years of a soft management of the decline, assisted by public resources, is responsible for the slow pace at which alternative solutions are being proposed and evaluated. New conflicts are now emerging. For example, there are differences of opinion over the management of derelict areas, abandoned by the steel plants and eligible for alternative uses, between the local managers (imposing high 'managerial barriers to exit'), the local policy-makers, willing to control the reconversion processes and the local community of real-estate entrepreneurs, in a context where public assistance reduces the social cost and blunts the necessity for innovative decisions.

In the European experience, a general philosophy of this kind, which we may label as 'assisted management' of the crisis is not the only possible solution. The 'neo-corporatist' model followed in the Federal Republic of Germany, which allows strong state intervention but requires more strict bargaining between the interests and forms of co-determination through consultative committees, was in fact followed by the European Community in its general strategy for the iron and steel sector. Two other possible models or philosophies rely more deeply on market criteria (or ideologies):

(a) the 'centralistic conservative' model, which is typical of the recent British experience, where a 'hard' management of the crisis was

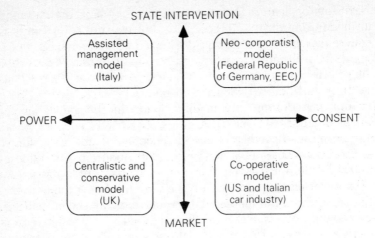

Neo-corporatist model: "soft" management of the decline, "three chairs bargaining", alternative job creation.
Centralistic-conservative model: "hard" management of the decline, clash with unions, low barriers to exit.
Assisted management model: rescues and public assiatence even in absence of social agreements; high barriers to exit.
Co-operative model: agreements on productivity gains, incomes and investments.

Figure 5.9 General philosophies in the management of declining industries (iron and steel).

characterized by a clash with unions, lowering of 'exit barriers' and privatization of restructured plants;
(b) the 'co-operative' model, based on internal agreements on productivity, incomes and investments. This last model is typical of other industries, where the problems are of a different intensity. An example may be found in some recent agreements in the automobile industry in the United States and in Italy (the Chrysler and Fiat agreements) (Fig. 5.9).

The choice between these alternative philosophies is not just a question of economics; it involves deep political, psychological and cultural elements, and depends on conditions in the local economy and labour market.

Conclusions

The concept of deindustrialization may be used in different ways. It may define the counterpart of a tertiarization process in the presence of a stable or growing total employment, as in the recent US experience (and I would call this a 'relative' deindustrialization concept). It may indicate a process of industrial job losses without the counterbalancing gain in tertiary jobs, as in the West German experience (Chapter 3) and in the Italian experience in the late 1970s (and I would call this an 'absolute' deindustrialization concept). It may be utilized to indicate a process of decline in international competitiveness, as in the traditional view of the Cambridge Economic School about the British experience. The indicators that may be used in these cases are very simple: a reduction in the share of industrial employment, a reduction of industrial and total jobs and a permanent disequilibrium in trade balance, respectively.

In this chapter, however, the concept of deindustrialization was used in a wider sense, with a deeper economic meaning. It refers to the inability of a regional production system to perform a process of industrial transformation successfully in the short and medium run. The focus of the statistical analysis shall be, therefore, on global industrial performance, synthesized by the combined indicators of relative productivity growth and relative employment growth with respect to the whole reference system. Intersectoral economic effects are indirectly considered, since the effects of advanced tertiary activities are primarily to enhance the effectiveness of the local industrial fabric, thus becoming incorporated into the indicator of industrial productivity growth.

Unlike other major economic transformations in history, for example the transition from agriculture to manufacturing, the present transformation from industrial to tertiary activities has not replaced industry as the main supplier of final goods – all production services function as intermediate goods, except for exports. The labour content of the final demand for consumer services is not expanding, and the main source for direct and indirect jobs remains the demand for industrial goods. Therefore, focusing on 'deindustrialization' processes, correctly understood, remains a relevant, sound and useful research programme.

Development always implies a process of 'creative destruction', at the aggregate as well as at the regional levels, particularly the substitution of declining or low value-added activities by new and dynamic ones. This process is seldom smooth, and it often implies accelerations, sudden jumps or transition difficulties; a crisis condition may easily emerge and prove to be long-lasting.

Research attention should focus on the insufficient or delayed adjustment capability of the *whole* regional productive structure, as revealed synthetically by the two combined indicators of *industrial* performance mentioned before. Nations usually adjust more rapidly to changing external and internal conditions, due to higher diversification of skills and activities, and due to the availability of a wider variety of economic policy instruments (macroeconomic policies and exchange rate variations). Regions usually do not have these tools at their disposal, and they may therefore suffer underdevelopment and industrial crisis for protracted periods, especially if they are in direct competition with more dynamic regions. In the theoretical literature they are therefore described as competing on an 'absolute advantage' basis, which may guarantee no specialization to the single region in the spatial division of labour, rather than on a 'comparative advantage' basis (a process that, after Ricardo, assigns 'paradoxically' a specialization even to the worst performing regions). Each local economy has therefore to define its 'vocation' actively, developing an economic 'advantage' and maintaining it in dynamic terms through effective reconversion and innovation processes.

The main conclusions of the empirical analysis of the Italian experience may be synthesized as follows:

(a) At the aggregate level, a condition of 'absolute' deindustrialization hit Italy in the early 1980s, in conjunction with a generalized slump in the world economy. Besides the weakness of the production sphere, an autonomous trend towards industrial rationalization took place at the same time, generating major employment cuts in manufacturing activities. Service employment, in particular in public administration, worked as a shelter and a shock absorber for the labour market.

(b) This condition was easily and rapidly overcome in the following years, from 1984 onward, when Italy developed at an average annual rate of 3.6 per cent. In these years industrial output grew at higher rates than the aggregate GNP (up 14.4 per cent overall in the period 1983–7, compared to 12.4 per cent of GNP),[22] led mainly by strong investment demand. These elements prove that it is impossible to speak of a structural deindustrialization in the case of Italy.

(c) In contrast, a condition of deindustrialization, with negative employment trends and lagging competitiveness, appeared during the 1970s in the three old industrial and 'core' regions of the Northwest, namely, Lombardia, Piemonte and Liguria. The causes were different: a general maturity of the economic environment, with a fall in industrial employment and a rapid growth in services in the case of Lombardia, where the Milano metropolitan area encompasses about half the regional product; a crisis centred on some big core firms (like Fiat and Olivetti) in Piemonte; and an exogenous crisis stemming from the worldwide slump in sectors in which the regional economy specialized: iron and steel and shipbuilding in Liguria.

(d) The ability of different regions to cope with deindustrialization was also completely different. Lombardia found important sources for revitalization in its diversified product structure and in a new linkage between advanced industrial production and advanced services. Piemonte followed some years later, through an important restructuring process of the main firms and their procurement context, while

Liguria, which was facing the most difficult task, failed to define a new strategy for reconverting resources towards new activities.

(e) Two elements greatly helped the revitalization process. First, the use of new production technologies, linked to new information and communication devices (flexible manufacturing systems, remote control over a dispersed production structure, etc.) relaunched the big companies through new technological and organizational flexibilities. Second, the revitalization of the big metropolitan areas of Milano and, to a lesser extent, Torino, was linked to a new role of the urban 'competence-and-control complex'. New economic synergies and new social solidarities eased this process, which was in some respects the feed-back effect of the previous crisis. In all cases industry rather than services was at the forefront of the revitalization process. Therefore, the term 'post-industrial' to depict this process may be rather obscure and misleading; 'meta-industrial' may be a better label, pointing out the importance of an industrial 'milieu' and of an industry–services linkage.

(f) Some regions of the South, characterized in the 1960s by an industrialization policy centred on the growth poles philosophy, are now facing serious development problems in the presence of a worldwide crisis in the sectors they specialize in (iron and steel and petrochemicals). In these cases, found in particular in Campania and Sicilia, the strategy of 'industrial conservatism' and of assisted development pursued in the 1970s only postponed any real solution to the problem.

6

Structural changes in the Spanish economy: their regional effects

JUAN R. CUADRADO ROURA

CASE STUDIES of specific economies have revealed complications in the use of the terms deindustrialization and tertiarization. Recent evolutions in both theory and data analysis appear to undermine a simplistic interpretation of how growth in services fits into change in the structure of economies at several levels.[1] The objective of this chapter is to study the deindustrialization and tertiarization of Spain, with special emphasis on its regional economies. This will be done by analysing the nation's economy from 1960 through to the present, by considering the effects of the economic crisis as manifested at the regional level, and by outlining the new economic profile of these regions that has emerged in recent years. Two 'macro-regions' will be analysed in detail, one in apparent decline, and another enjoying a period of economic growth. Some final remarks arising from the preceding sections will close the chapter, pointing out the differences between Spain and the more mature economies of the United States, the United Kingdom, the Federal Republic of Germany and France.

Recent developments and changes in the Spanish economy: an overview

From 1960 to the present the economy of Spain has passed through two very clear stages and is entering a third. During the 1960–73/4 period, its dominant features were transformation and strong economic development. The second stage, 1974–85, was characterized by adjustments to the international economic crisis. Most recently, four years of strong recovery (1985–9) have marked a new phase of high growth rates and new productive changes which will probably continue.

The development phase, 1960–73/4

In July 1959, Spain was suffering a marked crisis brought on by a dramatic lack of foreign exchange reserves (only US$65 million, a figure that virtually corresponded to immediate suspension of payments) and growing inflationary tendencies. Faced with these crises and the failure of the policies and the interventionism practised since 1940, the Spanish government, supported by the OECD and the IMF, devised a Stabilization Plan oriented to establish an internal–external balance. This emphasized flexibility, removing administrative and economic controls, and an international market orientation. The objective was to end Spain's isolation from the western world and to abandon autarky and internal interventionism.

At the time, Spain's per capita income was a mere US$374, 7.5 times less than that of the United States, 3.7 times less than the United Kingdom's and half Italy's. Of the employed population, 38.7 per cent worked in agriculture, 31 per cent in services and 30.3 per cent in construction and industry (where the productive limitations were important due to the international trade restrictions).

Four important trends were established as of 1960 and gained momentum around 1964. First, Spain experienced high and sustained GNP growth: an average of 7.4 per cent between 1960 and 1973 compared with 3.9 per cent in the United States, 5.6 per cent in France, 5.1 per cent in Italy, 3.1 per cent in the United Kingdom and 4.9 per cent in OECD countries. Second, extensive investments in productive capital and in infrastructure produced a mean Gross Capital Formation rate of 27.3 per cent of GNP during this period. Needless to say, these investments incorporated new technologies. Third, the Spanish economy was progressively integrated into the international economy. Foreign commerce led to foreign investments and the normalization of exchange rate policy. Fourth, this growth was accompanied by rapid structural changes. In 1973, employment in agriculture and fisheries had fallen to 24.3 per cent, while industry rose to 26.9 per cent (36.7 per cent if construction is included). Service employment rose to 38.9 per cent (see Appendix Table 6A.1). These changes simultaneously brought about a tremendous movement of the population towards the more dynamic regions (mainly to the Basque Country in the north, to Cataluña in the north east and to Madrid).

Industry was the great protagonist and the major beneficiary of this stage of development. The sector's mean growth rate during the 1962–73 period (excluding construction) was 10.1 per cent. This change affected all branches of activity, contributing to a diversification of the production system. As can be seen in Table 6.1, some sub-sectors underwent quite spectacular expansion: iron and steel (including automobiles and machine tools), chemicals, paper and graphic arts, construction materials, footwear and clothing. However, other sub-sectors registered far more moderate increases. Appendix 6A.2 gives more comparative data by branches of production.

There were five additional characteristics of this industrial expansion. First, there was a huge increase in the capital requirements of those sectors in which the economy was developing. Second, there was a decreasing demand for labour as the capital : product ratio increased. Third, intermediary inputs in the greater part of the manufacturing sector varied widely. This meant very different improvements in the added value of some sectors due to the great increase in value of the product and the decrease in the total cost of inputs. Fourth, Spain began to export more and more manufactured goods. In 1970, industrial exports already represented 65.5 per cent of the national total, which was radically different from the industrial expansion of the much more internal-market-oriented 1950s. Fifth, imports of industrial products also grew, reflecting the type of industrial structure. Unlike other European Community countries, Spain had an economy centred more on production of goods that were in low demand and mainly required by the internal market, and less on goods of strong and medium demand,[2] as reflected in the foreign trade

Table 6.1 Growth of industrial activities, 1960–73 (constant values, 1970)

Sub-sectors	1973 GDP as a proportion of 1960 GDP	Percentage participation of each sub-sector	
		1960	1973
Construction materials	5.29	3.56	5.31
Iron, steel, and metal transformation industries	5.18	23.44	34.27
Paper and graphic arts	5.16	3.75	5.46
Leather, clothing, and footwear	3.97	7.68	8.58
Chemical industry	3.54	12.74	12.73
Wood, furniture, and cork industries	2.99	6.37	5.37
Energy, gas, and water	2.98	8.49	7.13
Food and beverages	2.52	14.49	10.27
Mining	2.30	6.76	4.39
Textiles	1.81	12.72	6.49
Total	3.55	100	100

Source: Calculated from Banco de Bilbao. Renta Nacional de España y su distribución provincial.

Table 6.2 Foreign trade coverage (exports as a proportion of imports) in some European Community countries and Spain, 1975

Sub-sectors	Federal Republic of Germany	France	United Kingdom	Italy	Spain	Belgium	Netherlands	Denmark
Strong demand	1.94	1.15	1.38	1.08	0.28	1.08	1.37	1.94
Medium demand	2.24	1.45	1.14	1.22	0.73	0.91	1.16	1.47
Weak demand	1.03	1.08	0.83	1.78	1.27	1.31	0.66	0.61
Total	1.67	1.25	1.08	1.36	0.70	1.10	1.04	0.97

Source: Myro, R. 1988. *La industria española. Información estadística*. Madrid: figures from Spanish Input–output table 1975 and *Economie Européenne* **25** 1985.

coverage rate for 1975, under one in the case of the latter and above this in the case of the former (Table 6.2).

The growth of basic and manufacturing industries was accompanied by growth in the construction sector (with rates of 6.8 per cent as the mean average for the 1960–74 period) and in services (5.9 per cent real growth as a mean average). Construction responded to the development of infrastructure; the urban development process and the upward surge in tourism explain the sector's behaviour. Tourism, health and education were prominent sub-sectors in the growth of tertiary activities. These sparked off the uninterrupted tertiarization process in Spain.

There were four major causes for strong growth in this phase of Spain's economic development. First, a new economic framework was created that enabled the country to take advantage of both the economic boom of the European countries and the powerful desire for development existing within Spanish society. Second, Spain had a great reserve of cheap and still unorganized labour, which made expansion and more competitive production possible. Third, the domestic market – clearly developing thanks to the increase in income, consumer needs and actual intersectoral relations – remained, through diverse forms of protectionism, the reserve of Spanish companies or of those with foreign capital involvement in Spain. Lastly, Spain developed important foreign financial resources that made it possible to import equipment, accelerate technical change and purchase raw and intermediate materials. Three specific sources brought growing volumes of foreign reserves: the remittances of Spanish immigrants from Europe, foreign investment and contributions from the tourist sector.

However, this growth period also produced serious problems. Regional imbalance increased, and this was to be a significant factor later on. The system was unable to absorb the labour force expelled from the agricultural sector, with much of this excess emigrating to other European countries. Rapid expansion and transformation of the economy created heavy dependence on energy and other raw materials.

Economic crisis and productive adjustments, 1975–85

Differential notes on the case of Spain

The international economic crisis of the 1975–85 decade hit a Spain that, in terms of income level[3] and especially productive structure, was very different from the Spain of 1960. The country had not experienced phenomena such as 'stagflation' (although it had registered frequent inflationary tensions). Nor had it suffered the 'deindustrialization' already noted in other European economies. However, as acknowledged in several international reports, the real impact of the crisis was even

Table 6.3 The evolution of industrial demand

	Annual rates of growth				
	1972–75	1976–80	1981–83	1984–85	1972–85
Federal Republic of Germany	3.5	3.7	−1.7	3.0	2.4
France	3.1	3.8	−0.4	1.1	2.3
Italy	5.3	5.8	−4.7	3.4	3.0
United Kingdom	5.8	−1.5	0.3	4.4	1.8
European Community	3.4	3.3	−1.2	3.1	2.3
United States	2.8	5.0	0.9	6.6	3.7
Japan	5.5	10.0	3.5	8.0	7.0
Spain	6.1	2.4	−2.7	0.9	2.1

Source: Economie Européenne **25** (1985); Myro, R. 1988. *La industria española. Información estadística*. Madrid.

stronger for Spain than for the majority of European countries. A variety of reasons are implicated.

First, Spain had to face three types of problem simultaneously: the actual economic problems, political problems deriving from the transition from a dictatorship to a democratic system,[4] and social conflicts linked to the birth of free labour unions and the heavy demand for salary increases and other social benefits.

The second reason pertains to the fundamental problems of the production structure created in the previous phase, especially in the industrial area. On the one side, the heavy dependence on the home market led to an intense deceleration of industrial demand in almost every activity when the latter began to fall around 1975–6. Even though domestic demand had grown at a faster pace than in other European home markets before 1975, Spain's market now grew at a much slower pace (Table 6.3). In addition, several of the industrial sectors most affected by the crisis had registered a strong growth rate in their production capacity in the preceding years (the iron and steel industry, shipbuilding, automobiles, machine tools, electrical household goods, etc.) and were facing not only a fall in national demand and problems with exports but also a major increase in production costs (energy, heavy salary rises, greater tax pressures and increased financial costs) that put many companies on a downward run.

Third, another special condition in the case of Spain was the delay in applying adjustment policies. Due to the political situation between 1973 and 1976, the appropriate steps were not taken, as they were in other countries. In 1976–7, the imbalance in Spain's economy became much worse: an inflation rate of 26.4 per cent, a heavy foreign trade imbalance and an increase in unemployment. In July 1977, a serious adjustment programme was at last proposed and agreed upon by the government and all the political parties.[5] Unfortunately, from the end of 1978 until nearly the end of 1982 a new difficulty arose, mainly due to the political transition and the impact of the second oil price increase, inhibiting correction of measures to contain the inflation rate and curtail unemployment. No official efforts at industrial restructuring got under way although in the unprotected sectors and companies the market did its own restructuring. Only when the first socialist government was constituted (December 1982) was a tougher anti-inflation policy applied (in which the labour unions collaborated) along with a series of industrial reconversion programmes. By the end of 1985 the annual inflation rate was down to 8 per cent, a business surplus had appeared (allowing investment to begin), and the public business sector had begun to recuperate.

Industrial recession and tertiarization
Several facts are critical to an understanding of this period. The first is that Spain's unemployment rate rose from 2 per cent in 1973 to 21.9 per cent at the end of 1985 (2,970,000 unemployed).[6] All sectors, except for part of the service sector, contributed to this, but unemployment in the industrial and construction sectors was particularly significant. Between 1977 and 1985 (peaking in 1979) 883,400 industrial workers were laid off, as were 465,000 construction workers. Employment in the service industries increased by 53,000 due, in the main, to an expanding public sector. Agriculture continued to lay off workers, with employment down 910,200 in this period.

The structural changes within the population in employment for this period are important, as the variations in Figure 6.1 and the data in Appendix Table 6A.1 clearly demonstrate. Employment in agriculture was on a continuous decline. The weight of unemployment in services increased yearly although notable differences can be observed when the performance of the different branches is examined in detail (Table 6.4). Industrial unemployment fell by almost three points between 1976 and 1986, although the number of working hours per person also decreased, especially between 1975 and 1980. The construction sector behaved very much like the manufacturing sector, with employment falling and a low number of total working hours.

From the point of view of production, GNP growth rates were very low as of 1976 (showing negative growth in 1979 and 1981), with a mean of 1.57 per cent for the whole of the 1974–85 period. Industry grew even more slowly (mean 1.14 per cent) while services advanced faster than the national rate (mean 2.45 per cent). In general, the process of structural change in the Spanish economy was characterized, in relative terms, by a fall in agriculture, industry and construction and an increase in the services sector. A certain 'deindustrialization' and a continued tertiarization of the economy became evident. At the end of 1985, services represented 60.36

per cent of GNP, with a gain of 12.1 percentage points over 1974. Manufacturing industry lost 4.8 percentage points in its contribution to GNP between 1974 and 1985, and agriculture and fisheries maintained their historic recessive tendency, falling by 3.76 percentage points in their contribution to GNP.

The conjuncture of the drop in employment and the increase in production contributed to important gains in productivity for this period.

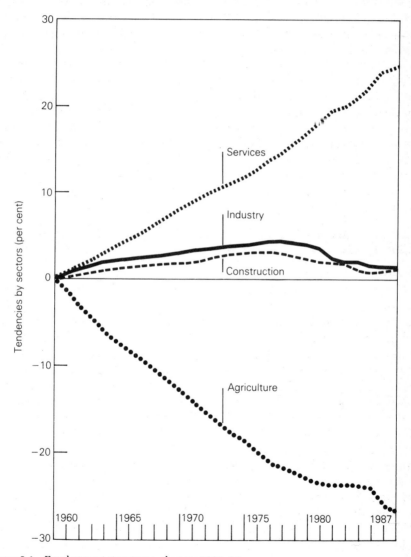

Figure 6.1 Employment structures, changes 1960–87.

The average productivity rate for the whole economy was about 3.1 per cent, for industry it was 4.7 per cent. This is compatible with a strong drop in investment, at least on a short-term basis. Gross Capital Formation dropped from 30.5 per cent of GNP in 1974 to 19.1 per cent in 1984, beginning gradually to recuperate in 1985.

As previously observed, data and percentages of an aggregate character conceal the internal changes to industry during the period under analysis. The economic crisis in Spain struck a sectorally structured industry very different from that of other European Community countries. As a result of the growth process, Spanish industry centred more on activities of low demand (several of which were also the most labour-intensive). Also, due to technological immaturity and a dependence on other industrialized countries, the strong demand sectors carried less weight in industrial production. At the same time, the lack of correspondence between the production structure and demand was translated into a foreign deficit, with the degree of import coverage being higher in the low-demand sectors mentioned above and the evolution of foreign markets forcing the introduction of production (and employment) adjustments leading to savings in intermediate consumption and a decrease in labour costs. The overall result was structural change that bolstered the strong-demand and

Table 6.4 Employment trends in the Spanish service industries, 1976–86

Employment increasing	Employment falling	Employment stabilized
Commerce intermediaries	Commerce (retailing and wholesale)	Hotels
Restaurants, bars and others		Railway transport
Communications	Repairs	Education and research (private)
Business services	Road transport	
Public education and research	Sea transport	
Public health	Air transport (private)	
Culture and leisure services	Real estate	
Air transport (public)	Social assistance	
Banking and financial services	Personal and home services	
Public administration, defence and social services	Insurances	
	Health services (private)	

Source: Author's calculations. Basic figures from National Institute of Statistics (INE).

Table 6.5 The balance of trade of Spanish industry, 1970–86 (exports less imports)

Sub-sectors	1970	1975	1980	1983	1985	1986
			(billions of 1980 pesetas)			
Strong demand	−150.3	−253.5	−250.4	−278.0	−265.7	−408.5
Medium demand	−61.3	−141.6	168.8	150.1	248.3	64.8
Weak demand	−4.0	65.7	185.6	456.3	485.5	244.8
Total	−215.6	−329.4	104.0	328.4	468.1	−98.9

Source: Figures calculated from Government statistics.

Table 6.6 Changes in the structure of Spanish industry, 1970–86

Sub-sectors	1970	1975	1980	1983	1985	1986
			(billions of 1980 pesetas)			
Strong demand	11.2	13.7	12.8	13.7	13.9	14.5
Aircraft	0.2	0.2	0.3	0.5	0.3	0.4
Information and office materials	—	0.2	0.2	0.4	0.4	0.6
Electrical machinery and materials	3.1	3.4	3.2	3.1	3.1	3.4
Electronic materials	—	1.2	1.3	1.8	1.8	2.0
Precision instruments	0.3	0.4	0.4	0.4	0.4	0.4
Pharmaceuticals	1.0	1.0	1.4	1.6	1.5	1.5
Chemicals	6.6	7.3	5.9	5.9	6.3	6.2
Medium demand	49.0	46.1	50.4	51.2	53.2	53.4
Rubber and plastics	2.8	3.7	2.9	3.0	3.2	3.1
Automobiles	4.8	5.5	6.7	7.1	7.6	8.2
Machinery and equipment	3.8	3.5	4.1	3.7	4.1	4.1
Railway materials	0.5	0.5	0.4	0.5	0.3	0.2
Other transport materials	0.4	0.3	0.3	0.3	0.3	0.3
Food, drinks, tobacco	21.4	20.7	21.5	23.2	23.8	22.9
Petrochemicals	10.4	6.2	10.1	9.0	9.5	10.2
Paper	5.0	5.8	4.4	4.3	4.4	4.5
Weak demand	39.8	40.2	36.8	35.1	32.9	32.1
Steel	5.7	6.7	9.0	9.5	9.8	8.5
Non-ferric metals	2.7	2.8	1.5	1.6	1.6	1.6
Shipbuilding	1.8	2.0	1.3	1.4	0.7	0.7
Metal products	4.5	6.9	6.9	6.6	6.1	6.0
Non-metal mining products	4.2	4.8	5.0	4.1	3.6	3.7
Wood, furniture and cork	4.3	3.8	3.6	3.2	2.7	3.1
Textiles	5.9	4.2	4.0	3.3	3.3	3.4
Leather	1.2	1.2	0.7	0.7	0.7	0.7
Clothing and footwear	7.9	6.3	4.1	3.8	3.5	3.4
Other manufactures	1.5	1.4	0.9	0.9	0.9	0.9

Source: Myro, R. 1988. *La industria española. Información estadística.* Madrid.

medium-demand industries (Table 6.6), reducing the importance of the low-demand industries, although the performance was not homogeneous in each bloc.

The new growth phase, 1985–9

Clear symptoms of the Spanish economy's recuperation were apparent by the second half of 1985. This tendency was maintained between 1985 and 1989, during which period Spain's GDP growth rates rose above those of its European partners (3.3 per cent in 1986, 5.5 per cent in 1987, 5.3 per cent in 1988 and 5.2 per cent in 1989, compared with European Community rates of between 2.9 per cent and 3.5 per cent, and the OECD's 3.3–4 per cent). This was accompanied by low inflation (5.5 per cent in 1988) and a high increase in investment, which increased Gross Capital Formation by over five points. At the same time, a rapid recuperation of employment had begun in industry (287,000 more jobs between 1985 and 1989), in construction (358,400 more) and in services (1,073,000 more), although many of these represented temporary contracts.

The economic upturn got under way between 1978 and 1985, along with the rebuilding of the business surplus, a climate of social accord and, of course, the stimulus of Spain's entry into the European Community as of January 1986. These bases of recuperation were supported by the parallel economic growth experienced by the western countries to which Spain is most closely linked.

From the sectoral point of view, industry has reached new, very high growth rates. Services have also contributed to growth, with rates very close to those of the whole economy (Table 6.7). Construction is experiencing a new boom.

The reactivation of the industrial sector has also been closely linked (together with the general causes listed above) to the rising expectations created by Spain's admission to the European Community which stimulated the acquisition of new equipment to modernize Spanish production processes and diversify its products. The machinery, transport equipment and chemical industries have experienced a more direct increase in demand. The food industry, with strong foreign participation, is undergoing very significant transformations. However, basic metals and textile industries continue to have a very weak growth rate and their reconversion has not yet finished, although some individual companies are doing very well.

Within the expanding service sector, an accelerated modernization process is under way that particularly affects commerce, banking and other financial intermediaries and services for companies. As in other countries, Spain is also developing services that require relatively unskilled workers, as well as rapidly growing public administration services and non-commercial services (mainly education and health, which require well-educated labour). This growth in the public sector has also

Table 6.7 Spanish GDP growth, 1986–89

	1986	1987	1988	1989
Agriculture	−11.0	9.5	3.9	−1.0
Industry	5.0	4.7	4.8	4.8
Construction	5.9	10.0	11.1	13.0
Services	3.7	4.9	5.1	5.0
GDP	3.3	5.5	5.3	5.2

Source: INE; 1989 figures are an estimate by FIES, Madrid.

been stimulated by the system of Autonomous Communities developed as a consequence of the new constitution of 1978.

Three major points outlined in this section are critical to understanding the regional effects of economic restructuring in Spain. First, the Spanish economy has been undergoing a radical transformation throughout the period under study. The structural change in terms of GDP and employment has been dramatic, with a constant fall in the primary sector, reinforcement and diversification of industry and a continued expansion of services. In fact, at least in terms of GDP, Spain's productive structure is very close to the European Community mean, with 6.6 per cent in agriculture, 27.7 per cent in industry, 6.6 per cent in construction and 59.1 per cent in services.

Second, the process of structural change continued during the central phase of the last crisis, although industry underwent a recession reflected in the GDP and, most of all, in employment. The recent recuperative years (1985–9) have, however, shown how industry has regained a certain leadership, almost completely rebuilding its share of output and, to a lesser degree, employment. At the same time, there have been important structural changes *within* this sector, as in the services (the relative increase of which has been continuous), particularly in regard to the relative significance of the high, medium and low demand industries. As changes continue to unfold in the current growth phase, the increasing complexity of the relations between industry and services clearly does not fit any simplified vision of a tertiarization of economies.

Lastly, these structural changes reflect many different positive adaptations, varying between sectors and periods of productivity. Agriculture has accumulated the most notable advances, mainly due to the drop in employment and the intensive mechanization. In industry, the improvement in productivity clearly slowed between 1973 and 1980, when once again it began to climb as employment dropped. In recent years there have been strong increases in output, with only a moderate high net creation of employment. The services have seen the lowest productivity increase in employment, as has been the case in other countries. Employment rose continually, as did production. However, some of the service branches were unable to absorb technical processes quickly. Also, public services expanded quickly, and their output is difficult to measure.

The regional impact of the industrial crisis

Some preliminary notes: regional trends and policy in the 1960s

It is too early to compare national deindustrialization in Spain to the experience of other European countries, since the Spanish process took a

Figure 6.2 Regions of Spain.

different form and has proceeded at a different pace. However, at the regional level there are more points in common with other European countries.

The Spanish regions to be analysed here are the current 17 Autonomous Communities (see Fig. 6.2), which are equivalent to European Community level 2 regions. In accordance with the 1978 Constitution, these Autonomous Communities have very considerable political and economic autonomy, including separate administrations and regional parliaments.

Several factors affected the regional economies. Development during the 1960–73 period reinforced the migratory movements from the agricultural and underdeveloped areas to the main growth areas which had begun in the 1950s (that is to say, Cataluña, the Basque country, Madrid and also the Communidad Valenciana). The population of these regions rose by 7.9 percentage points of the national total, while regions like Extremadura, Andalucia, Castilla–La Mancha, Castilla y León, Galicia and Aragón lost population.

Production also became more concentrated. Cataluña (particularly

Barcelona and its metropolitan area), the Basque country (and especially the Gran Bilbao area) and Madrid have been beneficiaries of this concentration in terms of GDP, although other provinces like Valencia, Alicante and Sevilla also shared this process.

From the territorial aspect, this strong expansion has followed a double line: on the one hand, it is concentrated in areas where a strong industrial presence and tradition that already existed before 1960 (Barcelona, Madrid and Vizcaya), now extends some 60–100 kilometers outside these; on the other hand, numerous other centres developed, either because they already had a base and locational advantages (Valencia, Alicante and Zaragoza), or owing to the support of activities from regional policies. The resulting distribution is reflected in Figure 6.3, simultaneously concentrated and dispersed.

Finally, rapid structural changes also occurred in all the regions, with an increase in tertiary employment (particularly in the tourist provinces

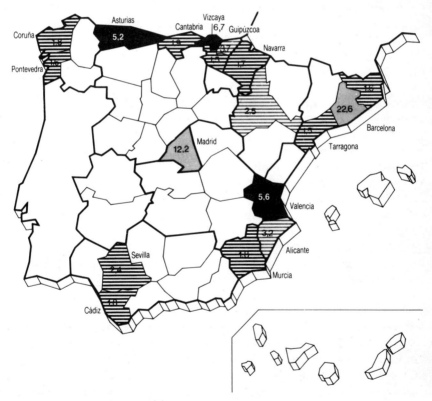

Figure 6.3 Geographical distribution of industry (1973 percentage of participation of industry at national level).

and industrial nuclei), a drop in primary employment and advances in industry and construction. Needless to say, these structural changes came at a different pace in each region, and exhibited remarkably different results at the end of the period.

From the point of view of regional policy, various aspects stand out, even in a very general form.[7] First, Spain was a highly centralized country in this period; between 1960 and 1975 all regional policy was programmed and directed from the central administration in Madrid. Even the very existence of 'regions' was not acknowledged, basically for political reasons, and the activities (growth poles, fiscal and credit incentives, agro-industrial development areas, etc.) carried out by the authorities were always of a local or, at the very most, provincial character.

However, official preoccupation with regional imbalances was far greater than it had been in the 1940s and 1950s, particularly during the three Development Plans carried out between 1964 and 1975. These efforts were designed to create or expand new industrial nuclei in the less-developed areas (mainly through a growth poles policy) and to promote particular areas of preferential industrial localization. Other instruments employed to develop some backward areas were: zones of interest for the development of tourism, new irrigation and the allotment of new infrastructure, and a few integral programmes for very depressed areas in Extremadura, Andalucia and Castilla, as well as measures favouring the diffusion of industry outside the main metropolitan areas. In one way or another, the intention of all these actions was to mobilize private investment toward certain places in Spain to develop their potential growth capacity or, at least, to improve the historic economic depression suffered by regions like Andalucia, Extremadura, Galicia the two Castillas and Aragón. The actual instruments used were very similar to those of other European countries (e.g. France and Italy), and included subsidies, credits, tax and export advantages and improvements in infrastructure.

Aside from a clear preference for highly localized improvements and an absence of plans for entire regions, the regional policy of this period had other clear characteristics: first, sectoral and national policies nearly always took precedence over regional objectives; and, second, the activities were geographically dispersed and, in general, the policy lacked continuity. Some results were obtained, e.g. the establishment of new industrial poles, and the regional policies were far more ambitious than any prior efforts and comparable with those of other European countries. However, the actual results were on the whole quite modest and did not counteract the effects created by market forces. Even the relative success of some of the industrial growth poles (Burgos, Huelva, Valladolid and Zaragoza) appear to be linked to developments of sectoral character and to the advantages that these cities' locations offered, rather than to the support measures granted, or the creation of adequate infrastructure which, in each case, was clearly deficient.

The effects of the crisis

The impact of the crisis is no longer felt in any of the Spanish regions, from the most depressed to the most industrialized and dynamic. Nevertheless, a careful analysis leads to two important conclusions. First, the regional effects of the crisis have been widely dissimilar in intensity and, most of all, in their consequences. Second, the economic–regional map of Spain is undergoing a series of changes that is altering the traditional growth areas.

The trajectories that the different regions are following are clearer if we divide the 1973–85 period at 1979, the year of the second oil price crisis. Growth rates dropped in the first sub-period, but industry was not always the culprit. The recession in tourism, one of the first sectors to be affected by the international crisis, was also strongly felt in the Canary and Balearic Islands.

In the second sub-period, the drop in the regional growth ratios is far more patent (Table 6.8) and extended into all parts of the industrial sector,

Table 6.8 Regional growth of GRP and industry regional growth, 1973–85 (real rates of growth per year)

Autonomous Communities (regions)	1973–79 GRP	Industry	1979–85 GRP	Industry
Baleares	0.1	2.7	4.1	2.5
Extremadura	1.8	4.4	2.3	0.7
Canarias	2.8	1.5	2.3	2.6
Comunidad Valenciana	3.0	3.8	2.0	1.8
Murcia	3.6	5.2	2.0	−0.3
Aragón	2.9	3.0	1.8	3.8
Navarra	2.7	4.9	1.7	1.1
La Rioja	4.3	7.1	1.6	2.4
Andalucia	2.6	2.8	1.5	0.0
Galicia	4.5	5.5	1.4	1.6
Castilla y León	3.0	6.0	1.3	0.6
Madrid	3.7	3.6	1.3	0.2
Castilla–La Mancha	2.0	3.6	1.1	2.2
Cantabria	2.0	0.7	0.8	−0.2
Cataluña	2.7	2.4	0.7	0.4
Pais Vasco	0.4	0.4	0.7	0.9
Astuarias	1.5	0.7	0.3	−0.5
Spain	2.7	2.9	1.4	0.8

Source: Calculated from Banco de Bilbao. Renta Nacional de España y su distribución provincial.

Table 6.9 Evolution of regional employment by sectors, 1973–79 and 1979–85

Autonomous Communities	Change of employed population (residents) 1973–79	Percentage employment change 1973–79				Change of employed population (residents) 1979–85	Percentage employment change 1979–85			
		Agriculture	Industry	Construction	Services	Total		Agriculture	Industry	Construction	Services	Total
Andalucia	−14.0	−19.1	−15.8	−11.2	11.6	−5.8	−11.9	−35.6	−15.7	−23.4	5.3	−12.5
Aragón	−7.1	−30.4	−6.4	17.3	9.2	−5.9	−14.0	−20.3	−12.1	−32.6	4.2	−8.8
Asturias	−4.5	−13.7	−5.7	−14.8	14.7	−2.8	−13.4	−30.7	−14.5	−11.8	2.8	−12.7
Baleares	1.1	−18.6	−18.7	−16.9	12.2	5.0	−10.8	−39.4	−7.5	−6.7	34.6	13.6
Canarias	0.2	−10.4	−14.3	−19.6	26.5	5.0	−10.5	−33.2	−11.7	−26.7	14.1	−3.5
Cantabria	−0.9	−13.3	−0.9	4.5	17.9	1.4	−13.6	−29.2	−22.1	−15.2	5.1	−13.6
Castilla–La Mancha	−15.9	−30.4	−4.2	−3.9	6.1	−11.7	−9.0	−23.5	−9.6	−18.4	−0.1	−11.5
Castilla y León	−12.6	−26.2	1.2	4.7	2.1	−9.1	−11.4	−27.4	−10.4	−19.0	0.2	−12.4
Cataluña	−4.6	−27.3	−13.0	−12.2	13.6	−3.9	−17.0	−22.7	−16.5	−32.8	5.9	−8.1
Comunidad Valenciana	2.0	−17.3	−2.6	−6.6	13.9	0.3	−14.5	−25.4	−10.2	−29.1	10.6	−5.5
Extremadura	−25.7	−32.8	−18.1	−10.6	2.6	−17.7	−10.7	−27.0	−19.6	−8.1	7.1	−10.5
Galicia	−6.8	−17.2	−2.4	8.9	9.0	−6.0	−9.0	−12.8	−12.7	−22.9	3.6	−8.6
Madrid	−4.9	−27.9	−5.8	−29.7	10.1	0.1	−8.6	−29.0	−11.2	−27.5	6.3	−1.6
Murcia	−3.4	−11.6	−0.3	−12.9	5.0	−2.7	−7.7	−26.5	−18.0	−32.0	14.7	−8.0
Navarra	−6.6	−33.3	1.3	−4.7	14.0	−3.7	−8.4	−26.5	−16.8	−17.7	8.9	−8.8
País Vasco	−8.2	−13.8	−10.1	−9.8	5.6	−4.6	−15.0	−31.1	−23.8	−24.8	−0.3	−14.8
Rioja, La	−6.2	−31.8	−3.6	−4.1	9.3	−9.4	−21.2	−33.6	−13.5	−26.7	11.4	−10.8
España	−7.4	−21.5	−8.3	−10.7	10.9	−4.4	−12.4	−25.9	−14.7	−24.8	6.5	−8.3

Source: Figures calculated from Banco de Bilbao. Renta Nacional de España y su distribución provincial.

with many rates close to zero and even negative (Asturias and Cantabria). However, the Balearic Islands recuperated as tourism revived, and some provinces in Cataluña, the Communidad Valenciana and Andalucia also felt these positive effects. The regions that best withstood the crisis with appreciable industrial growth were Aragón (3.8 per cent), La Rioja (2.4 per cent), the Communidad Valenciana (1.8 per cent) and the rest of Cataluña (2.5 per cent), if Barcelona is excluded. The existence of an industrial network composed of many small and medium-sized manufacturing companies with a greater capacity for adapting to the crisis, insulated from the downturn in the home market and having export possibilities, explains this improvement. In the case of Aragón and the Comunidad Valenciana, the recent implantation of large automobile factories and more auxiliary industries also had a positive effect.

From a global viewpoint there were several other trends. The first was the almost total and immediate stemming of inter-regional migratory flows, which halted the demographic concentration process under way before the crisis. Only short-distance population movements continued, from the large cities towards their immediate surroundings, and from the small towns and villages to the provincial capitals and some of the intermediate centres. The second is that the industrial recession continued, although employment rose at a somewhat slower pace than before. The services reaffirmed their historic growth, at lower rates than in the past, but higher than those of the other sectors. This, along with the recession in industry and construction, led to clear structural changes (see Appendix Table 6A.3) in which tertiarization appeared as a generalized phenomenon and deindustrialization, in terms of regional GDP, was more evident in some regions (Asturias, the Basque country, Cantabria, Cataluña and Madrid) than in others.

Taking employment as a reference, the effects of the crisis are perhaps even clearer. Within the sub-periods described above, employment throughout Spain fell first by 4.4 per cent and then by 8.3 per cent. However, the differences between regions (Table 6.9) are extremely important in both phases. Only the services nearly always gave positive variation rates. The drop in agriculture and in construction was general and, together with industry (a further 883,400 unemployed between 1976 and 1985), contributed to the relatively large losses in the Basque country, Cantabria and Asturias, followed by Andalucia and Castilla y León.

The deindustrialization effect at the regional level

Although virtually all industrial activities were affected by the crisis, it is also true that in some specific branches – mining, basic metal industries, transformed metals, the textile sector and transport equipment – these

Figure 6.4 Main areas affected by the industrial crisis.

effects were exceptionally grave. In some regions (Asturias, the Basque country, Cantabria, Cataluña and Madrid) the special concentration of these types of industry created an even greater impact, while in others, either because the weight of these industries was less or because they had greater product diversification, the results were less traumatic.

However, when we talk about affected regions we are giving a false picture of reality. The industries are localized in a concentrated form in many cases and the recessive effects end up being far more local than regional. Sometimes they affect virtually an entire city if it is tightly linked to a specific activity, as in the case of the shipbuilding industry in Spain, where recession heavily affected areas like Cádiz, Ferrol and Vigo. In other cities, crises in the iron and steel and other metals industries added to the problem. Figure 6.4 shows the main areas affected by the industrial crisis. In some cases such as Barcelona and its surroundings, Madrid and Vizcaya-Guipúzcoa, somewhat wider territories were involved but, in others, the effect was highly localized in one city and some nearby towns, even though negative effects were provoked in the entire region.

In any event, the industrial crisis has been the key to the evolution of the regional map of Spain. The application of shift–share analysis techniques to the regional figures of gross added value to the 1973–85 period, broken up into sectors and branches, makes interpretation easier.[8]

Table 6.10 shows some of the results. The positive or negative net total effect now gives us a first indication of the evolution of the different regions during the period under analysis. However, these figures only reflect the overall performance of each region in aggregate form, i.e. including agriculture, industry, construction and services. If we centre specifically on the case of industry, the only regions that obtain positive net total effect (NTE) values are Galicia, La Rioja and Cataluña (excluding Barcelona). To these the following could be added: Aragón, the Balearic Islands, Castilla y León, Extremadura, Murcia, Navarra and the Comunidad Valenciana, whose negative values are quite low. In a first approximation, these would seem to be regions that felt the effects and adjustments provoked by the crisis less strongly.

The breakdown of the NTE in industry into structural effects (SE) and differential effects (DE) permits us to say something else in this respect. Two effects are very clear. First, industry would have had negative

Table 6.10 Shift–share figures by regions, 1973–85

	All sectors			Industry	
	Structural effect (SE)	Differential effect (DE)	Net total effect (NTE)	Structural effect (SE)	Differential effect (DE)
Andalucia	−82,504.6	28,115.9	−54,388.7	−135,354.9	−38,281.0
Aragón	−32,708.6	46,684.9	13,976.3	−52,143.7	50,311.1
Asturias	−36,292.8	−102,814.8	−139,107.5	−77,616.1	−71,366.9
Baleares	67,039.3	−15,594.0	51,445.3	−14,895.6	6,719.1
Canarias	42,013.2	60,464.9	102,478.0	−21,240.5	2,269.0
Cantabria	−10,855.0	−22,972.3	−33,827.3	−28,203.5	−24,128.0
Castilla–La Mancha	−106,454.6	−14,976.1	−121,430.7	−40,275.0	24,978.6
Castilla y León	−78,054.8	27,687.7	−50,367.1	−78,857.3	70,203.6
Cataluña	17,209.0	−195,799.9	−178,590.9	−396,684.4	−110,280.0
Extremadura	−35,562.7	17,724.4	−17,838.3	−15,734.8	6,082.5
Galicia	−52,197.8	177,690.7	125.492.9	−65,282.6	69,804.8
Madrid	391,544.4	106,470.4	498,014.2	−180,054.2	1,329.3
Murcia	−6,024.0	44,891.7	38,867.7	−27,656.6	9,221.2
Navarra	−19,449.3	7,941.7	−11,507.6	−25,888.9	17,868.1
Pais Vasco	−36,285.8	−354,326.7	−390,612.5	−178,497.5	−115,598.2
Rioja, La	−11,321.4	20,432.8	9,111.4	−8,954.0	17,331.5
Communidad Valenciana	−10,094.5	168,378.8	158,284.3	−147,122.1	83,535.4

Source: Author's elaboration of figures from the statistical series *Renta Nacional y su distribución*. B.B.

performance (SE) in all the regions in the period under study, even though its relevance would have been different in each individual case. Second, industrial performance was even more negative (DE) in some specific regions (Andalucia, Asturias, Cantabria, Cataluña, nearly exclusively Barcelona, and the Basque country) as a result of the localization of industries that were particularly badly affected.

A further breakdown indicates fairly clearly which branches caused the displacements and, where relevant, the localized effects in each region. The diagnostic figure thus appears much more exact with respect to what could be classified as the specific profile of the sectoral adjustment at the regional level. The Basque country, Madrid, Barcelona, Asturias and Cantabria show the most intense locational disadvantages and, within these, the recession in metals and minerals, mineral and non-ferrous products, metal products and machinery, chemicals and textiles (in the case of Barcelona), sectors which showed major industrial adjustments during the 1973–85 period.

In those regions with a relatively high level of industrial development (the rest of Cataluña and the Communidad Valenciana) as well as in various less-industrialized regions, the good performance of some traditional activities (food and drink, metal products, clothing and footwear, furniture and other manufactures) and the presence of sectors with a future (electronics, optics, chemicals and pharmaceuticals) permitted much better results during the critical period. It should be remembered that services played a relevant compensating role, mainly in the many provinces along the Mediterranean coast.

The previous analysis and the data on growth rates, and the evolution of employment and unemployment by regions allows us to define the regions in which the deindustrialization effect has been present. They are, essentially, the three in the so-called 'Cantabrian Cornice' (the Basque country, Cantabria and Asturias), Madrid (which, however, has amortized its effect through the expansion of services), Barcelona and some more isolated industrial nuclei (Ferrol, Vigo, Cádiz), the latter being situated in less-developed regions.

Reconstruction of the industrial structure and areas of attraction

As explained above, aggregate data on the Spanish economy show an increasing recuperation of the industrial sector after 1984, which dovetailed with that already under way in services and construction. However, in the last few years not all the regions have reacted in the same way.

The study of the sectoral and spatial location of investments is an excellent tool for defining the changes under way in the more dynamic

Table 6.11 Location of new industrial investments, 1980–87
(Provinces with higher intensity indicator)

Province	New investments indicator	Total investments indicator[1]
Barcelona	4.152	3.738
Tarragona	2.647	1.554
Zaragoza	2.523	1.682
Madrid	2.395	2.967
Valencia	2.128	2.816
Alicante	0.945	1.143
Castellón	0.493	0.470
Baleares	0.396	−0.204
Navarra	0.238	0.608
Cádiz	0.087	0.490

Source: Author's elaboration. See Cuadrado, J., and Aurioles, J. 1989. *La localizacion industrial en España*. Ed. FIES: Madrid.
Note: 1) Including not only investments in new firms but substitution and enlargement investments. See note 11.

areas. In some ways, the investments, especially the 'new' ones (those determined by the opening of new factories) also anticipate the future. Therefore, an analysis of industrial investment (reinvestment, amplification and actual new ones) in Spain with data from the 1980–7 period allows us to draw clear conclusions.[9]

First, it has been proved that the case of Spain roughly coincides with other industrial trends in Europe in two ways.[10] Metropolitan areas with diversified industry have managed to regain their appeal to investment and to renovate their industrial network once over the crisis phase. This has not occurred in the old industrial regions dominated by more traditional products in recession (steel, coal, shipbuilding, etc.). Also, those regions that appear to be the most attractive to new investment and industrial plants are those that are well situated geographically and have a mixed industrial composition, although not yet highly industrialized. In certain cases, the companies are born and even look for nuclei and areas with low industrial tradition.

The second important conclusion to be derived from an analysis of the investments is that the provinces that obtain the most prominent position with respect to the creation of new companies and/or the attraction of others to these, form two virtual axes: that of the Mediterranean coast, with very high intensity indicators (Table 6.11) for Barcelona, Tarragona and Valencia, and that of the Ebro Valley, running from the Mediterranean to Navarra, with its main centre in Zaragoza.[11] Aside from these axes, only Madrid, Cádiz and the Balearic Islands stand out. Figure 6.5 illustrates both axes when we note those provinces with the highest new investment indicators.

In the same way, analysis of plant location of high-tech industries can give us an indication of how investments in these peak sectors are reflected at the regional level. In this respect, we will follow the definitions and results of Giraldez.[12] Activities such as high technology are included: pharmaceuticals, computers and office equipment; electrical material for measuring, signs, programming and control; telecommunications; aeronautics; precision instruments; medical equipment; and optics and other related materials. The study covers investments within the 1975–85 period and the main results are as follows.

First, the sectors that received the largest investments in this entire period were pharmaceuticals and electrical goods and equipment. As of 1980 activities linked to electronics and telecommunications also received growing investments. In a lower position were the optical, precision instruments and electro-medical apparatus and equipment industries. Investments have also improved and amplified already existing industries (an average of 75 per cent in this period) aided in many cases

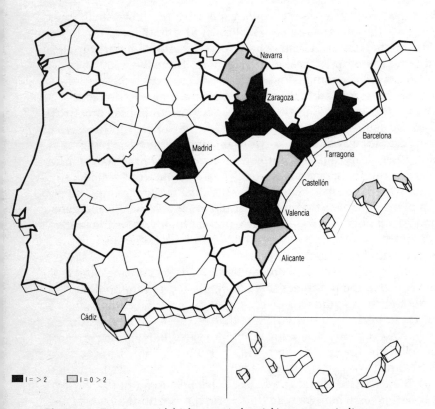

Figure 6.5 Provinces with higher new industrial investments indicator.

by foreign investment from the European Community countries, the United States and, more recently, Japan.

It is not easy to determine exactly where investments went (partly because, as we have just pointed out, the factories were already in existence) but, taking them as a whole, we can talk about a strong spatial concentration. Four provinces received more than 7 per cent of the period's total investments in high-tech activities: Barcelona, Madrid, Valencia and Vizcaya, which together reached 61 per cent of the national total. To these, some of the provinces close to Madrid should be added (Segovia, Avila, Toledo). Valencia, Zaragoza, Valladolid, León and Cantabria had somewhat lower percentages. Málaga, to the south, captured some important investments in electronics and telecommunications. Valencia is also in a good position, but Madrid and Barcelona easily dominate the field of electronic components and those of controls, electro-magnetic material and pharmaceuticals. The final conclusion about high-tech activities is that they tend to strengthen the role of the aforementioned large metropolitan areas, even though a search for new areas of localization can be observed, some of which lie on the Mediterranean coast.

To complete the picture, it is important to consider the varied success of the 'Urgent Reindustrialization Zones' (ZUR) created by the authorities to try to compensate for the negative effects of the industrial reconversion policy in some parts of the territory. Through grants and incentives, each ZUR was to attract new industries to the areas considered as those most affected by the reconversion.

The incentives for industry to locate in a ZUR have not acted in the same way in all regions. While Barcelona, Madrid and Sagunto (Valencia province), for example, have already met their objectives, investments are beginning to turn towards Bilbao (Nervion), Ferrol and Cádiz. Second, a substantial number of companies or multinationals might have settled there anyway, without any special incentives, due to the favourable conditions in the areas. ATT's case in Madrid is a good example. In Barcelona, the presence of German, Italian, Japanese and US companies is very prominent, but this is because today Barcelona occupies, without the ZUR, one of the first and most prominent positions with respect to foreign investment in Spain.

Recent economic recovery and the new economic–regional map of Spain

Even though the crisis has affected various zones and points of the territory, some have remained clearly damaged and show difficulty in recuperating, while others have either managed to recover well, or appear to be in even better positions to meet present conditions and anticipate the future.

The first case can be clearly observed in two of the northern regions, Asturias and the Basque country, to which a third, Cantabria, can be added in part. Together, these form a large industrial area in decline. In the second bloc we have the regions that showed the best performance during the crisis and seem to have overcome it. These include the Mediterranean Axis, from Girona to Murcia; the Ebro Valley, from Navarra to Tarragona, where they meet with the Mediterranean Axis; and the two archipelagos (the Canary and Balearic Islands), which have benefited mainly from tourism and a production structure with no grave problems.

Madrid, whose industrial base was badly damaged by the crisis, appears to be an attractive centre for new plants. In recent years, its role as the capital of the service industry has grown, a tendency that Spain's integration into the European Community has particularly favoured.

The data relating to regional conduct during the 1986–8 triennium, with its strong economic recuperation, have confirmed these trends and given greater form to Spain's economic–regional map. The indicators contained in Table 6.12 illustrate this. The Canary and Balearic Islands are two singular cases. They have undergone tremendous expansion since the beginning of the 1980s. The two axes continue to do well economical-

Table 6.12 GRP growth by Autonomous Communities, 1986–88

Autonomous Communities	GRP growth (percentage)		
	1986	1987	1988
Andalucia	1.8	6.4	6.4
Aragón	4.3	6.4	5.3
Asturias	1.8	3.5	3.6
Baleares	6.3	5.9	7.7
Canarias	5.7	6.7	7.4
Cantabria	3.6	4.9	5.7
Castilla–La Mancha	0.4	5.6	5.2
Castilla y León	0.4	6.4	5.7
Cataluña	3.7	5.6	5.4
Comunidad Valenciana	4.6	5.6	6.5
Extremadura	2.4	6.2	5.3
Galicia	1.9	4.2	5.1
Madrid	3.9	4.7	4.0
Murcia	3.4	5.9	6.2
Navarra	3.7	6.2	5.2
Pais Vasco	1.6	3.1	4.5
Rioja, La	2.2	5.6	6.2
España	3.0	5.4	5.4

Source: Fundación Fondo para la Investigación Económica y Social. Madrid.

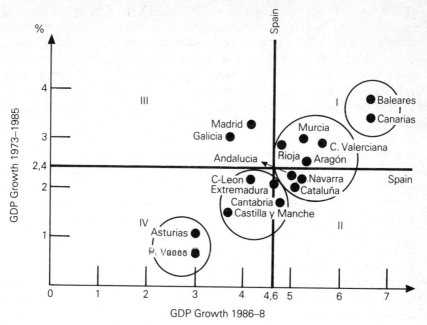

Figure 6.6 Regional growth, 1973–85 and 1986–8.

ly and it is quite clear that although Asturias and the Basque country, the two industrial regions with problems in 1988, have obtained very positive regional growth rates, this does not imply that the north is not still in recession. Between 1985 and 1988, this area continued to lose economic weight within Spain, as did the less-developed regions situated in the interior.

Figure 6.6 expresses the growth tendencies of the Spanish regions in the 1973–85 and 1986–8 periods in both comparative form and as a whole. Provisional figures for 1989 do not substantially change the tendencies described, in spite of the less favourable results of tourism activities which negatively affected the Balearic and Canary Islands.

What happened to the regional policy?

For the 1975–85 period, it is easy to answer this question: regional policy was virtually non-existent. While it existed officially, its real significance was negligible.

Throughout these years, some forms of regional policy inherited from the previous period continued, sometimes simply because their legal validity was extended but, in fact, all traditional regional development policy either lost importance or dropped out of sight, for several reasons.

This was, first, because the serious problems constituting the economic crisis (inflation, foreign trade imbalance, generalized unemployment, growing public deficit) demanded macroeconomic intervention, leaving no margin for regional nuances, nor many funds for them. At the same time, the industrial reconversion programmes were tackled from a sectoral and national angle. Specific regional measures (i.e. the aforementioned ZURs) were only taken when strong social protest arose in very specific locations where reductions in the labour force were imposed and companies were closed down, some of these being state-owned.

It should also be remembered that Spanish regional policy had been linked to indicative planning and, on abandoning this and substituting short-term for long-term and medium-term economic objectives, the regional objectives and instruments were left with practically no support whatsoever. All this also took place in circumstances in which the traditional problem regions (the most backward) were joined by others that had been until then the richest and the most industrialized.

Lastly, Spain differed from other European countries during this period because the establishment of democracy went hand in hand with the new political and administrative reorganization of the state. A redistribution of power was initiated as of 1978, linked in turn with the birth of the Autonomous Communities. The central government must now share its power with the regions or transfer it to the latter in various sectors, such as agriculture, industry, transport, tourism and education.

By the end of 1985, there was a regional policy with a very different profile from that of the previous efforts. One of its basic components derives from Spain's integration into the European Community, which has allowed it to benefit from the structural community funds (mainly the FEDER), and from the requirement to design medium-term regional plans for the ten Spanish regions that qualify as backward at the European level, and the declining industrial regions.

Regional activities have been reinforced from two angles at the same time. First, an Inter-regional Compensation Fund was established by the new Constitution to support greater economic equality among the regions. Second, because the powers now ceded to the regions (Autonomous Communities) give a wide range of action to regional governments. Today Spain is virtually a federal state. The regions have even more capacity to perform than their equivalents in some federal countries. Some regional governments are taking advantage of this autonomy and developing innovative programmes in support of industry and services, for the diffusion of new technologies (i.e. technology parks and institutes, enterprise centres, etc.) and for promoting and developing new markets for their companies. The case of the Communidad Valenciana, Cataluña and the Basque country are good examples.

Figure 6.7 The Cantabrian Cornice and the Mediterranean Axis.

The contrast between two macro-regions

In the previous section we examined Spain's recent evolution from the regional point of view: the incidence of the industrial crisis, the dynamism of services and the more recent phase of economic evolution are clear. It is impossible to analyse each individual region, but we will examine the main features and trends of two 'macro-regions' (on a national scale): the one denominated the Cantabrian Cornice and the other beginning to be called the Mediterranean Axis (see Fig. 6.7).

These areas are not completely homogeneous but, as we shall see, there are points in common with similar 'regions' of Italy, France, the United Kingdom and other European countries.

The Cantabrian Cornice: deindustrialization and decline

The whole of the Cantabrian Cornice can be seen as a declining region, although one of the integrated Autonomous Communities, Cantabria, is faring better because of its more balanced economic structure. Three Autonomous Communities are included in the region (Asturias, Cantabria and the Basque country), covering 23,115 square kilometers (4.57 per cent of Spain), a total population of 3,780,000 (10.24 per cent) and a 1988 GDP that is 10.36 per cent of the national total. The greatest economic and demographic weight is represented by the Basque country, one of the main industrial areas of Spain.

As in other European cases, the main production activities, such as mining, iron and steel, transformed metals, machinery and chemicals, have been overtaken by transformations in the world economy. There are three essential features of the current socioeconomic structure of these industries. First, they have a high degree of specialization, particularly in basic sectors such as coal and steel. Second, the production systems involved in these industries have not been adapted to other activities, and possibly cannot be. Third, traditional infrastructure is somewhat inadequate for new industries and there are insufficient services of the kind required to support new production systems or to perform the research necessary for technological development. Fourth, there has been a progressive loss of vigour and excessive rigidity in some of the institutions and directive groups. These have therefore become estranged from economic decision centres.

These socioeconomic features, which are fairly common to all declining industrial regions, usually provoke a relative decline in production levels and a loss of employment. This contributes to a loss of economic dynamism, compared not only with past eras but also with other areas of the country, especially with the new growth axes that are being considered in investment and the creation of new activity. The result is a decline in the region's economy, perceptible in the reconversion processes of the basic industries, the entrepreneurial closures registered and in the general phenomena of deindustrialization.

All this is happening in the Cantabrian Cornice, one of the traditional areas of Spanish economic growth[13] and a region that played a leading role in prominent episodes of Spanish industrialization. In 1960, four of its provinces were among the first six in Spain in their per capita income. Today, however, all except one are below the tenth position in per capita income and two are even below twentieth. In value-added terms, the area has lost importance within the nation as a whole, particularly as of 1974. Employment did grow until 1974–5, mainly due to an influx of immigrants to the Basque economy but, from 1975 until 1986, over 251,000 jobs were lost there, with more intense drops than the rest of Spain. These have particularly centred on industry and services, which have been unable to function as a buttress, being barely developed in terms of economic levels (37.1 per cent of employment in the service activities in

1975 in the Basque country; 32.6 per cent in Asturias and 33 per cent in Cantabria). In relative terms, the tertiary sector was much larger in 1985, but total employment varied little.

Different factors reflect the sector's industrial drop. The evolution of product demand is of course one of the keys, but the loss of competition and the growing inefficiency of the basic industries (some supported by the domestic reserve market, as indicated above), helped render productivity in these sectors below the national average, with a strong increase in labour costs. Furthermore, no new markets have opened abroad and investment rates have fallen.

None of the regions of the Cantabrian Cornice appears among those prominent in new industrial investments. If total investment is taken as a reference (including the investments in renovation and amplification), the levels are higher, but lower than those of the country's more dynamic zones, and especially Cataluña. In this context, the Basque country is the more prominent, despite everything else, while Asturias and Cantabria occupy very low positions at national level. The fact that the whole region seems to hold little attraction for foreign investors is also very significant. In 1986, 1987 and 1988, respectively, only 2.8 per cent, 2.2 per cent and 0.8 per cent of direct foreign investment was channelled into this zone.

From many points of view, this area is similar to other European regions that have suffered industrial deterioration. Cantabria, Asturias and the Basque country appear in high positions (28, 23 and 22 respectively) of the synthetic index of the 160 European regions. All three have been categorized as areas in decline. However, compared with others in Europe, they have some additional disadvantages. The withdrawal from the European 'centre' is one. The problem of transport along the Cantabrian Cornice is another, and the political problems in the Basque country do not favour a recuperative climate either, even though the growth rate jumped considerably in this region in 1988 and 1989.

A series of regional policy initiatives is under way and could already show positive results, particularly in the Basque country. But the dynamic Spanish economy is travelling in other directions for the moment, and the immediate neighbouring area (southwest France) cannot support a rapid recuperation.

The Mediterranean Region, an expanding axis

We include in this region Cataluña, always one of Spain's most developed regions, the three Valencian provinces which, aside from their agricultural wealth, have a long industrial tradition and, more recently, Murcia. If we take the territorial administrative limits as a reference, this 'region' covers 66,552 square kilometers (13.18 per cent of Spain) and a total population of 11,000,000 (28 per cent of the national total). It contributes 32 per cent of Spanish GDP, and this is on the increase.

Several features underlie the high growth rates in this area. First, agriculture is generally intensive, with small-scale farming predominant. The business is highly capitalized, the products are exportable and it has mean production levels considerably higher than the national average. Among the advantages of the region are its close proximity to Europe and the Mediterranean micro-climates. The region's diversified industry is dominated by small and medium-sized companies, many of these centred on the production of final assets and with an important export capacity. This region exports over 44 per cent of Spain's total agricultural and industrial exports. From a geographical point of view, an appropriate network of urban nuclei of intermediate character makes a decentralized and fairly balanced distribution of enterprises possible. Localities along the whole coast are linked by a fairly good infrastructural platform and communications network.

Service activities – commerce, transport, hotels and restaurants, trade, etc. – have always occupied a prominent position in the region, but it was the development of the tourist industry in the 1960s and 1970s that consolidated the Mediterranean axis. The relationship between those activities directly linked to tourism, industry and agriculture have sometimes been undervalued, but their multiplying effects are evident.

The entrepreneurial base, with proven capacity for demographic renovation and notable innovative tendencies, has been a considerable factor in the region's development. Its traditional openness to other countries has facilitated contacts, made new markets available and eased the introduction of new technologies. The dynamism of investment, once the first phase of the crisis was past, constitutes proof of the vitality existing in Cataluña, the Comunidad Valenciana and Murcia. Foreign investments have also been added. Between 25 per cent and 38 per cent of Spain's total foreign investments were made in Cataluña in the 1986–8 period and between 2 per cent and 7 per cent in Valencia–Murcia.

Finally, another characteristic worth consideration when we look at this region and its economic resurgence is the interdependence between its integrating economies. An analysis of the interchanges between Cataluña and the rest of the region demonstrates the precise amount the Communidad Valenciana and Murcia contribute to this effect (24.65 per cent entries into Cataluña and 19.7 per cent exits).[14]

The crisis had an intense effect on a series of industrial nuclei in the zone, including Barcelona and its metropolitan area, Valencia–Sagunto and other smaller cities. Practically all the employment rates in industrial activities fell; at the same time, the services sector continued to increase, but it could not absorb the labour force available. The changes contributed by the recent crisis (technological changes, an increase in production costs, market reorientation) have considerably affected some labour-intensive industries that have lost competitiveness and found themselves facing new competitors (mostly from other European countries and new industrial countries). This has forced them either to make changes and adjustments that are still under way and involve new investments and production techniques, with additional loss of industrial employment, or to look for a transitory outlet in the informal economy (with many examples in textiles, footwear, clothing and certain services).

From the standpoint of Europe as a whole, Spain could classify itself as a peripheral country. However, its integration in the European Community could modify this status. The Spanish Mediterranean coastline meets the Rhône axis on one side, and northern Italy on the other, thus connecting Spain to the European economic centre (the Hamburg–Stuttgart–Lille triangle). The Mediterranean axis can, at the same time, integrate with what is called the 'West European Euroregion', which covers large areas of France, Italy and Spain, with such important cities as Lyon, Milano, Barcelona and Roma, as well as Marseille and Valencia. This is a macro-region with considerable economic weight and important dynamism.

Whether this western Euro-region consolidates itself or not, two prominent facts are clear. First, the Spanish Mediterranean axis is a reality. It is extending toward the south and shows singular economic recuperation in both industry and services. Second, Spain's European Community membership offers new opportunities and resources to this zone and, although this demands structural changes, the diversity of production in the region itself appears promising. In fact, the current industrial renovation and the modernization and amplification of the services already prove this. The locational advantages, a favourable socioeconomic climate, the physical attractiveness, and the availability of qualified labour and ample subcontracting possibilities, are other factors supporting the continuity of the process.

Some final remarks

The evolution of Spain's economy over recent years has points in common with other European countries but also presents certain peculiarities. Various points should be noted.

First, in Spain's case, one could not talk about deindustrialization before the international economic crisis. There were practically no symptoms. Once the crisis hit the country, the facts had unique characteristics, on the one hand because the authorities tried to delay its effects, and on the other because, when the situation grew worse, the stagnation or decline in some industrial branches depended, most of all, on the evolution of the domestic market (although exports continued) and the rapid increase of production costs.

Second, if we qualify deindustrialization to mean a drop in employment and exports and a low industrial growth rate, we must conclude that the phenomenon was probably less important here than the slow, but very clear, renovation of the industrial production structure. The evolution of the weak, medium and strong industrial sectors demonstrates this. In Spain's case there are, above all else, productive changes that cannot be attributed, as in the United Kingdom for example, to the process of international production specialization.

Third, at the same time, the Spanish economy has travelled unimpaired towards tertiarization (53 per cent of employment and approximately 60 per cent of GDP at the end of 1987, the European Community average being 57.3 per cent and 62 per cent, respectively, in 1985). However, this process also includes structural changes within the services sector that should not be undervalued. Production and, most of all, employment changed appreciably in composition. The role of the public sector has been important in creating employment in the 1980s and has also influenced the expansion of consumer services, initiated in the 1970s. Only more recently have services to companies expanded, partly in response to growing relationships with other countries.

Fourth, at the regional level, the effects of the industrial crisis have been clear. The most obvious quantifiable impacts were felt in the three great urban industrial areas that have supported strong growth rates in the past (Madrid, Barcelona and their surroundings, and Gran Bilbao) but, at the same time, deindustrialization effects were registered in other places (mainly cities and towns, not regions) where very specific groups of companies were localized.

When we look at what has occurred within the phase of economic recuperation (1986–9), we can now see a clear redefinition of Spain's economic–regional map due to the effects of the crisis. The north is a large declining region characterized by deindustrialization where services have not played a supportive role. Two large expansion areas or axes – the Mediterranean coast and the Ebro Valley – which demonstrated higher growth rates in the 1973–89 period, are seeing enhanced industrial activity and receiving more production investment than the others. In the Mediterranean axis, services and the agricultural sector are playing an important supporting role in this recuperation. Lastly the rest of the country offers a heterogeneous panorama but with darker hues. The main exceptions are part of Andalucia (Almeria, Málaga, Sevilla, Cádiz) which may eventually link up with the Mediterranean axis, and some more isolated centres with a certain industrial level and supply of services (e.g. Valladolid).

On the whole, Spain's experience calls the deindustrialization concepts developed in other countries into question. In fact, this term is barely applicable to Spain's recent economic evolution. The only exception is perhaps the so-called Cantabrian region, clearly a declining industrial area. However, it is totally erroneous to apply this idea to other regions that are characterized dramatically by positive structural changes in industry and the services sector.

7

Regional perspectives on the deindustrialization of Sweden

FOLKE SNICKARS

The issue of regional deindustrialization

Sweden as a core and periphery

SWEDEN is both geographically peripheral and rural in relation to the international scene. This does not mean that it is economically peripheral or a region with strong specialization in agricultural production. On the contrary, its extremely steady economic growth since the 1930s is intimately tied to its industrial specialization.[1] It has attained economic strength from transforming its domestic natural resources into products to be used in the early stages of the international value-added chains. Its main trading partners are the nations of Western Europe. The market in these nations for intermediate products has been stable over time.

During recent decades the Swedish economy has been moving rapidly away from routine manufacturing to high-value production further along the value-added chains. Using coarse indicators of structural change such as the share of the labour force engaged in primary, secondary and tertiary activity, it can be demonstrated that overall, transformations in the Swedish economy follow about ten years after those in the United States.[2]

Deindustrialization[3] is a likely social and economic problem in any nation that has its economic potential dispersed among many small and isolated single-plant settlements. The problem is exacerbated in Sweden by major public sector investment in these settlements to upgrade and compensate for income differences. The backbone of Swedish regional policy is the principle of equal access to services and work opportunities in all regions. The Swedish system is vulnerable because it is difficult for economic transformation to proceed without migration from the stagnating industrial settlements to those where the service sector is expanding. The risk is that both private and public capital in the dispersed society will have to be depreciated faster than is technically required. Transport and communication become crucial factors in the process of deindustrialization in such a nation.[4] A natural question to ask is whether it will be possible to maintain economic strength in peripheral and rural Sweden in an international economic environment where comparative advantage seems to be shifting toward contact-dense metropolitan cores.

Forces of economic transformation

There is no non-controversial way to define regional deindustrialization. Any definition refers to a particular context. The theoretically proper regional context in a study of the economic transformation of Western Europe is that including the economic cores and peripheries of the European economies. Regions in this analysis will consist of economic conurbations extending over national borders, covering only parts of national territories. Regions may be areas of similar specialization, such as agricultural plains, mining districts, communities housing heavy industry, or cities specialized in scale-economic industrial activity. Deindustrialization in terms of the decline of leading industrial sectors seldom hits nations uniformly. Changes will occur first in regions with unfavourable resource endowments and obsolete economic specialization in all nations. Deindustrialization in terms of the growth of private service activities similarly does not occur at once on a national scale. Instead, the shift in economic structure is concentrated in the few regions in each nation endowed with growth-inducing infrastructure elements. Also, the expansion of population-related services tends to follow the distribution of the population.[5]

There is no obvious way to measure the direction and pace of deindustrialization. Obsolete production capacity will constantly be replaced with modern in any sector of the economy. This may be the result of a decline in established markets and the rise of new ones as demand changes. As the demand for services overtakes the demand for manufactured goods, manufacturing firms must be replaced with service enterprises.

Other factors may stimulate the replacement of production capacity. Even if demand is stable, markets may be served more efficiently by production establishments of different vintages. The obsolescence process may proceed with different speed in different countries and regions. At any geographical level smaller than the total market area of a commodity, deindustrialization may result from changes in the spatial division of labour.

The direction and speed of these processes can be measured in a variety of ways. In the strictly sectoral mode of analysis, border-lines are defined between industrial and service activities. The rate of deindustrialization is then the net flow of activity in some measurement unit over this border. In the strictly regional mode of analysis, border-lines are defined between geographical areas either in an administrative or a functional sense. The rate of deindustrialization is found by applying the sectoral definition to these regions. Adoption of the regional functionality concept implies that the regions in question may not necessarily be spatially contiguous.

Neither of these definitions is readily applicable in practice. A more realistic mode of measurement is to accept that deindustrialization is a relative concept. The issue then becomes whether the direction and speed of change towards deindustrialization fall within acceptable ranges in relation to international or national averages over space and time. At the core of the issue is the fact that processes of change have different speeds, leading to tensions between volatile and inert trajectories. There would be no need for worry if structural change proceeded without causing unemployment in some regions and an inadequate labour force in others. Negative net migration is only important in so far as it affects individual welfare and the need to allocate social resources.

Regional requisites and adjustments

Large portions of the Swedish economy are very sensitive to price competition and business-cycle fluctuations.[6] Firms in these markets are over-represented in some parts of the country with a clear concentration outside the metropolitan regions. Economic policy aimed at allowing high rates of return on private capital may be the best way to help these industries weather downturns in the business-cycle. However, it is unlikely that such policies can be maintained for long in Sweden with its strong distributional element in all fields of public policy. A more realistic scenario is that policy packages will build on an accelerated pace of structural change, increased investments in new technology and infrastructure, new market and R&D investments, cost pressures on obsolete industry and a subsequent increased rate of elimination of unproductive capacity. Such a choice implies a shift of large portions of Swedish industry away from price-competitive environments to environments characterized by dynamic product competition. Scope rather than scale will become the criterion for economic efficiency.

The traditional explanation for the existence of establishments with high productivity in a sector or region is capital intensity, implying a strong relation between the amount of fixed capital per unit of labour input and productivity. At present, this is less than half the truth. Even in industries not exhibiting high intensities of capital use, there are large differences in productivity among establishments. For example, in parts of the equipment industry, high productivity is likely to be related to non-material investments and economies of scope in the form of dynamic product competition. In such a competitive environment there is no established market price. Firms are price-makers rather than price-takers and anonymous auction markets are basically non-existent. The heterogeneity of the products offered to the market is so high that the buyers must always incur substantial search costs. The buyer may decide to buy at a price higher than the lowest given in another simultaneously cleared sub-market. This gives the seller a certain temporary monopoly even if there are several sellers in the same market. When services are involved the transaction can only take place as co-production via face-to-face contacts. In markets exhibiting rapid structural change there is also reason for the buyer to stick to the same seller to avoid further search costs. Lasting links will be established that reduce costs for both parties until the temporary monopoly is removed through new general market information. The markets thus become short and products run through their life cycles rapidly.

Table 7.1 contains a set of shorthand characterizations of the development conditions typical of the two opposite competitive environments, price and product competition. The idea behind the table is to point out the substantial differences that exist between the production conditions and requisites for different segments of the economy. This is a schematic view. In reality, no particular industry is entirely positioned in either of the environments. There may even be differences within individual firms in relation to the development conditions depicted in the table.

The prospects for any region or nation are tied to how well it is positioned in either of the two environments described in Table 7.1. Two ways emerge for national and regional development under the deindustrialization process, concomitant with a shift from price to product competition. First, different phases of the process indicated in the table may proceed in any region over time. At any given point in time, different segments of a national or a regional economy may belong to either of the two categories. The process of deindustrialization is actually one of continuing renewal or reindustrialization of any given region.

Table 7.1 might also be seen as a picture of two stable regional environments. The metropolis is the cradle of innovation and retains this position over time. Economic activity continually diffuses from the

Table 7.1 Traits and development conditions of industries operating under price and product competition

Characteristics	Price competition	Product competition
Sensitivity to world market prices	High and uniform	Low and variable
Market orientation	Global and saturated auction markets	Differentiated and volatile customer markets
R&D orientation	Low and process-directed	High and product-directed
Technology trade	Small and sector-specific	Expansive and sector independent
Use of information technology	Process guidance and control	Design, construction and logistics operation
Causes of productivity differences	Capital intensity and cost effectiveness	Non-material capital and temporal monopolies
Development strategy for firms	Trade with market shares	Trade with knowledge and patents
Production technology	Scale economic	Scope economic
Labour qualifications	Stable supply of skilled labour	Mobile supply of unique competence
Infrastructure needs	Bulk goods transport networks	Logistic and person transport networks
Regional milieu	Functionally separated physical environment	Contact-intensive sociocultural arena
Policy levers	Industrial and monetary policy	R&D and technology policy
Relevant economic theories	Vintage, trade and general equilibrium theory	Vintage, game, product cycle and innovation theory

Source: Snickars, F. et al. 1989. Chances for North Bothnia. Luleå: North Bothnian Learning Association.

innovative metropolis to rural and peripheral regions. The question for Sweden is whether the whole nation, except perhaps for the Stockholm region, can choose to be other than one of the price-competitive environments of Western Europe.

Structural change in the Swedish economy

Economic structural change

The Swedish economy has gone through substantial changes during the most recent decades. It is still a strong actor in the world market in economic segments close to the commencement of the value-added chains. The main problems that have persisted in the Swedish economy are related to production costs. It has been difficult for Swedish firms to compete in pure cost terms in the international environment. Many Swedish business leaders aspire to operate in a dynamic product-competitive milieu and to strive for quality in production and products. Public policies have helped to give Swedish products a competitive advantage through the imposition of quality and environmental standards.

One of the most important aspects of regional development in a modern economy is the growth and decline of different sectors, exhibiting different patterns of resource use and requisites for efficiency in production. It is instructive in an international comparative context to illustrate the change in the sectoral mix. Ideally, the structural change should be represented in terms of production. Employment figures can be used when technical change has been relatively uniform. The issue of deindustrialization is also intimately tied to labour use.

Table 7.2 gives an overview of changes in the use of labour for three broad categories of production. The share of the labour force engaged in manufacturing, including the primary sector, has declined steadily since 1950. At the same time employment in the production of public services has increased its share. The share of the labour force involved in the production of private services has remained relatively constant over a long period.

Total employment in the Swedish economy amounts to some 4.2 million persons. One out of three of these, some 1.4 million, is engaged in goods production. Forty years ago total employment was some 3.1 million. Three out of five of these, some 1.8 million, were engaged in goods production.

The restructuring of the Swedish economy from goods manufacturing to service production is comparable to the international average. However, in Sweden the growth of service jobs has been concentrated in the public sector. Another difference is that the labour force is divided evenly between men and women. The large share of female labour can be explained partly by the structure of demand but also by the fact that care

Table 7.2 Employment in three parts of the Swedish economy, 1950–80 and a forecast to 1990

Economic sector	Percentage share of employment in the sector				
	1950	1960	1970	1980	1990
Manufacturing	59	57	48	38	33
Private services	31	31	32	33	35
Public services	10	12	20	29	32
Total	100	100	100	100	100

Source: Regional development and interregional equity. 1984. Proposal from Government Committee on Regional Policy. Stockholm: Liber. p. 74.

and educational systems have been organized so that barriers to female labour participation may be lower than in other countries.

It might be added that the manufacturing sector makes up around one-third of the value added to products by the private sector in Sweden. The share has been stable during the last 15 years. However, the share contributed by the private services sector has been increasing. In 1984 it created half of the value added by the private part of the economy. The sectors in which the contribution to GNP has declined are the primary and construction activities.

The big job losers in the Swedish industrial sector during the last 20 years have been mining, textiles, stone and clay, iron and steel works and shipyards. Sweden's industrial sector does not differ in this regard from other economies in Western Europe. Within a pattern of total deindustrialization, a rapid restructuring is taking place among industrial branches. It is so marked and persistent that simple indicators such as changes in employment figures reveal it. However, such figures conceal even more substantial changes within manufacturing. These changes have occurred across sectors, as could be shown by contrasting employment changes with production or export changes. They have also occurred within sectors as forecast by the theoretical divide between price-competitive and product-competitive environments.

Structural change never hits a sector evenly but erodes the profit margins of obsolete establishments and boosts profits in modern highly productive establishments. When demand is sluggish, the competitive environment of the establishments in any industrial sector will shift from product-competition to price-competition. An internal restructuring of the production capacity will take place to keep establishments productive through cost reductions. Economic and industrial policy can be designed both to speed up this process of capital replacement and to slow it down. Swedish labour market policy has had both these faces. The aim of the so-called solidary wage policy[7] has been both to equalize worker earnings and to promote a steady pace in the shutdown and replacement of obsolete production capacity.

The rank order diagram shown in Figure 7.1 has been one of the tools used in the analysis of Swedish industrial change during the latest decade. It is a variant of a so-called Salter diagram in which the heterogeneity of establishments in any industrial sector can be shown.[8] The idea is to employ economic information as close as possible to the establishment level to characterize differences in productivity, gross profit shares, or any other indicators of performance relative to other establishments.

The data supporting Figure 7.1 have been collected at the lowest level of detail allowed by Swedish confidentiality laws concerning economic

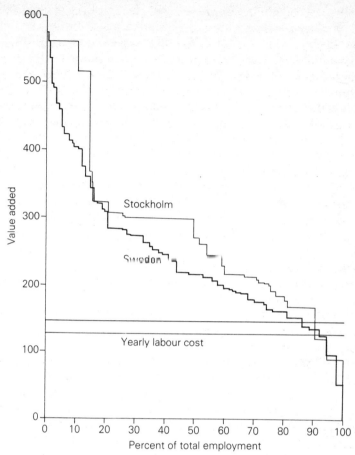

Figure 7.1 Rank order of labour productivity in manufacturing industry in Sweden and Stockholm, 1984 (value added in thousand Skr per employed).
Source: Johansson, B. & U. Strömquist. 1988. *Technology diffusion and import substitution: the role of the Stockholm region in technical change.* Stockholm: The Office of Regional Planning and Urban Transportation, Stockholm County Council, and Regional Economic Department, Stockholm County Board.

statistics. A step in the staircase corresponds to a small group of establishments of similar productivity. The establishment groups have been ranked by decreasing labour productivity, with the width of the steps equal to the total number of jobs pertaining to that establishment group. The horizontal lines in the diagram refer to the labour cost per person employed in the establishment group. It is partly a consequence of Swedish wage policy that this cost curve can be quite accurately represented by a horizontal line. This means that gross profits also differ substantially between establishments in the same industrial sector.

The diagram given in Figure 7.1 can be applied to any economic sub-sector as well as any region. Each staircase step in the current figure contains establishments from different sub-sectors sharing similar productivity properties. The figure shows that manufacturing industry in the Stockholm region almost completely dominates the industrial average in Sweden in regard to productivity. In 1984 only a few of the obsolete establishments had lower labour productivity than the Swedish average. These unproductive establishments are evidently on their way out of the Stockholm economy as well as the total Swedish economy. Their value added to products per employed person is lower than the wage cost, implying that their activity runs with a negative gross profit. The chances are high that these establishments will leave the market within the year.

In 1984, less than 10 per cent of Swedish manufacturing workers were employed in establishments with less productivity than the labour cost. The most productive establishments had a productivity advantage five times greater than the least productive ones. Even if this partly reflects technological differences between sub-sectors it is evident that manufacturing establishments with widely differing efficiencies co-exist in the market. The effect of the solidary wage policy is to increase the pace at which capacity falls below the break-even point for the gross profit margin, and to increase the gross profit margin for the most profitable establishments.

It turns out that the relationship between productivity and market exit and entry is not quite as simple as this analysis indicates. Empirical investigations have shown that shutdowns are convex decreasing functions of gross profit shares and that entries via new investments are concave increasing functions of gross profit margins. The production efficiency in a sector can also be enhanced through reinvestments in existing plants.

The productivity conditions are very different in different vintages of establishments in the Swedish manufacturing sector. Older firms use more persons to produce a given output than the emerging ones. While technical change reduces the demand for labour in manufacturing, the production growth in the service sectors is increasing the demand for service labour. A further increase in labour demand may be associated with generally lower labour productivity in this part of the economy. The labour released by technical change is one of the crucial factors behind the labour market and regional balance problems tied to rapid deindustrialization. In Sweden, the balance problems have been partly mitigated through very active retraining and matching schemes. These programmes have decreased some of the long-distance migration but they have failed to prevent unemployment and social problems in some of the core industrial regions.

Another consequence of the differentiation of production capacity according to the vintage of plants within the manufacturing sector is that product and process innovations, and the associated R&D efforts, are quite unevenly distributed. Figure 7.2 shows the share of turnover set aside for R&D in Swedish manufacturing firms at the end of the 1970s and the beginning of the 1980s. The information base for the picture is the annual economic reports for manufacturing firms with more than 50 employees, i.e. medium-sized and large firms.

The figure shows that substantial R&D efforts are being made by relatively few large firms. Except for the least R&D-intensive firms the share of R&D compared to turnover was higher in 1984 than in 1978. This is most evident for the research-intensive part of the manufacturing sector. One-third of the total manufacturing employment is found in those industrial firms that have intensified their research and innovation activities. Around 10 per cent of the employees work in firms that spend

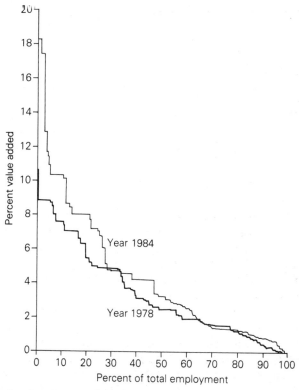

Figure 7.2 Rank order of R&D costs as a share of turnover in all Swedish manufacturing firms with more than 50 employees in 1978 and 1984.
Source: See Figure 7.1.

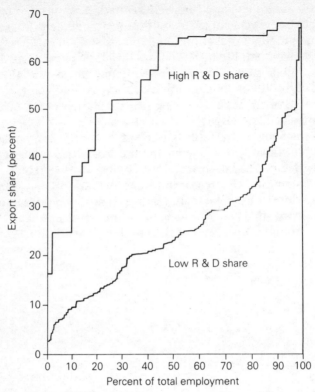

Figure 7.3 Relationship between export shares and R&D shares of turnover in two parts of Swedish manufacturing industry, 1984.
Source: See Figure 7.1.

more than 10 per cent of their turnover on research and development. These firms are over-represented in the pharmaceuticals and chemical industries as well as in sub-sectors such as instruments and electrical machines within the equipment industry. More than one-third of the manufacturing jobs in Sweden belong to long-term growth firms measured by R&D-intensity. This division of the manufacturing sector cuts through the established statistical reporting systems. Sunrise firms co-exist with sunset ones in a wide range of product markets.

The Swedish economy is highly oriented towards international markets, a pattern that will become even more pronounced in the future. The international orientation is explained in a revealing way by separating firms according to their level of R&D effort. Figure 7.3 shows this separation for the whole of Swedish manufacturing industry in 1984. The dividing line between the two groups of firms in the figure has been placed at 10 per cent of turnover devoted to R&D.

In 1984, some 80 per cent of total manufacturing employment in R&D-intensive firms was in firms that exported more than half of their turnover. Less than 10 per cent of the employment in the research-extensive industry belonged to firms that exported more than half of their turnover. Export orientation and R&D-intensity thus seem to be closely linked to one another. One way of interpreting this observation is that for the new industrial firms, markets tend to become international at an earlier stage in the product cycle than before. In a market environment characterized by price competition, exports tend to increase towards the final stages of the product cycle. The export campaigns are then conducted in terms of decreases in relative prices.

The combination of export surplus values and price increases can be employed as an indicator of early stages in the product cycle.[9] Firms operating in the environment of product competition can be identified on the basis of this criterion. This kind of analysis has been used at the regional level to trace economic potential across industrial sectors. The analysis shows that there is no clear regional concentration of the most dynamic industries. It also shows that productivity is different for single-plant firms than for establishments located in the same region but belonging to more comprehensive corporate structures.

The connection between R&D efforts and export achievements in the Swedish manufacturing sector has important implications for the demand for transport and communications, and in the long run for regional development.[10] Figure 7.4 brings out the secular trend in the value

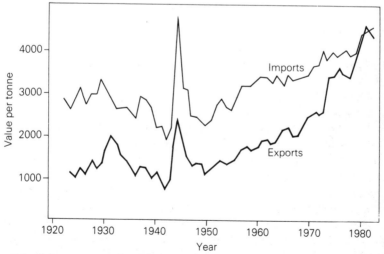

Figure 7.4 Value per tonne of Swedish exports and imports, 1920–85 (1985 price level). Source: Johansson, B. 1988. *The future labour market: dynamic competition and regions.* Stockholm: Council of Job Security, Swedish Employers' Association.

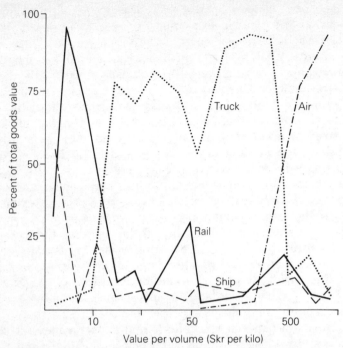

Figure 7.5 Relative importance of different modes of transportation in commodity export markets with different value contents in Sweden, 1984.
Source: Andersson, Å. E. & U. Strömquist. 1988. *The future of the K-society.* Stockholm: Prisma (in Swedish).

content of Sweden's foreign trade. The value content of Sweden's total exports has become higher than for total imports. The change occurred in the first part of the 1980s. The rise in the value content of exports has not meant that the export markets have changed. Sweden still exports most of its goods and services to Western Europe.

The deindustrialization process has led to a new international pattern of comparative advantage. In this new economic order, Swedish industrialists have become more and more international. Sweden represents a small export market for other countries as well as a small domestic market for the firms based in Sweden. In the domestic market Swedish firms face massive import competition. Import substitution has become a lever towards technical change with an ensuing increase in the competitive edge in the high-value segment of the export markets.[11] As a consequence of this process, technology trade has become an increasingly important sector of the economy. The metropolitan region of Stockholm plays an especially important role in this context as an import node for new technology. Recent empirical evidence has shown that the Stockholm region is also a market for new technology. The most recent technical equipment traded in the world market and brought into Sweden via Stockholm is purchased by Stockholm-based firms. Older products are more often sold to firms in other parts of Sweden. Some two-thirds of the value of information technology products sold to Stockholm firms comprised equipment that was less than two years old in the world market in 1986. The corresponding share for the whole Swedish market was about 30 per cent. The majority of the rest of the products was younger than five years.

The value content of goods shipped defines sub-markets dominated by different means of transportation, as indicated in Figure 7.5. Low-value commodities are ship-loaded, medium-value goods are loaded on trains, moderate-value merchandise is brought in ever smaller shipments to the markets by truck and high-value products are distributed globally as air cargo. Human beings can be seen as medium-value or high-value commodities in this regard. In recent decades, firms have been willing to spend more on transport and communications to save other production and distribution costs. Examples from the telecommunications sector indicate that half of the production, inventory and distribution costs can be saved by doubling the transport costs. Since the latter make up only some 5 per cent of the total commodity value, the result is a massive cost-saving. As a consequence, the market orientation has become stronger and production schedules tighter. Inventories have been centralized and brought down to a minimum capacity.

Twenty years of regional change in Sweden

In what sense is there a regional problem?

Deindustrialization can have either a positive or a negative connotation. The industrial way of organizing production has led to specialization, both in manufacturing and consumption; division of labour, both between persons and locations; and division of time, from both daily and life-long perspectives. For many persons working in tedious jobs, spending substantial amounts of time in travel between home and work and having little time for leisure, deindustrialization would come as a relief. For other persons who have difficulties in adapting to new technology, are too old to upgrade their skills, or lack income and social resources, deindustrialization simply means that they lose their jobs and disappear from the labour market for good. For them, deindustrialization means only a more subtle form of exploitation.

For other persons, the new economic order is a challenge. They learn to make new technology work for them to enhance their welfare. They are constantly renewing their personal competence and making use of greater flexibility in the labour process to combine work and leisure. For these people deindustrialization means becoming part of the knowledge society. They have left the worker form of living and replaced it with the career or independence forms. They have no single life-long occupation but may change occupations several times. Flexibility and preparedness for change are traits not only of the expansive firm in the knowledge society but also of the successful individual.

Table 7.3 gives an overview of regional development tied to the national structural change discussed above. The regional sub-division has been made on a functional basis, containing the three metropolitan regions, regional service centres in southern and northern Sweden, local service centres in southern and northern Sweden and four basic industrial regions with different specializations.

The employment change has been most rapid in the larger regions. The growth pattern has been similar in the two smaller metropolitan regions and the regional centres. The growth in the Stockholm region has been exceptional and is forecast to remain higher than elsewhere in the future. Metropolitan growth has been as rapid in the mid-1980s as it was during the phase of urbanization during the mid-1960s. The fact is that job growth in the Swedish metropolitan regions has been very stable during the last decades. Population growth has been more variable, with a decline in the growth rates for all metropolitan regions during the 1970s,

Table 7.3 Development of population and employment in functional regions of Sweden, 1975–84 and forecast for 1985–95

Region	Employment		Population	
	1975–84	1985–95	1975–84	1985–95
		(percentage changes)		
Stockholm region	9	14	5	9
Göteborg region	7	9	1	5
Malmö region	5	9	0	5
Southern regional centres	7	6	2	3
Northern regional centres	9	7	2	3
Southern local centres	3	−1	0	−2
Northern local centres	3	−1	0	4
Western textile region	−1	−4	−1	−4
Central steel region	−7	−1	−5	−5
Southeastern forest region	1	−6	−3	−8
Northern forest region	1	−1	−8	−5
Sweden	5	6	2	2

Source: Regional analysis. 1987. Long Term Economic Report, Appendix 24. Stockholm: Liber. p. 4.

partly as a consequence of the intra-regional increase in the supply of female labour. The regions housing universities have had very stable job and population growth.

For both employment and population, and for both past and future, with only two exceptions, the larger regions exhibit growth rates above the national average. Evidently, both jobs and population diffuse upwards in the settlement hierarchies. The job losses have been quite dramatic in some industrial regions. The steel belt in middle Sweden has lost both jobs and population at large scale and for a long time. One reason for the population decline is that the attraction points for young persons in the metropolitan regions are not that far away. A move is more likely to be permanent if the home region is located in the northern or southern peripheries.

In the most recent years, there have been indications that the most intensive pressure on the metropolitan regions is fading away. One reason for the slowdown of growth is that the building sector has not been stimulated sufficiently to build more urban housing. The mechanism seems to be the classical one. Regional growth is curtailed as high demand for personnel and residences, and concomitant wage and housing price increases outstrip labour resources and housing supply. Under these circumstances it is difficult to keep regional inflationary tendencies within acceptable ranges.

The traditional analysis of Swedish regional development performed above indicates that different regions have quite different relationships to the emerging knowledge economy. The Stockholm region has deindustrialized most rapidly, yielding its position as the most important national centre of the equipment industry. At the same time the industrial establishments in the Stockholm region have changed in character. Only a minority of the industrial workers in the Stockholm region now perform manual work. The industrial jobs in this part of Sweden have been replaced by service jobs. The more peripheral the region, the more rapid has been the job growth in its manufacturing sector. Also, the relative growth of business-oriented service jobs has been higher outside the metropolitan regions than within them. The reason is that in the more rural regions, the number of these traditional city jobs was low.

The regional distribution of decline

The perception of which regional problems are associated with deindustrialization has changed over time. It has taken well over a decade for industrialists, employees and trade-union leaders, as well as public policy-makers, to recognize and take action to mitigate industrial decline and prepare for growth in the service sector. The 1970s was a decade of recurrent crises and conflict between actors engaged in those crises. The loss of jobs seems to have surprised industrial leaders, trade unionists and politicians. The increasing inter-regional and international dependency, which was seen as an avenue towards higher prosperity by some actors, was regarded as a fundamental threat by others. One common argument was that the core exploited the periphery. The national and multinational corporations showed no responsibility for the welfare of regional and local labour and for infrastructure capital. Industrial power relations extended beyond the control of local political leaders, and not even national policy-makers could influence shutdown decisions. Projections included persistent unemployment in declining regions, coupled with outmigration, loss of real estate value and deterioration of public services.

Figure 7.6 provides an example of a type of analysis performed to assess local vulnerability to unexpected industrial decline. It shows the 74 Swedish municipalities which were most heavily dependent on a single industrial plant in 1980. Dependency is measured in two ways. The first criterion is the ratio of jobs in selected manufacturing sectors to all jobs in the region. The second measures the dominance of one plant, or firm, in the selected manufacturing sector. Dependency is high only when the sectoral composition is skewed, and the distribution of capacity among establishments is biased. The cut-off points in Figure 7.6 are 40 per cent of total employment in one leading sector and 30 per cent of employment in one dominant plant, respectively. Using these limits, one out of four Swedish municipalities was shown to have a vulnerable industrial mix in 1980.

Other studies have indicated that in the early 1980s, over one-third of the obsolete industrial production capacity ought to be replaced by more effective new establishments. During this process, economic decline will prevail in the municipalities affected. Government will have to augment aid, both to the crisis industries in their internal restructuring and to people directly or indirectly affected by job losses.

The information contained in Figure 7.7 further supports the notion of inequality in the regional distribution of industrial decision-making power. The figure shows the levels of national and regional remote control of workplaces. Each point in the figure indicates the conditions in one of the 24 Swedish counties in 1985.

Firms with their head offices in the Stockholm region employ 45 per cent of their total workforce in establishments in other counties of Sweden. Only some 8 per cent of the employees in the private sector establishments in the Stockholm region work in firms owned externally. The situation for most of the basic industrial regions is exactly the opposite. In some counties there is virtually no external control in this

Figure 7.6 Municipalities in Sweden dominated by a single industrial firm or establishment in 1980 (indicated by shading).
Source: SOU 1984:6, Appendix 9.

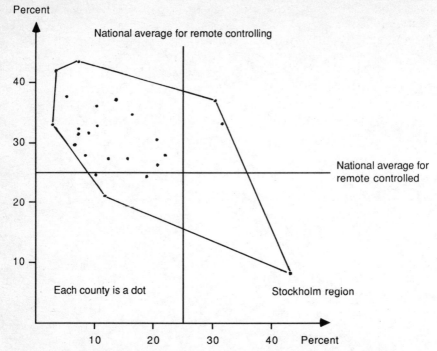

Figure 7.7 Inter-regional ownership patterns in Sweden's private sector, 1985. Share of employees working in firms with head office outside the region, and share of employees in firms with head office in the region working elsewhere in the country (percentage).
Source: Snickars, F. et al. 1989. *Chances for North Bothnia*. Luleå: North Bothnian Learning Association.

sense. The dependency indicated in the above discussion does not refer to ownership ties among firms, but to connections between the plants and the firms using them for production. Most establishments are single-plant firms.

The corporate structures above the firm level have also changed markedly during recent decades. It has become common for a corporation to restructure through sales, acquisitions and mergers rather than by investing in new capacity. Jobs at the local level have become increasingly dependent on external financial markets of a speculative nature. At the same time, local capital markets have emerged, both for long-term infrastructure projects and in the form of venture capital for risky projects. A major political debate has taken place in Sweden over whether firms should place profits into so-called wage-earners' funds controlled by the labour unions. Such funds have been set up as regional capital suppliers in six parts of Sweden.

As in several other countries the industrial crises have been primarily associated with large firms and large establishments. Small firms have generally experienced industrial growth, albeit at a moderate volume of overall activity. In recent years much effort has been devoted in Sweden to the encouragement of regional self-reliance through local entrepreneurship to counter the decline in basic industries. Rather than subsidizing large basic enterprises to maintain inefficient capacity in the price-competitive markets, the public sector has supported small-firm formation, and diverted funds to regional investments in material and non-material infrastructure.

The regional distribution of growth

The new division of labour within the economic sectors, illustrated through the twin concepts of price and product competition, divides the Swedish regions. The dividing lines often cut through the established concepts of administrative and functional regions. The new economy has a network character. In this network economy, physical distance has been transformed through new technologies of transport and communications. It has become more natural to think of a city or region as the smallest link in a network of larger and more trade-intensive cities and regions than to regard it as the central place of a market hinterland. The traditional notion of comparative advantage seems to be outdated, making way for advantages created by temporal monopolies and technological, rather than physical factors of location.

It is important to note that in Sweden regional employment growth is not congruent with the notions of competitive environments and comparative advantage discussed above. Such analysis suggests that persons in knowledge occupations are becoming crucial factors for growth, and

Table 7.4 Sectors in the Swedish economy ranked according to employment growth for the period, 1970–83

Rank order	Economic sector	Index value (1970 = 100)
1	Social care	205
2	Health care	183
3	Recreation and culture	174
4	Interest organizations	161
5	Consulting	156
6	Education and R&D	155
7	Public administration	152
8	Insurance companies	150
9	Sewerage and cleaning	150
10	Jurisdiction	146

Source: Author's computations from census data.

that ever larger portions of the labour force will be engaged in work with an international orientation.[12] The actual nature of development is depicted in Table 7.4.

Most of the rapidly expanding employment sources in the economy are to be found in the area of personal services, both private and public. These jobs are necessarily local in nature. Thus, the shift towards public service employment in Sweden is regionally neutral. There are also clear tendencies toward concentration in the fields of consulting, banking and other business-related services (see Figure 7.8).

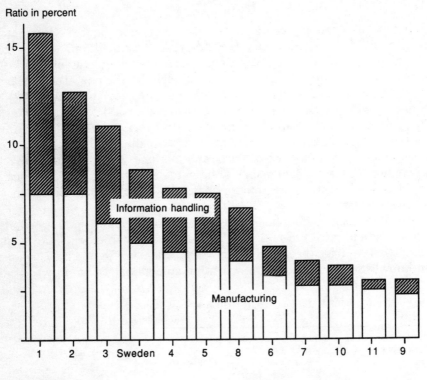

Figure 7.8 Employment in business services in relation to total population aged 16–64 years in different regions of Sweden, 1984.
Source: Ståhlberg, L. 1986. *The private service sector: regional development possibilities.* Report 1986:8. Stockholm: Swedish Industrial Board.

Table 7.5 Share of R&D personnel of total employment in private service sectors in four parts of Sweden, 1970–84

Region	Share of R&D personnel		
	1970	1978 (percentages)	1984
Stockholm	29	26	22
Göteborg and Malmö	21	24	28
Forest industrial regions	13	13	13
Other regions	37	37	37
Sweden	100	100	100

Source: Axelsson, S. 1988. *New traits in innovation work.* Stockholm: Department of Regional Planning, Royal Institute of Technology.

The regional pattern of employment in business services illustrates the sub-division of expansive activity among Swedish regions. The fourth column is the national average of jobs in the business service sector in relation to the active population. The provision of jobs is higher than this average in the three metropolitan regions and lower everywhere else. It is especially low in the sparsely populated regions.

The unevenness is more marked for those business services that involve information handling activities. It is less pronounced for the firms providing services to the goods manufacturing sectors. As noted earlier, the regional distribution of jobs in the sectors engaged in population-related services is quite even when compared to the distribution of population among Swedish regions. Currently, the number of public sector jobs is not growing as fast as it did during the latest 15-year period. Instead, business service jobs are multiplying at a rapid rate.[13]

Although business service jobs are concentrated in the metropolitan regions of Stockholm, patterns of growth are not reinforcing this pattern. As illustrated in Table 7.5, the opposite seems to be true. There has been a steady decline in the Stockholm region's share of national employment in occupations with a distinct R&D character over the latest fifteen-year period. At the same time, there has been compensating growth in the other major metropolitan regions. The division of labour in industrial R&D between metropolitan areas and smaller settlements has been very stable over a long period. The growth occurring in the Stockholm region has to do with the role of the metropolis as a market and meeting place for the exchange of concepts and ideas, and for financial transactions.

Changing regional policy perspectives in Sweden

The persisting goals and the changing means

The goals of Swedish regional policy have remained basically unchanged since the 1960s, as a fundamental balance between efficiency and equity was sought at both the national and the regional level. Regional development policy in Sweden aims at the most efficient use of resources available in a region to attain growth and welfare goals. Special means have been introduced to influence the development of each separate region. The regional equity issue has been the other cornerstone of Swedish regional policy. Equity has been operationally defined at different levels of ambition at different times but the goal has not changed. The objective of Swedish regional policy is to create conditions for balanced population development in all parts of the country, and to give people in all regions equal access to work, services and good environmental conditions.

It may be interesting to note that the goal formulation does not specify the kind of work that should be provided across the nation's settlements. For this reason, deindustrialization is not regarded as a critical issue in itself. During two decades of active regional policy, questions concerning the composition of the job supply at the regional level have not been crucially important. For single-plant settlements, this implies that what is important is not the type of industrial activity but whether there are no alternative job opportunities. One-sidedness or regional overspecialization can mean that most of the jobs in one locality are associated with a growth industry.

In Sweden, as in most other countries in Western Europe, regional policy proper has come to be associated primarily with the equity issue. Even though the Swedish goal is to level welfare differences between individuals across the regions, many of the regional policy instruments have been directed towards the manufacturing sector. The rationale behind these support schemes has been that job security will be maximized if the prerequisites and conditions for production are made equal across the country. This equality or neutrality has been expressed in terms of compensatory schemes to reduce differences in production costs. Much of the debate in the 1970s referred to whether the funds used up in direct subsidies to remove regional differences in production costs have been effective. Some major investigations performed in the late 1970s indicated that the schemes had saved or generated very few jobs, though the results were quite uncertain due to major difficulties in making the measurements.

The provision of public services has been one of the cornerstones of regional policy work. Most of this activity has dealt with the equity issue, i.e. equal access to public services in all regions. Less attention has been given to the job opportunity aspect. An exception is the scheme for moving some of the national public agencies away from Stockholm to other parts of the country. Although the results of this relocation of government offices have been hotly debated, similar new initiatives are currently under way. In the late 1980s, regional policy support has also been given to private service industries in Sweden, for example through the recent introduction of schemes to reclaim some of the costs of business travel for peripherally located firms.

From the above discussion it should be clear that the goal of bringing jobs to the people has primarily been fulfilled through the adoption of production cost compensation schemes. As a matter of fact, there is also a cost compensation scheme for local government in Sweden. It is one of the most important transfer schemes in terms of the size of the funds involved. The basis for the redistribution is partly local differences in the potential of the local labour force to generate income, and partly the differences in infrastructure needs according to the geographical traits of the municipality at hand.

The changes in the delineation of areas where industrial enterprises are eligible for regional policy support are depicted in Figure 7.9. Only when an establishment is located in these parts of Sweden may a firm ask for regional employment premiums when it has increased its labour force, or for transport cost support. A firm may also receive investment subsidies and other special funds if it locates in a rural area.[14] In recent years, there have also been experiments in applying general, rather than selective support policies to firms in peripheral areas. An example is the overall reduction of the wage tax in the northernmost parts of Sweden.

Over time, criteria determining an area's eligibility for support have changed from general to specific considerations. It is now possible to acquire eligibility for municipalities in any part of Sweden. The basis for the decision is the severity of the labour market problem and the imbalance in population growth. Interest in using the traditional cost-reduction policies appears to be declining both at the national and at the regional level. The background for this is the shift in the major guideline for regional policy away from equity between regions toward self-reliant intra-regional development programmes.

In the 1960s regional development planning was initiated as an independent activity. The regional administration was to provide the national government with actual and forecast information to mitigate existing development problems and to avoid future ones. Schemes were developed in which cities, towns and settlements were classified for eligibility in public investment planning. The schemes were co-

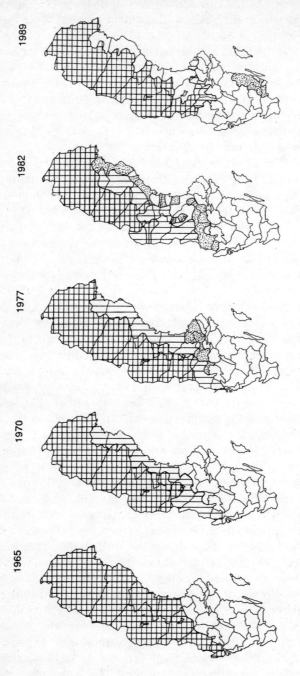

Figure 7.9 Delineation of the geographical areas for eligibility of regional policy support in Sweden, 1965–82.
Source: Törnqvist, G. (ed.). 1986. The Swedish economy in a geographical perspective. Stockholm: Liber, and Government proposal in 1989.

ordinated at the county level. In Sweden, a politically elected county council is in charge of health care planning and comprehensive development planning at that regional level. Concurrently, there exist several agencies of the national government in charge of supervising the fulfilment of national goals, also at the county level. The county board encompasses some of the regional agencies while others, such as the one associated with labour-market policy, are external to the county board. The regional decision-making situation is further complicated by the fact that municipalities and county councils together collect major revenues, taxing the average citizen some 30 per cent of gross earnings.

The national government transfers substantial resources to the municipalities and to the regional agencies. There has been a long-standing conflict between the planning and operational activities of the sectoral agencies and the co-ordinating activities of the county board. During the latest decade, the national government seems to have accepted the fact that the co-ordinating activities of regional development planning can be performed more efficiently at the regional level than at the national. A so-called package policy to stimulate regional development in crisis regions has been initiated. The idea behind this policy is that national sectoral ministries should each contribute resources to a common pool to be used in a better co-ordinated way for regional development. The down side of this approach is that there are smaller chances that continuing policy support can be obtained at the same level of funding. Packages of this kind have been launched for several of the regions in Sweden hit by industrial crises, e.g. for the northern mining region, for the steel belt in middle Sweden, for the shipyard cities on the west and south coasts of Sweden and, most recently, for the weakly industrialized southeastern part of the country.

The package policy can be seen as a response from the national level to regional initiatives in the area of development planning. The largest change in Swedish regional policy is the increased level of ambition of the municipalities in economic and employment planning. It is actually outside of the law for municipalities to act directly in co-operation with industry in job creation and infrastructure projects. Still, a number of such joint initiatives have been launched in the latest five-year period under the general heading of local self-reliance. The role of the county board has thus changed from that of a co-ordinator of national government activity to a participant in project-negotiation activities involving public and private actors. Consequently, it is likely that the county boards will be substantially reorganized. Also, the issue of deindustrialization has been transformed from a conflict-ridden problem area to an arena for project initiatives in partnership between the private and the public sector. In terms of the distinction between price and product competition, the conclusion is that public policy actors are working to encourage development by taking risks rather than striving to attain stability by removing risks.

Contemporary policy issues in Sweden

Two subject areas dominate the current discussion on regional policy in Sweden. One relates to projects in the field of transport and communications infrastructure. The other refers to the regional distribution of higher education, R&D activities and knowledge diffusion. There is a consensus that policies aimed at providing underpinnings for the sustainability of regional systems are more efficient than policies aimed at levelling out the production and factor cost differentials.

The discussion concerning infrastructure investments may be illustrated by comparing the two parts of Figure 7.10.[15] The figure shows the total investments made for a 15-year period in the road and telecommunications sectors, respectively. It also shows the total investments made in infrastructure in the two sectors as well as the demand for financing claimed within the government budget process. Telecom Sweden is in charge of both the investment and maintenance of the nodes and networks for telecommunication and the operation of the telecommunication services under government monitoring. For this reason it is difficult to separate the part of telecommunications investments that is infrastructure-oriented. The Swedish Road Authority, on the other hand, is responsible for the bulk of the investment and maintenance in the road networks.

A comparison reveals that the investments are of comparable size. The road investments have declined steadily while the telecommuncations investments have fluctuated over time. The infrastructure component has exhibited larger swings than the total. Total investments in infrastructure have been higher than the demand for funds via the ordinary budget channels, a sectoral difference that has increased over time. The explanation for the deviation is that road investments have been made for labour market reasons, as well as for reasons of economic performance. An increasing share of the investments in telecommunications networks is financed outside of the government budget.

There is currently a question whether resources should be diverted to the road sector again. Strong demand from trucking interests is meeting energetic environmental opposition. The telecommunications sector is also being considered from a regional perspective. A more detailed analysis of the figures shown in the graph reveals that the Stockholm region dominates these investments. It is not self-evident that a regional dispersal of investments in telecommunications networks would spur

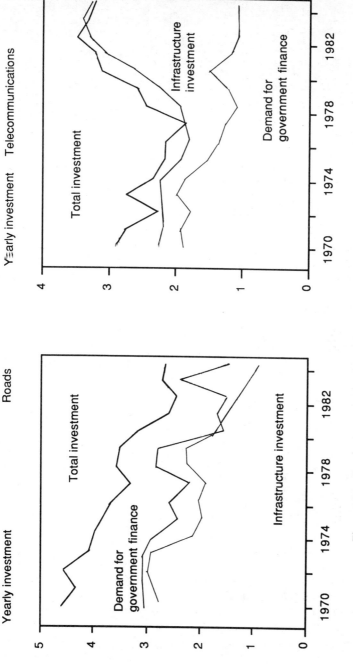

Figure 7.10 Total gross investments, total infrastructure investments, and claimed demand for government-financed infrastructure investments in Sweden's road and telecommunication sectors, 1970–84 (million Skr in 1984 prices).
Source: Sundberg, L. & G. Carlén. 1989. Allocation mechanisms in public provision of transport and communication infrastructure. *Annals of Regional Science* **20**, (3).

Figure 7.11 Location of national and regional universities in Sweden, 1985.
Source: SOU 1984:74.

economic growth in the regions receiving the investments. The spin-off effects are most likely to occur in those regions where the R&D necessary for the infrastructure technology is performed.[16]

The establishment of universities and other centres of higher education can be efficient levers for regional development. At the beginning of regional policy proper in the 1960s, a policy of establishing universities in the northern part of Sweden was employed as a means to promote long-term growth. It can now be said that the establishment of the northern universities was a successful policy. This can also be said for most of the regional university centres formed during the 1970s and in the early 1980s (see Figure 7.11).

There are six national university regions. In most regions there are also technical universities in the cities that house national universities. The exception is the northern region, where the technical university is located in Luleå and not in Umeå, which is the university centre. There are regional universities in 18 cities and towns. In some cases, two cities share the same regional university. Each of the regional universities is connected to a national one. In the latest decade, local politicians have started seeing universities as major assets for the future, much in the same way that the county board has become a regional co-ordination centre.

The current debate focuses on whether university education can be extended in the regional universities to encompass doctoral programmes and long-term R&D. It seems very likely that such initiatives will be taken. The problems of critical mass and research quality will most likely be addressed by linking the regional universities into networks with connections to the established national universities. The distribution of funds between university regions is given in Table 7.6.

If the population distribution is used as a norm, the table shows that the regional distribution of educational capacity is uneven between subject areas. It may be possible to increase the total capacity of the system for

Table 7.6 Funds for education and R&D in five subject areas allocated to six national university regions in Sweden in 1984.

University region	Technical	Business	Care	Teaching	Culture	Total	Population
			(percentage of national total)				
Stockholm	30	19	38	20	47	28	19
Uppsala	7	21	12	18	9	15	19
Linköping	11	5	5	10	2	7	8
Lund	19	21	19	13	19	19	19
Göteborg	23	19	16	25	22	20	23
Umeå	10	15	10	14	1	11	12
Sweden	100	100	100	100	100	100	100

Source: Regional development and interregional equity. 1984. Proposal from Government Committee on Regional Policy. Stockholm: Liber. p. 74.

higher education and R&D by moving supply closer to the places where potential students live.

It has generally been regarded as self-evident that young people will have to move to find the right education and that people having a higher education will have to move often during their careers. The Swedish regional policies for higher education appear to be related to the needs of a deindustrialized society, in which continuing education is a fundamental part of work and social life. In this way, regional policy can also move beyond the industrial way of thinking to a stance more appropriate to the knowledge society.

Following ten years of regarding deindustrialization as a problem, there are signs that policy-makers are seeing in this process some challenging prospects for the Swedish regions, even for those with obsolescent industrial cores.

Discussions of Swedish regional balance have characteristically seen development in the metropolitan regions as inevitably linked to decline in the industrial regions. The most interesting fact concerning this discussion is that it is based on largely faulty information. Since the 1960s moves to the Stockholm region have not been forced. Instead, people migrate to the metropolis to engage in challenging work tasks and to minimize long-term risks, as shown by several surveys of migrants. Politically, there has been weak acceptance that the metropolitan economy is crucial to the nation as a whole, or that its environment is so unique that growth could not materialize elsewhere. According to this view, Sweden deindustrialized is not Sweden urbanized. There is considerable political feeling and scientific evidence that a knowledge society could thrive in a regional structure not necessarily dominated by metropolitan regions, as in the industrial era. The unprecedented metropolitan growth that has occurred in the OECD countries has been intimately tied to industrial production. Organizing mass transit of persons to the Central Business District (CBD) to work might not operate for the knowledge society, not only because the transport systems of the metropolitan regions are congested, but also because the CBD may not be the most efficient milieu for the production and transfer of knowledge.

Regional analysis of deindustrialization enforces this conclusion. In the industrial society productivity was increased via more intensive use of material capital. Labour was used to serve the capital equipment at fixed sites where efficiency required the accumulation of material capital. Knowledge is produced by combining human capital resources with a much smaller quantity of material capital. The productivity of human capital can be obtained by moving people to the right environments for appropriate periods. It is not self-evident that this can be most efficiently achieved in a functionally specialized core area of a metropolitan region.

If this were the case, there would be far fewer demands placed on meeting places in resort environments of different kinds, and no specially built high-tech cities. A possible reason for the continuing pressure on metropolitan regions in the deindustrialized society is again the minimization of risk. Profits from service production can be placed in the stock of real estate in a region, since this stock, if it has been properly planned, functions as long-term infrastructure. If this hypothesis is valid, the conclusion is that single-central metropolitan regions might well be economically inefficient in the long run. It would be better to place some of the profits from economic growth in transport and communications infrastructure to connect smaller cities and regions into a multiple core network region. From a Swedish perspective, the question is whether appropriate regional coalitions can be formed to implement such an idea politically and economically.

III
Perspective studies

8

Problems of regional transformation and deindustrialization in the European Community

PAUL CHESHIRE

IF UNEMPLOYMENT is a symptom of deindustrialization, the existence of concern in the Chukyo region of Japan and worry in the Midwest of the United States might suggest that there should, in Europe, be panic and dismay. The large industrialized countries of the European Community (see Table 8.1) have populations in the 50 million to 60 million range, comparable to that of California or some of the other US Census regions. In recent years (1986–7), whole countries in the Community had mean unemployment rates of 11.2 per cent (United Kingdom), 10.6 per cent (Italy), 9.9 per cent (Netherlands) or 11.2 per cent (Belgium). In those two years half the member states had unemployment rates of more than 10 per cent. The precise relationship of these unemployment rates to deindustrialization is not immediately clear. It is true that the highest rates of industrial job loss generally occurred in those countries with the highest unemployment rates. However, the countries with the highest unemployment rates, Spain and Ireland, were not those with the oldest and most mature industrial economies, which are now showing symptoms of decline. Both countries had proportions of employment in industry below that of the Community as a whole. Within the Community context, they appeared, rather, to be low-wage countries that still had large traditional and low-productivity agricultural sectors.

Table 8.1 Summary data on the experience of deindustrialization at national level, 1977–86

Country	Total population (thousands)	Proportion of employment in industry relative to EC 12[1] (percentages)	Annual rate of industrial employment change 1977–86[2] (percentages)	Unemployment rate 1986–7 (percentages)	GDP per capita at ppp[3] (thousands of ECU)
Belgium	9,862	96.0	−3.1	11.2	12.8
Denmark	5,121	87.9	2.3	5.9	13.4
Federal Republic of Germany	61,062	123.4	−2.2	6.6	14.1
France	55,392	96.6	−2.7	10.3	13.2
Greece	9,966	81.2	0.4	7.4	6.8
Ireland	3,537	89.0	−1.4	18.8	7.6
Italy	57,247	—	−3.0	10.6	13.1
Luxembourg	367	109.3	−2.1	2.9	15.4
Netherlands	14,567	87.4	−2.2	9.9	13.2
Portugal	9,716	105.0	−1.5	7.9	6.1
Spain	38,688	93.0	−4.2	21.4	9.0
United Kingdom	56,763	—	−4.2	11.2	12.1
European Community	322,268	100.0	—	10.7	12.2

Source: Eurostat. *Indicators statistiques pour la mise en oeuvre de la reforme des fonds structurels* October 1988 and *European Economy.* February 1989.
Notes: 1) The highest value 1977–86 where data available but Belgium = 1977 and 1985; Denmark = 1981 and 1985; Federal Republic of Germany = 1980 and 1985; France = 1981–5; Italy = 1982–5; Spain = 1980–5; United Kingdom = 1982–5. 2) Annual rate of change from highest value available to most recent date for which there are data. 3) Purchasing Power Parities.

Europe is still a complicated and highly diverse area. It is arguable that the regions of the United States are no more diverse than the regions of France, a country of roughly the same size as a large US state. Yet France contains within it old, declining regions such as the Nord–Pas-de-Calais and expanding sunbelt regions such as Midi–Pyrénées. Even this statement conceals complexity. The Nord–Pas-de-Calais itself is diverse. It contains both the stagnant urban region of Valenciennes, with an ageing economy based on coal and declining industries, and the far more prosperous Lille. Lille is an administrative centre of growing importance and long history; its prosperity over the next ten years will be greatly enhanced by the construction of the London–Paris–Bruxelles TGV (train à grande vitesse) system. Lille will be the key junction on this system linking Paris, London, Bruxelles and Amsterdam. The recent unemployment rates of the respective Functional Urban Regions (FURs) of Valenciennes and Lille reflect this diversity. In Lille the unemployment rate in 1985–7 was less than half that of the urban region of Valenciennes. The 'sunbelt' region of Midi–Pyrénées, the expansion of which is centred on

Toulouse and the rapidly growing aerospace and high-tech industries of that city, is equally diverse. The growth is occurring, however, not in the context of a modern region, with a highly capitalized agriculture and modern infrastructure, but in an area that is probably still one of the most backward in France, where oxen were still a common source of agricultural motive power until 25 years ago and which most urban French would still dismiss as a rural backwater.

This variation is within one country. The other large countries of Europe are as, or almost as, internally heterogeneous; Italy and Spain perhaps more so, the United Kingdom perhaps less so. What a non-European reader must bear in mind, therefore, is that there are many regions in Europe with populations of 2.5 million that are in every way as diverse as a US census region 15 times as large in terms of population and 100 times the area. Indeed, there is one entire country, Luxembourg, with a separate language and national administration, but with a population of only 367,000.

Europe thus varies from modern, service-oriented, world metropolitan regions, such as London or Paris, to backward regions where more than 40 per cent of the population still derive their livelihoods from peasant agriculture. This heterogeneity provides both a strength and a weakness for the purposes of the present study. It is difficult to find large enough samples of comparable regions for rigorous statistical analysis. This difficulty is made infinitely greater by the fact that the national, and even regional diversity is carried through into variations in statistical definitions that make the collection of comparable data an extremely difficult and often impossible task. There tend always to be too many independent variables, not all of which can be measured. On the other hand, diversity naturally produces genuine variation. To some extent one can hope to identify, at least in a qualitative way, the influence of policy or institutional differences because there is genuine variation. The problem is that the variables in which one might be interested often vary together, with the result that it is hard to identify clearly the influence of, say, explicit policy compared to institutional or cultural differences. It may be possible, for example, to argue that industrial policy can be successful in offsetting the undesirable effects of deindustrialization because Europe has a successful example in Sweden; but it is not necessarily possible to conclude from this that a similar industrial policy would work in, say, Lorraine in France or Wallonnie in Belgium.

Some definitional issues

There are important economic and social changes in progress in the world at present. These manifest themselves in a number of dimensions; but change in each of these dimensions interacts with change in others. There is an *organizational* dimension to change. Enterprises are reorganizing themselves and the relationships between each other. Thus large Japanese firms are establishing semi-independent facilities in Europe. This applies to both manufacturing and service firms. Major European firms are buying or forming alliances with complementary companies in other European countries. Major retailing firms – of which British-based Marks & Spencer is perhaps the best-known example – are emerging as quasi-manufacturing companies. They specify and market, with more and more precision, goods (and now services) produced by subcontractors. At the other extreme some traditionally manufacturing companies, Fiat for example, are moving more towards subcontracting production to specialized, large-scale firms and concentrating on design, assembly and marketing.

There is a *sectoral* dimension to change. This is manifested most obviously in the growth of employment in the service sector and the decline of employment in industry. The composition of and relationship between the service and industrial sectors is itself changing so that increasingly people are searching for new definitions that might capture more accurately the shifting reality: people, for example, talk of knowledge-based activities, or the information-intensive sector.

There is, then, a *spatial* dimension to this process of transformation. This occurs on at least three levels. First, manufacturing of more labour-intensive and less technologically sophisticated products has been systematically shifting to the newly industrializing countries and to lower-cost countries in the more developed world. In Europe, for example, Spain and Portugal have been recipients of some of this activity. This process may sometimes smother indigenous firms. Second, there is an inter-regional dimension to transformation. Within countries activities are specializing and relocating, most frequently by means of differential rates of growth. Third, there is an intra-regional level to spatial transformation. This, at least for industry, has taken the form of decentralization, usually again more by differential growth than by direct relocation. Continuing decentralization of industrial employment, however, is not inconsistent with emerging recentralization of employment overall. Service employment is growing faster than industrial employment and it is a large proportion of total employment. If it continues to have an urban orientation, therefore, net recentralization of employment can occur even though industry (and even sections of services) continue to decentralize.

This is not an exhaustive list of the dimensions within which transformation is occurring. Overall the process is probably the most fundamental economic and social change that has occurred since the Industrial Revolution. It is intensified in Europe by the process of European integration, which is accelerating change in both the organizational and spatial dimensions of transformation. In this perspective deindustrialization appears as just a special case of one dimension of transformation. The definition of deindustrialization proposed by Camagni puts deindustrializing regions into a broader context of change and such regions appear as a special case.[1] Still, however, that definition may be too confining since it focuses only on aggregate employment and productivity changes in regions compared to larger units. It abstracts from the interrelated spatial, organizational and other dimensions of transformation.

It follows from the perspective outlined above that a proper understanding of the ways that regions are transforming, and of the problems that arise, must be based on an appropriate definition of regions and an appropriate measure of change.

One of the important lessons of the European experience is that the conclusions one reaches about both the incidence and extent of transformation and deindustrialization and the judgements one forms as to the impact and policy implications of the process depend critically on how regions are defined and how the process is measured. Analysis of the experience of deindustrialization in Europe can throw light on the role of policy, on the process of transformation and on the varying capacities of regional economies to adapt to change. But in the process it can also throw light on certain important definitional and conceptual issues because we find we only get consistent and intelligible results if we measure variables in consistent and appropriate ways and use sensible definitions. Fortunately, the regional delimitations of the European Community are in principle capable of yielding a fine-grained spatial analysis.

The spatial definitions used in this chapter draw on those of Hall and Hay,[2] Cheshire and Hay[3] and Eurostat. Hall and Hay defined a system of Functional Urban Regions (FURs) for Europe that included all countries of the present European Community except Greece. Cheshire and Hay extended this system to include the two major urban regions of Greece, Athens and Salonika, and produced a partial integration of the system of FURs with the set of nesting regions (Nomenclature des Unités Territoriales Statistiques or NUTS) defined by Eurostat for purposes of statistical reporting. Detail of the NUTS Level 1, 2 and 3 regional system is provided in the Appendix. Further work by Cheshire and Bevan[4] has now fully integrated the Hall and Hay FUR system with the NUTS system.

All FURs consist of a core, defined in terms of a concentration of jobs, and a hinterland or ring, defined in terms of commuting relationships.[5] All 229 FURs with a total population of more than one-third of a million, which together accounted for 69.8 per cent of the population of the European Community in 1981, have been disaggregated into a series of 'building blocks'. Each building block falls exactly into either a core or a hinterland of a major FUR and also into a single 'Level 3' NUTS region;

that is, the smallest European statistical reporting unit, corresponding to a *département* in France or a county in the United Kingdom. Where data exist for the building blocks, these are aggregated to give totals for the FUR and its components and can be compared with each regional level of the Eurostat system. Where data are not available for the FUR building blocks, the most appropriate available data – in most cases either population or employment – are used to weight the relevant Level 3 data to yield FUR estimates.[6] Cheshire and Hay provide maps showing the spatial boundaries of these FURs and how the FURs relate to the Level 2 regions.[7] The relationship between FURs and Level 2 regions varies considerably, both according to the degree of urbanization and the administrative conventions of different countries. In heavily urbanized regions, especially where Level 2 regions are smaller, FURs may be spatially quite small and there may (for example in the Ruhr in the Federal Republic of Germany, or in the English Midlands) be several separate FURs within one Level 2 region. In less urbanized areas, in Ireland, France and parts of Spain and Portugal, FURs are extensive and may be larger than individual Level 2 regions.

The use of FURs has many advantages. First, they largely abstract from national differences in regional definition. There are inevitable differences, but in general FURs will be consistently representative of the spatial economic sub-systems of each country. We can compare the Paris region with the London region knowing that, as far as possible, differences that only reflect spatial definitions have been eliminated. Second, they have the great advantage, in the present context, of permitting the important distinction to be drawn between decentralization and deindustrialization. As has already been indicated, deindustrialization, however defined, reflects a complex set of inter-related causal mechanisms. Partly it can reflect, especially at the regional level, a change in the structure of comparative advantage of which the most obvious and extreme manifestation is resource depletion. It can also reflect, again in terms of comparative advantage, changing relative factor costs compared to transport cost. It may also (and this interacts with changes in comparative costs) reflect technical change in production systems. These may merely be an extension of the secular trend toward more capital-intensive methods of production or they may reflect more dramatic shifts in production organization – one of the dimensions of transformation – whereby production activities are disassociated from 'service' type activities allowing relocation of each type of activity. A further, but inter-related, causal mechanism involves changes in ownership and organization of large, usually multinational, corporations. These changes reflect, however, the economic forces already alluded to. Yet a further causal mechanism, again related to factor cost differentials and the changing structure of transport costs (including communication and data transmission) is relocation of activities that handle goods in bulk including, but not confined to, industrial production.

As a result of these various forms of change, what appears as deindustrialization from one spatial perspective may appear as relocation from another. Outsourcing from the United Kingdom appears as deindustrialization from the perspective of the United Kingdom, but from the perspective of a newly industrializing country it may appear as industrialization and economic growth. Deindustrialization from the mature regions of the industrialized economies may be seen, in part at least, as a manifestation of global relocation or transformation.

This problem is even more marked at a regional level because there may be deindustrialization and decentralization together. From the viewpoint of the FUR core there may be deindustrialization, but to the extent that this reflects decentralization from core to hinterland, there may be no net deindustrialization of the FUR as a whole. Even this is likely to depend on how deindustrialization is measured, however, since both the process of direct relocation and of differential growth (the most important cause of core employment loss found in British studies[8]) is almost always accompanied by reorganization of plant and new investment. As a result, it is usually associated with reduced employment but higher output per worker and often higher total output. As measured by FUR industrial employment change there is likely to be apparent deindustrialization, while as measured by industrial output change there may be industrial growth. Measured in Camagni's way, by relative productivity and employment change, there may not appear to be any problem of deindustrialization at all.[9]

There is yet a further factor to consider and that is the extent to which the observed change reflects what might be called a 'squeeze' phenomenon or a 'pull' phenomenon. In fast-growth regions, with rising real per capita incomes and land prices, there may be rapid rates of change of the comparative advantage of different types of activity. In, for example, Berkshire (a Level 3 region in the NUTS system), west of London, there has been rapid population and employment growth,[10] rising relative wage levels and rising land prices. This reflects both decentralization of service activities from London, indigenous growth, particularly of high-tech industries,[11] and the effective constraint on land supply enforced by the operation of the British land use planning system.[12]

British Aerospace, which has a major establishment in this region, provides an example of the results this has. Each year the activities of this establishment have been routinely reviewed to ensure that the value added at the site justified the opportunity cost of the site and labour costs. Over a period of years higher value-added activities, human capital, and higher salaried employees became concentrated there and lower value-added activities were relocated to establishments in lower-cost regions. This inevitably produced apparent deindustrialization. In 1989, as part of a wider reorganization of British Aerospace, it was announced that the Berkshire facility would close. The primary manufacturing – the actual metal cutting – had already been moved to an existing facility near Manchester, and by 1989 the Berkshire site was used only for managerial and design functions and for assembly. The assembly was to be transferred to existing facilities in the South West, mainly in Plymouth, where there would be retooling investment as well as new jobs. Most managerial and design functions would also move to the plants in the South West, but a few of the highest-level functions were to be concentrated on the remaining divisional plant elsewhere in the South East.

Similarly, as Table 8.2 shows, there are complete FURs (the FURs are located on Figure 8.1) that exhibited large percentage losses of manufacturing employment between 1971 and 1981 but had low unemployment and high income. There were others, however, with large percentage losses of manufacturing employment that had low – or relatively low – per capita income and high unemployment. Thus Frankfurt, a rapidly growing financial centre in the Federal Republic of Germany, lost both residential population from its core and industrial jobs from the whole FUR (and Level 2 region) but had the highest per capita income of any major FUR in the European Community and among the lowest unemployment. Although the population of the FUR itself was growing slightly (the growth of the hinterland more than offsetting the core loss), the wider Level 2 region containing Frankfurt was growing more rapidly. This implies that there was a faster rate of growth in the area beyond the hinterland than within it. What appears to have been happening in Frankfurt, therefore, is a classic manifestation of the predictions of urban economic location theory. Growing service sectors and their office buildings outbid residential and manufacturing uses for land; this, in turn, was associated with relocation of population and industrial employment beyond even the FUR hinterland.

If Frankfurt represented an extreme case of the squeeze forces, Liverpool, a declining port and industrial city in northwest England, represented an extreme case of the opposite form of causation, 'pull' (or sink) forces. Industrial employment fell by over one-third during the decade of the 1970s but total employment also fell. The loss of manufacturing employment represented 79 per cent of total change. Since the boundaries of the FUR and Level 2 region were coincident, the level of spatial analysis here did not affect the apparent loss, although the rate of loss was worse in the core than in the hinterland. Frankfurt and Liverpool are the polar extremes of a continuum; Madrid, the capital of Spain, represents an

Table 8.2 Manufacturing employment loss from selected FUR cores, FURs and Level 2 regions 1971–81,[1] FUR unemployment 1985–7, and FUR per capita GDP 1985

FUR	Manufacturing employment change			Level 2 region[2] as percentage of manufacturing employment, 1971	FUR population as percentage of Level 3 region	FUR mean U rate 1985–7		FUR per capita GDP 1985	
	FUR core as percentage of manufacturing employment, 1971	FUR as percentage of manufacturing employment, 1971	As percentage of total employment change, 1971–81			Rate	as percentage of nation	as percentage of EC 12	as percentage of nation
Amsterdam	−30.0	−29.3	227	−10.0	234.5	10.2	102.3	127.8	118.8
Barcelona	−22.7	−12.4	94	+5.4	99.9	23.1	106.9	81.7	112.9
Belfast	−40.6	−33.0	2.098	−17.7	66.0	14.9	131.9	86.4	83.1
Bilbao	−12.1	−4.5	26	−1.8	92.9	24.2	111.9	85.4	118.1
Bochum	−7.0	−5.5	73	−6.3	304.3	10.9	160.3	132.1	114.0
Bruxelles	—	—	—	—	338.8	11.3	100.6	156.9	154.5
Frankfurt	[−14.3]	[−7.7]	[−276]	[−6.3]	364.5	4.6	68.2	330.4	285.1
Liège	+11.1	−20.6	−602	−20.6	137.0	14.6	130.3	101.7	100.1
Kobenhavn	−49.3	—	—	−30.3	303.3	6.2	90.6	136.8	117.4
Liverpool	−39.5	−36.2	79	−36.2	91.6	18.2	161.2	97.2	93.5
London	−35.6	−31.9	122	−36.0	133.0	9.7	85.9	145.5	140.0
Madrid	−30.9	−4.8	−35	−5.8	102.5	19.6	90.9	82.2	113.7
Manchester	−23.0	−16.6	189	−15.8	75.9	13.0	115.3	103.8	99.9
Rotterdam	−28.8	−12.4	−46	−11.7	145.6	9.8	97.7	121.7	113.2
Torino[3]	−28.0	−8.4	−156	−4.2	87.4	10.7	105.7	124.0	120.1

Notes: 1) Years vary for some FURs. 2) In which highest proportion of FUR core population resided in 1981. 3) Data for Italy are difficult both because of reporting problems and because public services were included in the census of employment in 1981, but not in 1971. These calculations assume employment in public services was the same at each date and therefore almost certainly overestimate 1971 total employment and underestimate total employment gain. [ ] = Provisional — = Not available.

Figure 8.1 Location of selected European urban regions.

intermediate situation. There was a net loss of manufacturing employment from the FUR as a whole; the core was rapidly growing and sqeezing out manufacturing employment from it. There was, in fact, a strong growth of the intially small manufacturing sector in the hinterland; manufacturing employment in the Madrid hinterland increased by 98.3 per cent (compared with an increase of 162.4 per cent in services). The hinterland extended beyond the Level 2 region, however, so manufacturing loss in the FUR as a whole was less than in the Level 2 region. In Barcelona, in northeast Spain near the French border, the Level 2 region, Cataluña, is, however, far larger relative to the FUR than is the case with

Madrid, so although the same spatial relativities applied, there was net growth of manufacturing employment in the Level 2 region as a whole.

The spatial definition of region, therefore, is clearly an important factor in determining the perception of deindustrialization. The more broadly regions are defined, the greater the proportion of the change that systematic patterns of intra-regional relocation have been causing will be netted out. From one perspective Barcelona, as a contiguously built-up metropolitan area, was suffering severe problems of deindustrialization during the 1970s. From another, Barcelona as Cataluña was an industrializing region. Frankfurt was a deindustrializing region, but apparently because higher value-added activities were outbidding industrial uses of land. Liverpool, Belfast in Northern Ireland, or, in a less extreme way, Bochum in the industrial Ruhr of the Federal Republic of Germany, were deindustrializing FURs in deindustrializing wider regions.

Industrial activity was not being priced out of Liverpool, Belfast, or Bochum because of the pressure from other, more profitable, activities. It was simply declining in regions that had lost comparative advantage over a wide range of activities. Indeed, in Liverpool and Bochum the other sectors of the local economy were declining too, so that the loss of manufacturing employment was less than the loss of total employment (79 per cent and 73 per cent, respectively). In Liverpool one-third of employment was lost from the port; even the two primary service sectors, distribution, hotels and repairs, and banking, financial and business services, showed employment losses. In Bochum there was a loss of 12 per cent from the energy sector – coal mining was still a significant local industry in 1971 – and other sectors showed only very modest growth. Belfast, in contrast to Liverpool, had growth in services. Reflecting its status as the regional capital of Northern Ireland, it had substantial growth in the two key service sectors (34.8 per cent and 91.6 per cent, respectively) and also in public administration. Its loss of employment overall was, as a result, very modest compared to its loss of industrial employment. Liège represented yet a further point on the contiuum. Like Belfast it was the administrative centre of a generally declining region (Wallonie, the French-speaking southern area of Belgium). As was the common pattern with cities in France, the core of Liège was a large centre before major industrialization so, although there was a concentration of older, heavy industry in the Liège FUR as a whole, it was concentrated in the hinterland (a core : hinterland location quotient of 0.56 in 1971 compared with 1.42 in Bochum, for example); only the newer, other manufacturing sector was concentrated in the core. The result was a very modest 1.2 per cent increase in overall employment, with the growth strongly dominated by banking, financial and business services and by public administration. The result of this overall pattern was that there was some growth of manufacturing employment in the core, serious loss in the hinterland and in the FUR overall, but the negative change in manufacturing employment in FUR was six times as great as the increase in employment overall. Although the spatial extent of the FUR was smaller than the Level 2 region of Liège, and its boundaries were not coincident with those of the surrounding administrative regions, the percentage change in manufacturing employment in both the FUR and Level 2 region was the same.

Measuring deindustrialization

It would appear that implicit in the use of employment change in industry as a definitional measure of deindustrialization, is a concern with the labour market implications, and perhaps with the impact deindustrialization may have on the distribution of income. If mean per capita income is the focus of interest, then measures based on absolute and relative output changes and measures that take account of productivity changes would appear to be more appropriate. The labour market impact of industrial employment loss is not, however, a simple problem. Definitionally the change in the number of unemployed in a particular labour market is given by

$$\Delta U = \Delta N + \Delta M - \Delta E_i - \Delta E_a - \Delta E_s + \Delta C$$

where ΔU is the change in number unemployed; ΔN is the net change in economically active as a result of natural change; ΔM is the net change in economically active as a result of migration; ΔE is the change in employment in industry (i), agriculture (a) and services (s); ΔC is the net change in commuters.

If labour market outcomes are the focus of interest, concentration on ΔE_i implies that this is the main source of variation. Such a view might be consistent with migration and commuting being constant, natural change being exogenous, agricultural employment being unimportant and industrial employment being the 'economic base'. These are, to say the least, strong assumptions. There is evidence that regional employment changes in one sector tend to be positively correlated with those in other sectors.[13] That an industry is 'basic', however, is only a minor part of the explanation, if any. The main explanations seem to be joint causation and sectoral diffusion effects. That is, because of mobility and relatively low attachments to particular sectors, fluctuations in employment and unemployment in one sector in a regional economy spill over into other sectors.

Agriculture is unimportant in employment terms in most of Northern

Europe or North America but that is observably not the case in Southern Europe (see below). For certain regions it is valid to assume that a change in industrial employment is a good indicator of total employment change, but these regions are unusual. In the short run, it may be reasonable to assume that migration flows, natural change and commuting are given. In the longer run, however, these factors, particularly commuting and migration flows, will adjust to employment opportunities.

The implications for unemployment of loss of industrial employment will depend, therefore, on the local economic context within which the reduction in employment occurs; on the adaptability of the displaced labour and on the degree of independence of a particular regional labour market from that of other regions; and on the economic context of those other regions. In principle – leaving aside any pro- or anti-interventionist predisposition – policies to improve labour market mobility, both geographic and occupational, may be as appropriate as policies aimed at reducing the rate of deindustrialization or at generating new jobs in non-industrial sectors. In the context of a strong regional squeeze on industrial employment, policies focused on occupational mobility and retraining would appear to be the most appropriate. In the context of general decline and deindustrialization – such as Liverpool – policies that attempt to increase rates of productivity growth and encourage job generation, and even geographic mobility, may be more appropriate. The argument is, then, that since causes vary, appropriate policies also vary. The spatial definition of the region is important because that, in turn, is closely related to the degree of independence of the 'regional' labour market and the extent to which quasi-automatic adjustment to job loss in particular sectors can be expected.[14]

How deindustrialization is defined and measured should reflect the focus of interest. This is true both for the spatial unit over which it is measured and for the indicator used. As the national data presented in Table 8.3 show, the results depend strongly on the measure used. There is no correlation (rank correlation = 0.02) between changes in industrial output and industrial employment over the period 1983–7, and the country that did worst on the employment measure, Ireland, did best on the output measure. The country that did best on the output measure did worst on the simple productivity measure.

The experience of the individual member states of the European Community between 1983 and 1987 represents an unexpectedly extreme situation. In the longer term a stronger positive association between output and employment growth might be expected, especially if the way in which these were measured was abstracted from cyclical factors. We may provisionally, if a little simplistically, conclude that observers whose primary concern is with unemployment and distributional issues will focus on employment-based measures of deindustrialization. However, if their interest is in issues related to aggregate income or to trading and competitive positions, then output and (relative) productivity changes are more relevant. In a regional context, adaptive capacity, mobility and the extent to which industrial activity is either being squeezed out or declining because of loss of absolute advantage, are highly significant. So, also, is the institutional and policy framework within which deindustrialization may be occurring.

One measurement problem, albeit a more practical one, remains to be considered. The period over which the measures are made has a strong influence on the apparent incidence of deindustrialization for two reasons. First, while there does appear to be a trend toward declining industrial employment share and declining absolute industrial employment in the economies of advanced regions and countries, it is not a linear trend. As is shown by Rowthorn and Wells,[15] the share of manufacturing

Table 8.3 Industrial production and employment for EC member states: cumulative change 1983–7 (1983 = 100)

		Change in output: change in employment		ΔEmployment	Rank on ΔOutput	$\frac{\Delta \text{Output}}{\text{Employment}}$
Belgium	Output	110.3	1.20	5	7	5
	Employment	91.8				
Denmark	Output	119.1	1.13	1	3	9
	Employment	105.7				
Federal Republic of Germany	Output	111.7	1.17	3	6	8
	Employment	95.7				
Greece	Output	103.0	—	—	—	—
	Employment	—				
Spain	Output	113.9	—	—	—	—
	Employment	—				
France	Output	106.1	1.20	7	9	5
	Employment	88.7				
Ireland	Output	139.0	1.58	9	1	1
	Employment	87.8				
Italy	Output	107.5	1.21	6	8	4
	Employment	89.2				
Luxembourg	Output	132.3	1.40	4	2	2
	Employment	94.2				
Netherlands	Output	113.8	1.18	2	5	7
	Employment	96.3				
Portugal	Output	121.8	—	—	—	—
	Employment	—				
United Kingdom	Output	116.3	1.32	8	4	3
	Employment	87.9				

Source: European Economy, February 1989.

in total employment follows a relatively predictable path once other factors are allowed for, tending to rise as real per capita incomes increase to about US$3,800 at 1975 values, and tending to fall thereafter. Since both the level of real per capita incomes and their trajectory through time, vary across the regions and countries of the European Community, we should expect to observe differences in their industrial employment change and changes in industrial employment share over any given period, even if the pattern of these were ultimately to be identical. The conclusions one draws, therefore, from any cross-sectional data with given start and finish dates have to allow for the level of development of the country(s) or region(s) concerned. We should be chiefly interested not in the decline of manufacturing employment share, but in departures from the expected rate of decline and the problems of adjustment.

In addition, measurements must allow for the incidence of the trade cycle. Since the 1950s there has developed both a theoretical[16] and an empirical[17] literature on the inter-regional transmission of the trade cycle and its sequential regional incidence and amplitude. The differential incidence of the trade cycle implies that measurement problems are much worse in the context of the regions of the European Community than they are in the context of US or Japanese regions since it is necessary to take account simultaneously both of national differences in the incidence of the trade cycle and regional differences. In principle the problems associated with the trade cycle can be overcome by taking longer runs of data and by averaging (problems associated with levels of development are less easy to deal with in this way), but data availability does not always permit this.

Measures of regional deindustrialization in the European Community

It is not surprising, given the brevity of its existence and the scarcity of data, that there has been comparatively little work done on problems of regional deindustrialization across the European Community as a whole. A study undertaken by the Centre for Urban and Regional Development Studies (CURDS) attempted simply to document industrial employment change for Level 3 regions of the then nine-member Community between 1974 and 1982 using national data sources.[18] Part of the differences they found reflected national differences in the incidence of the trade cycle. The United Kingdom, which entered the recession of the early 1980s first and most sharply, tended to dominate the list of deindustrializing regions and relatively few problems of deindustrialization appeared to exist in France, a country that entered the economic recession late.

A more significant problem resulted from the level of analysis used.

Since the administrative Level 3 regions vary in their relations to functionally defined regions both in size and coincidence of boundaries, there was a severe problem of confounding decentralization and regional deindustrialization. As is shown below, the wider region of Wallonie, in southern Belgium, is by any standards a deindustrializing region, with declining heavy industries, including resource-based industries such as coal and also steel and heavy engineering. The decline is concentrated in the belt of medium-sized industrial towns, such as Mons and Charleroi, which grew up on the coal measures stretching from northeastern France through Hainaut. Charleroi, if defined as a FUR, lost 63 per cent of its employment in energy and water (NACE 1) and 13 per cent of its employment in manufacturing industry between 1971 and 1981. Since the Charleroi FUR was of rather similar size to the Belgian Level 3 regions (66 per cent of its population being in the Charleroi administrative region), the Level 3 analysis described its deindustrialization with reasonable accuracy. However, the CURDS study also found very strong deindustrialization in the Level 3 region of Brussel–Hoofd. In reality the case of Brussel[19] was quite different from that of Charleroi. No individual Level 2 region – much larger than the Level 3 regions used – contains more than 61 per cent of the population of the Brussel FUR. Its population is split between 16 Level 3 regions with the most important, Brussel–Hoofd, the region which happens to bear the name of the city, and which historically represented the city, entirely within its modern core. Brussel–Hoofd in 1981 contained less than 30 per cent of the total population of the FUR of Bruxelles. In terms of the Level 3 analysis, therefore, the discovery that Brussel–Hoofd had lost more industrial jobs between 1974 and 1982 than Charleroi is likely to have said more about the intra-regional change in progress in the FUR of Bruxelles than about problems of deindustrialization. In the period 1977–86, Brussel–Hoofd had the second lowest concentration of industrial employment of any Level 3 in the European Community for which data were available whereas Charleroi had a concentration of industrial employment of 1.22 compared to the Community as a whole. This is an extreme example, but it illustrates an endemic problem of using the individual, fine-grained administrative regions as the basis for analysing deindustrialization or, indeed, any other aspect of regional transformation.

Among the few other studies of regional deindustrialization on a Europe-wide scale are those of Keeble and his various associates.[20] These made use of data from the Labour Force Surveys and so, because of problems associated with sample size, their analyses were confined to Level 2 regions. This meant that the problem of confounding intra-(functional) region relocation with deindustrialization was considerably reduced, but other limitations remained.

One reasonably clear conclusion did emerge, however. Keeble et al.[21] were the first to document, on a European scale, the general pattern of decentralization of industrial activity from more urban to more rural regions. Other conclusions that were drawn seem to have been less reliably founded. No allowances were made, for example, for the influence that the particular start and end dates used should have been expected to have had; nor did the studies allow for the general relationship (documented by Rowthorne and Wells[22]) between the share of industry in total employment and the level of economic development.

The start and end dates analysed in the earlier Keeble study were 1973 and 1979; those in the more recent study, 1979 and 1983. The conclusion of the first study was that

> The Community's weakest and most vulnerable manufacturing plants and firms would appear to be those located in its peripheral regions. Secondly, central region employment decline may reflect capital–labour substitution and increasing labour productivity, coupled with actual disposal of manufacturing capacity to adjacent intermediate regions.

This was a widely quoted and influential conclusion. It not only influenced academic thought but was seen by policy-makers to justify and so reinforce the focus of European regional policy on the problems of 'peripherality'. It fitted neatly into an emerging political and academic conventional wisdom of the relative positions of the 'core' and 'periphery' regions of the European Community. Although, as shown in the analysis in the next section of this chapter, European economic integration has had and is having a systematic influence on regional fortunes, the actual position is more complicated. The changes in industrial employment found in the Keeble studies are reproduced here as Figures 8.2 and 8.3. From Figure 8.2 it can be seen that the first study's conclusion with respect to peripheral regions was largely determined by the experience of Italy. This mainly reflected the failure of the traditional regional policy of the 1960s, which had involved the transplanting of heavy industry to the Mezzogiorno. Elsewhere many peripheral regions – the southwestern regions of France, the southeastern regions of the Federal Republic of Germany, Ireland and southwest England – exhibited some of the most rapid rates of manufacturing employment growth.

The second period, shown in Figure 8.3, illustrates the problem of the national incidence of the trade cycle. All United Kingdom regions were in one of the three categories of greatest loss, and eight of its regions were in the category showing the greatest employment loss of all. Italian and French regions generally showed gains, or only small losses, reflecting the

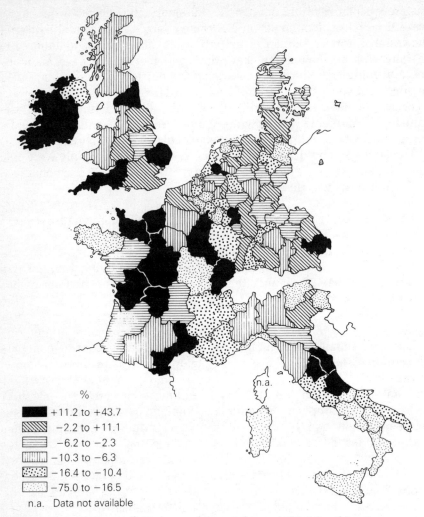

Figure 8.2 Percentage manufacturing employment change, 1973–9: Level 2 regions.

later incidence of the trade cycle in those countries. But the regional incidence in both countries was markedly different to the earlier period. The Italian south and northeast did relatively, and in some cases absolutely better. In France the previously strongly growing regions – Normandy, in the northwest, Champagne–Ardennes and Franche-Comté in the southeast and Poitou–Charente and Limousin in the southwest – had turned to sharp manufacturing employment loss.

In fact the results of Keeble *et al.* suggest that the regional pattern of manufacturing employment change was negatively and significantly related between 1973–9 and 1979–83; the correlation coefficient across all regions was r = −0.224 (t = −2.26 N = 102) and the only country exhibiting a significantly stable pattern was Belgium (r = 0.762, t = 3.33). The United Kingdom was the only other country where regional manufacturing employment change in the two periods was positively associated, but the relationship was not significant. In all other countries analysed, there was a negative relationship. In view of this general reversal of pattern, Keeble *et al.*'s more recent conclusion, that there was 'better regional manufacturing employment performance in the 1980s in the

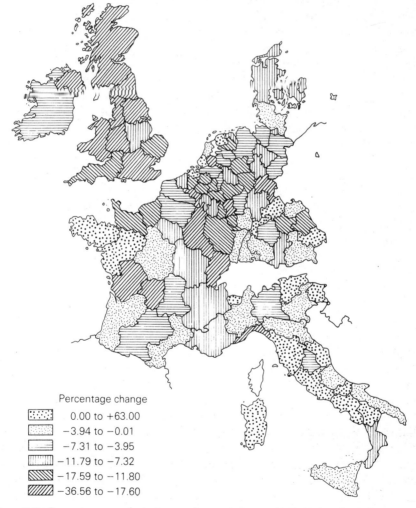

Figure 8.3 Percentage manufacturing employment change, 1979–83: Level 2 regions.

peripheral regions than elsewhere in the Community' is to be expected; but it is not clear what significance can be attached to it. Because both studies failed to allow for the differential national incidence of the trade cycle or the varying degree of economic development, the conclusions that can be drawn with respect to general patterns of deindustrialization are weak.

If we examine the two sets of data carefully, we do find some stabilities, however. Some regions exhibited industrial employment loss over both periods and others exhibited industrial employment gain. These patterns do not abstract from levels of national or regional economic development but, because they reflect two separate time periods, they do largely abstract from the differential incidence of the economic cycle. They are shown in Figure 8.4. It will be seen that no region in the Federal Republic of Germany appears in the deindustrialization set despite the problems that are known to exist in the Ruhr. There are various possible reasons for this, including the relatively strong performance of the West German economy over the period in question, but an important reason seems simply to be one of regional definition. It happens that no Level 2 region includes the whole Ruhr area; regions that contain a part of the Ruhr also contain other areas that had strong growth.

Even these more stable patterns represent a varied range of causation and experience. Some of the regions that were peristently losing manufacturing employment are 'classic' cases of declining, mainly resource-based industrial regions in deindustrializating national economies. The regions of southern Belgium – Hainault, Liège and Luxembourg – fit this pattern; so do those of the United Kingdom: Northern Ireland, Scotland, the North West and Wales. Lorraine in France and Liguria in northern Italy are other examples, although each has its own special features. Other regions persistently losing manufacturing employment represent, to a significant extent, the accident of regional definition. The cases of the Ruhr and of Brabant, in the context of decentralization from the Bruxelles FUR, have already been commented on. The Grand Duchy of Luxembourg which, as was already seen from Table 8.3, has performed outstandingly well in manufacturing output and productivity terms, appears to represent a different explanation, that of a strongly restructuring region in transition. Labour costs have risen rapidly there, as has employment in services. It would appear that loss of employment in manufacturing is primarily the result of a 'squeeze' effect. The differential growth of the service sector, mainly as a result of the location of important institutions of the European Community in Luxembourg, is a significant source of this 'squeeze'.

Patterns of manufacturing employment growth similarly reflect a range of causal factors. The strongest concentration of growth has been in the emerging industrial regions in the southeast of the Federal Republic of

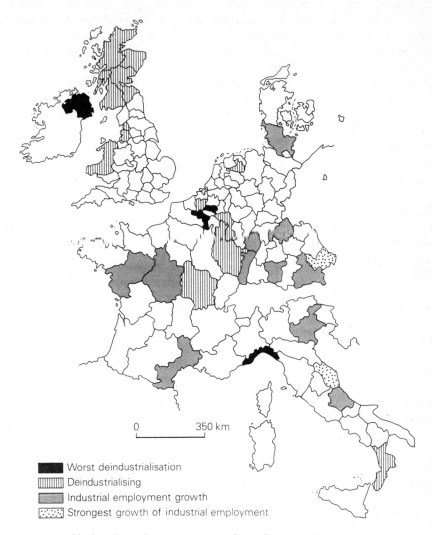

Figure 8.4 Stable deindustrializing regions or industrializing Level 2 regions 1973–83.

Germany, centred on München. But the decentralizing shift in France, towards the south, the west and the east, is also persistent, as is the growth of some of the previously less industrialized regions in Italy. Schleswig-Holstein appears to be the most likely example, the inverse of Brabant, of a recipient region of manufacturing activity from a decentralizing urban region, in this case, Hamburg.

Regional transformation as a source of adjustment problems

Given these difficulties of measurement what can be done to attempt to answer the question implicitly posed at the start? To what extent is regional transformation, particularly as it manifests itself as deindustrialization, a source of problems? To what extent, and in what circumstance, can it be successfully accommodated? To answer these questions it is necessary to try to avoid the analytical and measurement problems already identified. In the analysis that follows, therefore, we use FURs and we analyse only the 122 largest urban regions. This makes it possible to distinguish deindustrialization almost entirely from intra-regional decentralization; only 'almost entirely' because the FURs were defined in 1970 so there are some, such as Frankfurt, Paris, London, or Madrid, that in recent periods may be decentralizing beyond their 1970 boundaries. Second, comparisons between single start and end dates are avoided. The indicators of problems used in this analysis are averaged over a number of years to exclude, so far as possible, purely cyclical movements. Those explanatory variables that are not averages relate to single dates to reflect some characteristic aspect of the FURs.

Transformation makes regional economies change. These pressures for change will vary from region to region depending on the characteristics of each. They interact with what we may call the adaptive capacity of each regional economy; its flexibility and responsiveness. Unemployment has been the problem cited most often as a consequence of one facet of transformation, deindustrialization. The question is whether changes in unemployment expressed as an identity can be converted into a set of explanatory relationships. A further reason for looking first at unemployment is that the data are available. A definitionally comparable measure of unemployment has been available for the Level 3 regions of Europe since 1977 (values for Spanish regions for the period before Spain joined the European Community can be computed from the Spanish Labour Force Survey[23] which used a survey question that for all practical purposes was the same as that used in Community countries).[24]

The dependent variable to measure changes in unemployment rates is the absolute difference in percentage points between the mean rate for the FUR in 1985–6–7 and the FUR mean for 1977–9–81. The change over this period ranged from an increase of 18.4 points in Cordoba in southwest Spain, to a fall of 2.9 in Aarhus, Denmark. The best results obtained, using ordinary least squared multiple regression techiques, are set out in Table 8.4. Given that there are 120 observations for a cross-sectional analysis, an overall adjusted R^2 of 0.92 is good.

The explanatory variables have been chosen to reflect the factors that in the European context may have generated forces for regional adjustment and that might be associated with the adaptive capacity of regions. The measure of economic structure used relates to the percentage distribution of the labour force, not in the FUR itself, but in the wider regional context – the Level 2 region – within which the urban region is located. As no direct measure of change in the size of the labour force induced by natural change alone is available, the overall rate of natural population change from 1971 to 1981 is included to reflect this supply-side variable.

Two sets of results are reported, one excluding natural population change, the other including it. The results suggest that there may have been an association between a specialization in industrial activities in the wider Level 2 region in 1975 and a greater increase in unemployment in the urban region over the ten years in question, but any relationship was significant only at the 12.5 per cent level and accounted for only a small part of the total variance. If the natural rate of change of population in the

Table 8.4 Dependent variable: change in mean U rate, 1977/81 1905/7

Independent variable coefficient	Estimated coefficient (excluding natural rate of population change)	Estimated (including natural rate of population change)
Intercept	−0.6021	0.9103
	(−0.44)	(0.23)
Change in economic potential with European Community integration	−1.2690*	−1.1621*
	(−2.36)	(−2.10)
Natural rate of change of FUR population 1971–81		0.1038*
		(2.05)
Dependence of local economy on coal	0.6967*	0.8167**
	(2.50)	(3.10)
Dependence of local economy on port	0.3507**	0.3437**
	(2.73)	(2.73)
Percentage of labour force in agriculture, 1975	0.0609**	
	(2.64)	
Percentage of labour force in agriculture, 1975, squared		0.0014*
		(2.09)
Percentage of labour force in industry, 1975	0.0359	
	(1.55)	
Mean unemployment 1977–81	0.1899**	1.1621**
	(2.98)	(18.83)
Country dummy: Spain	8.1418**	6.8903**
	(14.08)	(9.67)
Federal Republic of Germany	2.2228**	2.7484**
	(4.91)	(5.45)
N = 120	adj R^2 = 0.81	adj R^2 = 0.92

Notes: 't' values in parentheses * = significant at 5%   ** = significant at 1%.

FUR itself was included (right-hand column), then specialization in industry ceased to be a significant variable at any reasonable level. A relative concentration on agricultural activity in the wider region was statistically far more closely associated with an increase in unemployment in the FUR than was a concentration on industry (there is a very low correlation between concentration on agriculture and concentration on industry). Although for obvious reasons there is a relatively low concentration on agriculture in most Level 2 regions associated with major cities – with less than 5 per cent of employment accounted for by agriculture in the great majority – in less-urbanized parts of Europe, FURs have extensive and quite rural hinterlands. The most extreme case is in northwest Spain where the Level 2 region containing the urban region of Vigo had nearly 41 per cent of its total labour force engaged in agriculture in 1975.

It seems clear that while regional economies more dependent on industry had some adjustment problems, they were considerably more serious for regions experiencing urbanization and agricultural transformation. This provides a different, but firmer, foundation for European policy concern with the problems of poor, backward regions; but not, as is argued below, justification either for existing agricultural policies or for regional policies that direct assistance to the rural parts of those backward regions. It should also be noted that in a period when growth was concentrated in the service sectors, regional economies with a specialization in that sector experienced fewest adjustment problems.

Considerably more significant, in statistical terms, than dependence on industry, was the presence of a port, a coalfield, or both. These explanatory variables are measured as dummies. The port dummy has a range of 0 to 4. Values are assigned strictly according to the volume of seaborne traffic handled at a date, 1966, shortly before that at which the analysis started.[25] The coalfield dummy ranges from 0 to 2 with 1 for FURs that are partly in the area of a coalfield and 2 for FURs wholly within a coalfield. The reasoning behind this formulation of the variable is that problems of adjustment might have arisen not only from the direct loss of employment in coal-mining areas but from their relative lack of attraction to new, decentralizing industries.[26]

Another systematic influence on regional unemployment change across all the major urban regions of the European Community was the impact on the region of European economic integration. The results of Clark et al.[27] supplemented by the results reported by Keeble et al.[28] for the regions of Spain make it possible to construct a measure of the change in regional economic potential resulting from the combined effects of eliminating tariffs within the European Community and from falling transport costs. The more an area had gained from these effects the less, *ceteris paribus*, unemployment rose. A further factor explaining the variance of unemployment changes was a strong positive association with the mean level of unemployment during 1977–81. Unemployment increased most in those urban regions where it was already highest.

The final explanatory variables are country dummies. The most important implication of these would seem to be that even using differences between averaged values of unemployment does not fully offset measurement problems associated with the differing timing of national economic cycles. Other chapters in this volume show that the economic downturn in Spain after 1974 was particuraly serious and that recovery did not start until 1984/5, and that the Federal Republic of Germany entered the recession rather later than other major European countries and has suffered a relative worsening in its unemployment rate (albeit from a very high standard) compared to other countries. These effects are reflected in the values of the country dummies. It seems probable that there is also some problem of colinearity between the measure of regional specialization in industry and the country dummy for the Federal Republic. It will be seen that in the results in Table 8.5, where Federal Republic values effectively act as datum, specialization in agriculture is rather less significant and specialization in industry rather more so. The implication may be, therefore, that there is some difficulty in clearly distinguishing the 'German' effect from the effect of specialization in industry.

Table 8.5 shows the results of a rather similar analysis using as the dependent variable the change in the value of the urban problem index[29] over the whole period 1971–87. Again urban problems are conceptualized as being problems of adjustment. The major difference in the pattern of results is that the log of the population is significant; allowing for other factors, larger cities performed better. Other minor differences are that the gain in economic potential was relatively more significant, and concentration on agriculture not so significant compared with the results obtained examining changes in unemployment alone. The difference in results partly reflects the longer period covered by the analysis – 16 years compared with 10 – but also the fact that more weight in the dependent variable is implicitly given to income and service sector growth. Larger centres have benefited more from service sector growth.

What do these results suggest about regional transformation in general, regional deindustrialization in particular, and the problems that are associated with it? First, concentration on industry was, by itself, a small contributing factor in increasing regional unemployment, although it was rather more closely associated with an increase in the more broadly based urban problem index. There seem to be statistical difficulties in distinguishing between the alternative propositions: that West German urban regions tended to have had a relative increase in problems or that regions

Table 8.5 Dependent variable: change in urban problem index, 1971–87 for FUR[1]

Independent variable coefficient	Estimated coefficient (excluding natural rate of population change)	Estimated (including natural rate of population change)
Intercept	16.3172** (3.62)	17.2253** (3.34)
Log FUR population	−0.8649** (2.75)	−0.9606** (−2.95)
Change in economic potential with European Community integration	−4.7410** (−5.27)	−4.8178** (−5.31)
Percentage of labour force in industry, 1975	0.0651* (2.20)	0.0673* (2.16)
Percentage of labour force in agriculture, 1975		0.1685* (1.86)
Dependence of local economy on coal	1.2245** (3.34)	1.2076** (3.42)
Dependence of local economy on port	0.6429** (3.80)	0.6295** (3.85)
Natural rate of population change		0.1742** (2.75)
Country dummies Spain	6.9051** (6.75)	4.5387** (3.55)
Italy	−2.3011** (−3.07)	−3.7437** (−4.50)
United Kingdom	−3.3858** (−4.48)	3.7495** (−4.62)
France	−2.1967** (−3.50)	−3.4958** (−5.12)
N = 117	adj R^2 = 0.78	adj R^2 = 0.80

Note: 1) Negative = improvement: therefore, since explanatory variables have positive values, negative coefficients 'better'. For definition of variable see text.
* = significant at 5%   ** = significant at 1%.

specializing in industry have had a relative increase in problems. Regional problems of adjustment were far more closely associated with a concentration on specific resource-based, older, industrial activities: coal mining and ports and related industries. Ports are an interesting case because the problems of local economies that grew up around them have received rather little attention. It would appear that it is not just that there has been considerable capital–labour substitution as containers and roll-on, roll-off ferries have replaced traditional cargo-handling methods but that ports have lost their locational advantages to processing industries as transshipment points. Containers can be unpacked anywhere.

Regions that were specialized in industry but where that specialization was skill-based rather than resource-based (the Birmingham or Manchester areas in the United Kingdom, Stuttgart in the Federal Republic of Germany, or Torino in northern Italy come to mind as examples) have adapted to capital–labour substitution and to loss of employment in their traditional industrial sectors more readily than the old, resource-based industrial regions. Torino and Birmingham emerged as new problem regions in the late 1970s and early 1980s. The case of Torino is analysed by Camagni.[30] It has recovered sharply in relative productivity terms since 1976–9 although these gains have still been at the expense of employment. Comparable data are not available for Birmingham, but employment appears to have grown strongly since 1985 and unemployment relative to London, for example, which was rising during the 1970s and reached a peak of 1.65 in 1983–4, fell to 1.26 the London rate, in 1985–7 and to 1.10 the London rate, in 1988. In the terminology used earlier in this chapter, their adaptive capacity was superior even though the forces generating the need for adjustment may have been as strong.

There remains the question of how deindustrialization affects the level and distribution of income. The evidence is not available to provide a systematic answer for European regions. It is known that the distribution of income in the United Kingdom has changed radically since 1975,[31] but to date only one study is known to the author that throws light on the pattern of change within a city region, London.[32] The distribution of household incomes became less equal in London between 1979 and 1985 than it did in the United Kingdom as a whole and, in London, the contribution of occupational change to the observed change in income distribution appeared to be greater than in the United Kingdom generally. This is one isolated and still preliminary finding. The Greater London area is certainly one of the areas within the European Community that has experienced the strongest and most rapid tertiarization; the United Kingdom has likewise experienced particularly strong deindustrialization.[33] The small amount of available evidence supports the view that deindustrialization generates increasing income inequality. However, it is not adequate to conclude that increasing income inequality is an inevitable consequence of deindustrialization.

There is evidence that across the major urban regions of the European Community there have been problems of labour market adjustment, and wider problems too, resulting from the decline of industrial employment, but that a concentration of regional employment in industry alone was a barely significant factor. The diversity of the Community's urban regions meant that urban centres in poor agricultural regions suffered worse problems. The major problems of adaptation to industrial change that we can measure were essentially features of resource-based, older industrial regions.

Some concluding remarks

What wider implications can be drawn from this experience of regional transformation and deindustrialization in the countries of the European Community? Such implications are necessarily speculative, but may be useful all the same. The first is that to focus on deindustrialization is at best misleading; at worst dangerous. Camagni's definition of 'deindustrialization'[34] is the most illuminating because it suggests that deindustrialization is just one special case in a wider process of transformation. It is important to remember that even in this definition the regional economies being analysed exist as but points in space and the process of intra-regional and inter-regional transformation is hidden from view. So, too, are other dimensions, such as the organizational and sectoral ones referred to above.

Analogy can be dangerous, but in the present context it seems helpful. Economic transformation is the most fundamental process of economic and social change since the Industrial Revolution. We do not, however, spend our time analysing 'de-agriculturalization' and 'de-agriculturalizing' regions. Yet if we look at the British Isles between about 1770 and 1850 we can apply the Camagni schema to regional change, substituting word for word 'agriculture' for 'industry'. We would have 'de-agriculturalizing' regions – much of Ireland, Scotland and the South West, for example – losing both employment and output from agriculture and with productivity falling relative to other parts of the United Kingdom. There would equally, in Camagni's terms, have been 'agriculturally restructuring' regions and 'virtuous circle' areas. During the Industrial Revolution the economic and social structure of agriculture as well as its regional geography was totally transformed. We do not talk of de-agriculturalization, although it was significant, because it was not the essential feature of change. The essential feature was industrialization. In Europe now deindustrialization is similarly only one feature of industrial change and we have seen that agricultural transformation still appears, in many parts of Europe, to be creating more serious problems of regional adjustment.

The spatial aspect of regional transformation is also important. There is long-distance inter-regional relocation, which can appear as regional deindustrialization, but intra-regional decentralization is spreading out to such distances that it is spilling over traditional regional boundaries. Evidence in Cheshire and Hay[35] shows that decentralization of population follows similarly predictable patterns and has moved south through the urban regions of the European Community since 1950, also tracking levels of economic development. One characteristic of economies undergoing advanced tertiarization, however, seems to be a form of recentralization. This can be seen in urban regions such as London.[36] Decentralization started early in London and so much industry had gone by 1980 that further loss, even at constant annual rates, had a falling impact on the local economy. The strongly growing service and information sectors had a stronger urban orientation and made an increasingly important contribution to total employment in London. The evidence is that employment increased in London with a turning date in 1982 or 1983 and that, reversing a trend of 40 years, there was also an increase in the number of households. These changes were accompanied by increasing income polarization.

The association of deindustrialization with counter-urbanization, therefore, may be for many urban regions only another temporary phase. The advanced stages of tertiarization may produce new forms of urbanization. These are likely to be at much lower residential densities expressed as people per hectare than in the past. It does not follow, however, that job densities will not return to those of the high point of industrialization because job densities in the service sector are so much higher than in industry, even nineteenth century industry. Not all urban regions are likely to experience strong reurbanization, however. Another characteristic of falling transport and communications costs and the decline of resource-based regions is that, although the new activities may be urban-oriented, they are not constrained to locate in one particular urban centre. In that sense economic activity is becoming even more footloose.

Another lesson to be learnt is that measuring and analysing these changes at a regional level requires appropriately defined regions. Those used for the present analysis have been functionally defined urban regions, but it is likely that regions defined in other ways could also be appropriate. They must be areas in which intra-regional decentralization is occurring as well as the centres from which decentralization is emanating and they must be defined consistently. While regions may be broadly defined to include several smaller but similar urban regions, it would make measurement difficult, and the results of analysis misleading, if regions were defined to sometimes include several smaller urban regions but tightly drawn about very large metropolitan areas. It is also important to be careful not to confound secular trends with cyclical variations.

Certain factors in the European analysis came out so strongly that it seems likely that they apply to regional adjustment and transformation in countries outside Europe, especially long-settled countries with an agriculture based on peasant traditions and with an historically long-established urban system. The first is that regional problems of agricultural transformation, which tend to be most acute in regions that also have other adjustments to make, seem to be every bit as severe as those associated with specialization in industry. Indeed, the evidence is that they are more severe and very much more significantly associated with problems of unemployment. Older, resource-based regions whose industrial development was based on coalfields or ports appear to have the most significant problems associated with deindustrialization. The implications of this are that industrial regions in which development was based on human skills or communications may lose industrial employment but they have less difficulty in adapting.

Although the European Community is unique in the extent to which it is a policy-assisted process, it is not the only case of increasing economic integration between countries. Falling transport and communications costs mean that countries are themselves becoming more integrated; globalization means that economic integration is increasingly a transnational phenomenon. Integration confers spatial gains in comparative advantage but also implies relative losses. The European experience shows those effects are systematically related to changes in the regional incidence of unemployment and to the changing incidence of urban problems. To some extent this supports the European policy concern with problems of 'peripheral' regions but it suggests that changes in peripherality are as, or more significant than absolute levels of peripherality and that the processes involved extend beyond simple issues of deindustrialization. The gains and the losses are experienced by the whole range of the economy. As an illustration, we can see that in the severest cases of loss from economic integration and transformation, represented by Liverpool, in northwest England, there is not only loss of employment from industrial sectors, but from all sectors. Equally, although they have gained greatly in terms of economic potential, the old, resource-based industrial regions of northeast France and southern Belgium have experienced severe adjustment problems. The gains may reflect the economic power of concentration, but it is a highly selective power, and to a significant extent produces decentralized concentration.

The problems resulting from the impact of agricultural transformation and increasing peripherality as a result of economic integration seem to provide a justification for the European Community's preoccupation with the problems of the poor, backward rural regions of southern Europe. However, regional losses from increasing peripherality appear to be just as great in the deindustrializing northern and western fringes of the Community as they are in the south. These two sets of problems emphatically do not provide a justification for directing aid to the rural parts of these backward regions or for the form that European agricultural policy has taken. The farmers that remain in the rural areas are more capitalized, have larger units and rising asset values and incomes. A consequence of this is that there is a flow of usually unskilled workers and low-income population into the urban centres of the poor, mainly rural regions. There is evidence of rising unemployment in the core cities of these types of FURs compared to their still very rural hinterlands. Although there is still too little research to be sure, the likelihood is that it is these urban centres in the poor, rural regions of the Community that are becoming the nodes of disadvantage, not their rural areas. There is, therefore, an equity argument for being concerned with the poor positions of such regions but for focusing help on the urban areas within them. There is also an efficiency argument for such a policy since successful adaption and transformation of these regions must almost certainly be urban-based.

One final point relates to income and income distribution. Here the evidence is quite inadequate, but there is an indication that deindustrialization and tertiarization are accompanied by increasing income polarization and inequality. It is interesting to note that in the European Community country where these processes are most advanced there is striking evidence of such changes in income distribution and also a change in the perceptions of policy-makers about the relative importance of spatial differences in welfare. In the United Kingdom since about 1977[37] attention (and funding) has steadily moved from inter-regional to intra-regional welfare disparities. There should be some scepticism as to the justification for this shift of focus since these types of intra-regional disparities, although they may be growing, seem to reflect growing social disparities that happen to have a spatial manifestation rather than disparities that truly represent spatial welfare differences. Given the political pressures on policy-makers, nevertheless, this reaction to the intra-regional spatial implications of transformation may be difficult to avoid in other countries. Spatially based policies to combat this new form of intra-regional disparity, however, are difficult, perhaps impossible, to formulate because the origins of this inequality are not spatial but sectoral.

It is interesting to note that a similar shift has not occurred at the level of the European Community. There, as was noted above, policy concern is still mainly focused on inter-regional differences, particularly those between poor backward rural regions and the prosperous and mature regions of much of Northern Europe. This emphasis has been strongly reinforced by the accession to the Community of Greece, Portugal and Spain, three Southern European countries with a strong representation of such poor regions. This has sharpened the Community's focus on inter-regional welfare differences. Partly as a result, Community spatial policy continues to give serious attention to the problems of declining industrial regions which, like the problems of poor rural regions, truly represent inter-regional welfare differences.

9

1957 to 1992: moving toward a Europe of regions and regional policy

PAUL CHESHIRE, ROBERTO CAMAGNI,
JEAN-PAUL de GAUDEMAR &
JUAN R. CUADRADO ROURA

Introduction

THIS CHAPTER describes the policies developed at the level of the European Community that have an impact on patterns of regional change. Within this description, particular attention is paid first to the evolution of policies aimed specifically at reducing regional disparities, so-called regional policies. Our second aim is to interpret this evolution of policy. The slow, and at times almost imperceptible, development of policies controlled by the supranational Community, rather than nationally by the member states, reflects the development of the Community itself and the impact that development has had on patterns of regional transformation. The effects of European integration on regional change have been slow and long-term. They will not come into existence overnight on 31 December 1992. Similarly, the significant reform of the Community's Structural Funds in 1988 did not represent a sudden emergence of Community regional policy; that has been emerging since the 1960s. Sometimes the balance has shifted to the Community and sometimes it has shifted back to member states as it did between 1973 and 1975; but the long-term trend is not only towards direct Community influence over European regional policy (and other policies too) but also towards a direct relationship between the European Commission and the regions of Europe independent of national governments. Further changes are in prospect as Europe approaches the symbolically important 1992; and other changes that may, or may not, be on the immediate agenda, as we note in Chapter 10, should be anticipated.

The European Community is still very far from being a single integrated sovereign state. Compared even to federal states its internal diversity is immense and its budgetary resources are feeble; about 1 per cent of total Community GDP. Within that budget there is an agricultural juggernaut (itself only a small fraction of the true cost of agricultural support to citizens of the European Community) that still consumes two-thirds of spending and incidentally increases regional disparities. After allowing for other commitments, that has historically left some 10 per cent (or 0.1 per cent of total GDP) for structural spending, of which a part only has been used to 'reduce regional disparities'. At the same time other aspects of Community policy, themselves designed to promote European integration, have both caused strong local economic effects and constrained national policy responses to those effects. The most important of these policies have been industrial and competition policy and macroeconomic policies designed to produce monetary union and economic integration.

Movement toward a single European market is likely to produce major economic benefits (and probably social and cultural benefits as well) to European citizens. The sources of the benefits are partly those identified by the conventional economic analysis of the effects of customs unions, but more of them stem from dynamic effects, which are difficult to model and quantify. Not all firms, all workers and all regions will be winners in this process, however. In recognition of this, each major move toward integration has been complemented by the development of structural policies aimed first at simple transfers to poorer regions and disadvantaged people, but increasingly at attempts to improve the adaptive capacity of regional economies adversely affected by processes of economic transformation or to increase the growth potential of backward regional economies. These structural measures, while significant in terms of absolute expenditure (ECU 13.35 billion is projected for 1993) are still trivial in comparison to the scale of the problems. As the Secretary General for Regional Policy wrote in 1989: 'Despite the increase in resources decided in Bruxelles in February 1988, assistance accounting for such a small percentage of total investment in the Community cannot expect to have any significant impact overall'.[1]

The agreements reached by European heads of state in 1988 that produced the Single European Act increased the speed of movement towards an integrated Europe but did not initiate it. Competition and industrial policies have for 20 years or more been influential in transforming Europe's steel, textile and shipbuilding industries and in eliminating national measures of protection. Steps towards monetary union have restricted exchange rate adjustments (even, to an extent, as they affect agriculture) as a means of offsetting changes in national competitiveness. Transport policies, communications and, more recently, environmental policies are having regional impacts. Their spatial impact, especially the development of a European high-speed rail network, is likely to increase over the next 20 years.

The main regional effects of 1992 will be in the service sectors, in hitherto protected sectors such as public procurement and in the movement of capital and people. It is also likely that the most important effects will be psychological and catalytic rather than direct. The European economy has been experiencing the most prolonged investment boom since the 1960s and new organizational changes are rapidly appearing. Firms are attempting to forge alliances or produce strategies to serve the whole European market; new international systems of subcontracting and marketing are appearing. In the public sector the role of the supranational European institutions is becoming inexorably more important but so also is that of regional authorities. A direct relationship is emerging between institutions at a European level and at a regional level. Even in the United Kingdom, where, against the trend in the rest of Europe, national government has been increasing in power at the expense of local and regional government, local authorities are building links with Bruxelles.

These changes will shift the focus increasingly to urban regions. The integration and transformation of the service sector together with its continued relative growth, will favour the largest metropolitan areas. So, too, will the need for large firms to have representation in national markets. At the same time, more specialized service centres will emerge in smaller cities and some urban economies will develop successful specialized integrated production, marketing and supporting service economies. These changes, partly because of the increasingly footloose nature of economic activity on a European scale, will accompany increasing competition between the economies of urban regions. Local economies in core regions, or with good communications to them, will have advantages in this process. Some local economies will lose out because they fail to specialize in appropriate sectors, others because of initial handicaps. The most important of these are likely to be a specialization in old declining industries – especially in smaller industrial cities that do not have a strong service sector or natural economic hinterland – and smaller cities in poor, backward and peripheral agricultural regions. Such cities not only tend to have primitive service sectors but will lose local markets, such as procurement, that had previously been protected. Their economies will have to absorb the rapidly increasing supply of largely unskilled labour created by the continuing transformation of peasant agriculture and by their demographic structure.

Although the main beneficiaries of European integration are likely to be the inhabitants of urban regions, particularly those in more central urban regions, the nodes of disadvantage created by this process are also likely to be increasingly urban. On grounds of equity, therefore, Community structural policies should develop effective mechanisms of urban intervention. There is also an efficiency argument for such a development. Policy has tended to shift from spatial redistribution to trying to facilitate 'regional convergence'. Successful adaptation and transformation of regional economies must almost certainly be urban-based, although not necessarily in historic urban cores.

Major features of the European Community

The European Community is a long way from the GDP budget ratio observed in sovereign states, even when they have a federal structure. The examples of the United States, the Federal Republic of Germany, or Switzerland, show that a federal budget, even in a very decentralized country, represents at least 10 to 15 per cent of the national GDP. The Community's general budget (excluding the European Coal and Steel Community) today amounts to about ECU 40 billion. The budget forecast for 1989 was ECU 44.8 billion, i.e. approximately 1 per cent of total Community GDP (ECU 3,700 billion in 1987). The difference between the Community and an integrated federal state is thus very great, even when indicators of integration other than the ratio of GDP to budget are considered.

The second fundamental budgetary fact relates to the structure of Community expenditures. The vast bulk of expenditure is linked to the Common Agricultural Policy (CAP). These expenditures are managed within the Agricultural Fund. The pattern of expenditure from the Community budget is shown in Table 9.1. In the early 1970s, before the first enlargement of the Community, agriculture consumed three-quarters of total spending. Even in 1980 it still accounted for nearly that proportion and was still almost two-thirds in 1988. The main part of agricultural spending (almost 95 per cent) is devoted to agricultural price support, the balance (only about 5 per cent) is reserved for 'orientation', that is to say for the restructuring of farms, farm modernization, or the training of farmers.

This progressive reduction of the relative weight of agricultural expenditure should not mask the fact that these expenditures are still growing strongly in absolute value (about ECU 28 billion in 1989 as opposed to

Table 9.1 Structure of the European Community's budget

		Expenditure 1986 (billions of ECUs)	Shares in Total Expenditure			Projected expenditure 1992 (1988 prices) (billions of ECUs)
			1972	1980	1986	
1	Agriculture and fishing, of which:	23.0	76.2	73.6	65.5	...
	1.1 Price guarantees	21.9	75.0	69.7	62.4	...
	1.2 Fishing	0.2	—	0.3	0.5	...
	1.3 Structural	0.8	1.2	3.6	2.4	...
2	Other sectoral policies (R&D, transport, energy and other industries)	0.8	3.6	1.9	2.3	...
3	Social Fund	2.5	2.9	4.7	7.2	
4	Regional Fund	2.4	—	6.7	6.8	12.9
5	Mediterranean programmes	0.13	—	—	0.4	
6	Development and co-operation	1.2	6.1	3.1	3.3	...
7	Administration, etc.	1.8	5.9	5.0	5.2	...
8	Repayments	3.3	5.3	5.1	9.4	...
9	Total	35.1	100.0	100.0	100.0	—
10	Structural spending (1.3+3+4)	5.7	4.1	15.0	16.3	13.35
11	EC Budget as percentage of EC GDP	0.97	0.56	0.83	0.97	...

Sources: Commission of the European Communities, 1987. Third periodic report on the social and economic situation and development of the regions of the Community. Annex Table 4.1; and Council Regulation (EEC) No 2052/88 of June 24, 1988 (The Reform of the Structural Funds), Article 12.

ECU 23 billion in 1986 and ECU 3.6 billion in 1973) and produce a serious conflict between their growth and the growth of the Community's own resources.

The European Council of February 1988 did, however, move to bring this situation under control. It agreed in principle to a growth ceiling for agricultural expenditure of the order of three-fourths of the growth rate of the GDP. Community spending remains burdened with this relative weight of agricultural expenditures, but the commitment is beginning to be controlled.

The background to this situation is a Europe that displays great regional diversity compared to a typical sovereign state. Figure 9.1 shows the variations in per capita GDP between the Level 2 regions of the Community in 1985. Even ignoring the tightly bounded city regions of Greater London, Paris and Hamburg, per capita income in the richest region in the Federal Republic of Germany was 3.5 times that of the poorest region, in Greece. Nor was regional variation confined to income. In 1985 the unemployment rate in the region where it was highest, in southwest Spain, was 12 times that of the lowest unemployment region, Luxembourg; three regions in Greece had more than 50 per cent of their labour force in agriculture, four regions in Spain (in 1984) had more than 30 per cent.[2] In the United Kingdom no region had as much as 5 per cent of total employment engaged in agriculture. The European Commission, in calculating the intensity of regional problems uses a 'synthetic index'. This arbitrarily gives equal weight to divergences of GDP and unemployment from the Community mean and has provided a first approximation of the map of regional aid; the opening offers from which political haggling starts. This synthetic index of regional problems for 1985 is shown in Figure 9.2.

Although committed to reducing 'regional disparities', the first call on the Community budget, agricultural spending, consumes two-thirds (and historically up to 80 per cent) of the total funds available. Moreover, the incidence of its expenditure runs directly against the aim of reducing regional differences. The mechanism of the Common Agricultural Policy is to raise output prices farmers receive by means of minimum import prices and levies. This creates surplus output that is brought into store or subsidized for export. Thus the main part of agricultural spending – the costs of buying up surpluses and export subsidies – which consumes so much of the Community's budget, is but a small fraction of the costs of the CAP to European consumers.

Since agricultural protection, via a system of supported prices, necessarily gives most assistance to the most productive farmers, the more highly capitalized farmers of Northern Europe, on more fertile land, receive most support. Numerous studies, including those of the Commis-

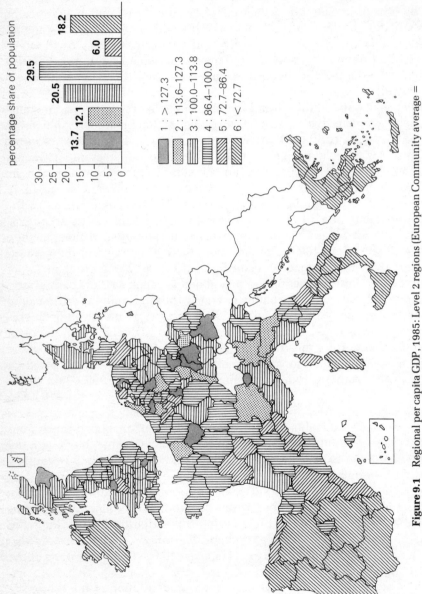

Figure 9.1 Regional per capita GDP, 1985: Level 2 regions (European Community average = 100).
Source: Commission of the European Communities, 1987.

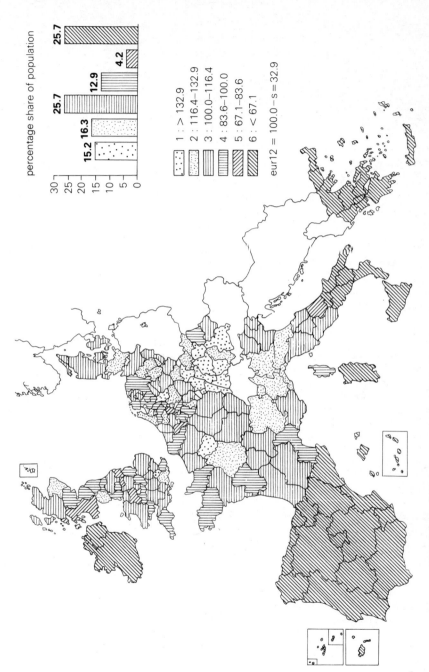

Figure 9.2 Value of 'synthetic index' for 1985.

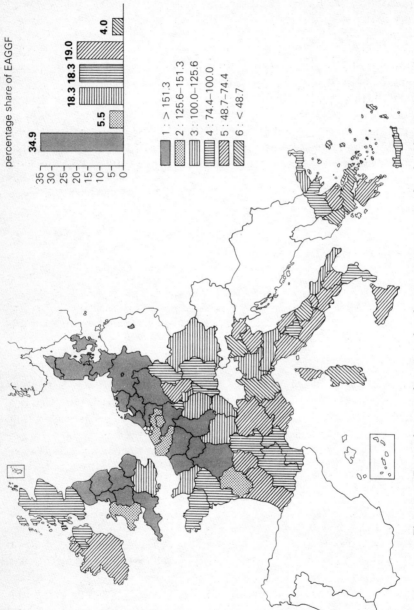

Figure 9.3 Regional receipts from agricultural price support per agricultural work unit, 1986 (European Community average = 100).
Source: Commission of the European Communities, 1987.

sion of the European Communities itself, document this regressive pattern.[3] Figure 9.3 shows the pattern for 1985 expressed as price support per agricultural work unit. Read together with Figure 9.1, it is possible to see how regional variations of expenditure on price guarantees is almost directly proportional to per capita incomes. In most regions of the northern Federal Republic, Denmark, England, the Netherlands, Belgium and the Paris Basin, southwest to the Atlantic, expenditure per person engaged in agriculture was 25 per cent or more in excess of the Community mean. In the Community's poor regions it was mainly 25 per cent or more below the mean. Since, as we can see from Table 9.1, the costs of price guarantees have historically swallowed 75 per cent or more of the Community budget, its regional impact in the context of the total regional influence of Community policy is highly significant. Read together with the synthetic index, shown in Figure 9.2, we can see how the regional incidence of agricultural price support spending per agricultural work unit failed to help those regions identified as needing assistance.

Structural policies

Over time the European Community has developed a set of what have come to be known as 'structural' policies. The first was the structural guidance section of the CAP. As can be seen from Table 9.1, this has always been a poor relation of the expenditures on price guarantees. The Social Fund was established before the Regional Fund but has some responsibilities that overlap with it, particularly in relation to unemployment and training. Together these structural policies still accounted for only 16.3 per cent of Community spending in 1986.

The importance of non-agricultural structural spending is growing, but it is inevitably small in relative terms. Table 9.1 shows that spending directly on regional policy, discussed in detail below, has risen to 6.8 per cent of the total and, as a result of the European Council's decision of February 1988, will be doubled in real terms by 1993. Despite the strong rise in unemployment since 1973, the share of social policy only grew to just over 7 per cent of expenditure by 1986. This is a good indicator of the real difficulties of building a 'social Europe'. The Social Fund, like regional spending, is set to double by 1993.

Finally, one should note the timidity of the Community's technological effort in spite of the many political declarations on this subject in the media. The Community's assistance for technical development represents approximately 2.5 per cent of the general budget. That is a minute proportion of the Community's GDP and indicates the still very national character of technological development policies (the developed countries devoting on average from 2 per cent to 3 per cent of their GDP to these policies). Actions speak louder than words, and this provides an objective indicator of the real reluctance of member states to commit themselves deliberately to a Community policy of research and development, in spite of the success of programmes like EUREKA. The reasons for this are discussed briefly below (p. 285).

These are the instruments of direct and structural intervention available to the Community. While they have a certain effectiveness, their effect is not as significant as sometimes claimed. In any case, they are, to date, trivial compared with those of the member states or of those at the disposal of, for example, the US federal budget. On the other hand, it would be a mistake to see the capacity for financial intervention of the structural policies as the only relevant mechanisms or instruments. Community policies are concerned as much with the co-ordination and control of national policies, notably with regard to competition and aid to industry (see Articles 92 and 93 of the Treaty of Rome) as they are with direct spending. If one considers this set of policies, the role of the Community has been important in pushing its member states into structural adjustment for the industrial sectors hit by deindustrialization and the restructuring of the world market for goods, such as the steel industry, textiles, shipbuilding, or even the automobile industry. It is interesting to examine in this context the way in which the Community has attempted to combine an energetic regulatory intervention, in the form of directives concerning the control of national aid policies, and programmes of aid to reconversion of those same sectors and the regions in which those industries are concentrated.

Community regional policy: the first moves

We may start with explicit regional policy, the area of policy most obviously designed to promote regional convergence and to offset the damage done to local economies and the goal of 'reducing spatial disparities' by other Community policies. The early days of the Community saw some discussion of regional problems and statements of broad principle in favour of developing policies to tackle them. The Treaty of Rome, for example, did mention the general aim of reducing regional disparities, and Article 130 included in the remit of the European Investment Bank the need to make loans for 'developing less developed regions'. There were two significant actions during the 1960s. The Commission pursued a general interest in regional policy culminating in a report on regional problems to the Council in 1965. In 1967, when the executives of the original three Communities[4] were merged, a Directorate General for Regional Policy (DGXVI) was set up.

The real situation amongst the original six member states, however, was that only Italy attached serious weight to regional policy, because only Italy had serious regional problems. However, in the 1960s these were not problems of deindustrialization. In Italy there was, and still is, an endemic problem in the South, the Mezzogiorno. Almost all indicators of economic development revealed two Italies, an advanced and dynamic North and a backward, peripheral South, highly dependent on peasant agriculture. In addition, partly because of its large, low-productivity peasant agricultural sector, Italy received relatively little from the Community budget in the early years. Over the whole period 1954–72, for example, Italy received less than 60 per cent of the French level of agricultural support per head and only 33 per cent of that going to the Netherlands. Italy saw the development of regional spending, therefore, not only as a mechanism to aid its own regional development but as a means of at least partly redressing the balance of financial flows within the Community as a whole. Other countries within the original six had regional differences but they were less serious than in Italy. At the Community level, therefore, nothing of substance was done with respect to regional policy before the first enlargement of the Community.

The enlargement that brought in Denmark, Ireland and the United Kingdom (Norway was originally among this group of new candidates for membership but withdrew following a negative referendum result) changed the political balance in two ways. First, the process of negotiation led to a rethinking of Community priorities and made it possible to re-open discussion on dormant issues. Second, it changed the political balance. Ireland saw itself in a position very similar to Italy, and could perhaps even be identified with southern Italy. It was a largely poor agricultural country and was peripheral to the rest of Europe.

The United Kingdom, like Italy, was concerned with regional differences. The 1960s had seen a series of important measures introduced to assist its Development Areas.[5] By 1971 these assisted areas covered virtually the whole of the north and west of the United Kingdom. British concern, however, was almost solely focused on the problems of declining industrial regions.[6] In addition, and perhaps equally important from a British viewpoint, regional policy was seen as a financial counterbalance to the Community Agricultural Policy (CAP). In whatever way the net benefits of British entry into the Community were calculated, there could be no avoiding the fact that the CAP would represent a major and continuing financial drain and distortion of resource allocation. The British perceived the development of Community regional policy as a way of directing funds away from agricultural spending and of partly redressing the drain on the national exchequer represented by the CAP. The first Commissioner for Regional Policy, Lord Thomson, who was British, looked forward to the day (he predicted it would come in the 1980s) when regional policy expenditure would rank on a level with agricultural spending.[7]

The result was that at the October 1972 Paris meeting of heads of state, where the seal was put on the agreements for the first enlargement of the Community, there was a clearly defined intention to establish a Community regional policy. The final communiqué ran:

> The Heads of State ... agreed that a high priority should be given to the aim of correcting, in the Community, the structural and regional imbalances which might affect the realisation of Economic and Monetary Union.

Two specific actions were proposed. The first was a co-ordination of national regional policies; the second, the establishment of a Community Regional Development Fund (ERDF) by 31 December 1973. The Fund was to be used to correct the main regional imbalances, 'particularly those resulting from the preponderance of agriculture and from industrial change and structural underemployment'. From its first inception, then, European regional policy identified two types of problem regions: those based on agriculture and those suffering the effects of industrial change.

The next step was the Thomson Report (1973).[8] This provided the first official analysis of the comparative standing of the regions of the enlarged nine-nation Community and of the nature and causes of regional problems. In its summary it identified two specific types of problem region:

> The agricultural problem regions tend to be situated on the periphery of the Community ... they usually have the characteristic of severe structural underemployment and in some cases there is also high, long term unemployment (these latter features are particularly significant in the case of Ireland and of the Italian Mezzogiorno). Whatever the variation in this respect, a common feature of all these areas is a relatively low income per head ... and a high dependence on agricultural employment.
>
> The areas suffering from industrial change have usually been those where there has been a high dependence for employment on ageing industries. Their problems of economic transformation are often underlined by a constantly slow rate of growth and by high levels of unemployment stretching over many years.[9]

There followed a period of inaction. A change in government in the United Kingdom led to a demand to renegotiate the terms of British entry and, in addition, the Federal Republic of Germany, as the main financer of Community spending, developed doubts about adding regional policy to the financial demands of the Community budget. In the economic climate of 1973 and 1974 these doubts were shared to some extent in other countries. The deadlock was broken by the initiative of Italy and Ireland. They both threatened to boycott the December 1974 Paris summit, which, among other things, was supposed to conclude the process of renegotiating the terms of entry for the United Kingdom. The ERDF was duly established, but with an initial budget of ECU 1,300 million over three years compared with the ECU 3,000 million proposed by the Commission. Not only was its budget restricted, but so was its 'Community' element. Its powers were restricted to co-ordinating national regional policies and to supporting the policies of member states. The original Commission proposals had been for an independent European policy that could largely have determined its own criteria and priorities. Thomsom saw it as adding a 'new human dimension to the Community. One which is essential to any degree of economic and political unity'.[10]

In the event, not only was the funding proposed by the Commission whittled away by national governments during the negotiations, but it proved impossible to agree Community criteria to designate specific regions that should be eligible for Community aid. Governments would only agree to Community aid being used to assist nationally initiated projects in regions they had designated themselves for national regional policy purposes. The remaining token of a genuinely European form of action (as opposed to expression of sentiment) was that money from the ERDF should be additional to funding available from national sources. This principle, so-called 'additionality', is the thread of continuity in Community regional policy. It is widely acknowledged to have been evaded by member states on a large scale but it has always provided a Community dimension on which the Commission could tug. It gave a specific role to the Commission that could be, and has been enlarged.

It is probable that this original formula allowed each country to sell the development of European regional policy in a way that related to national perceptions of regional problems. The Commission, however, appears to have sold the policy in a quite even-handed way. Thomson, in virtually all his public statements, stuck closely to the format of the Thomson Report.[11] There were two basic types of regional problem in the Community; a Community regional policy would not only help those poorer regions but also the congested central regions and was also necessary to secure a basis for European unity.

National debates were, not surprisingly, more self-interested. In the debates during the campaign for the referendum on Community membership in the United Kingdom, between late March and mid-June 1975, the emphasis became more and more nationalistic. So far as regional policy was discussed at all, the emphasis seemed to be on the fact that the

United Kingdom could make its own decisions with respect to regional designation without 'interference' from Bruxelles. In addition, although Thomson might talk of aid for 'poorly productive agricultural regions and where there was an overdependence on ageing industry'[12] everyone in the United Kingdom was happy in the knowledge that in the United Kingdom he meant the Development Areas. These, since before the Second World War, had been identified as the areas of ageing nineteenth-century industries in the North of England, Scotland and South Wales. When, however, Thomson told a meeting in Venezia that the 'way to give a fair deal to Italy ... is by a vigorous regional policy of the kind the Commission has been proposing ... Italy has received few financial advantages from membership ... in the agricultural sector, for example, total Commission subsidies in Italy between 1954 and 1972 were £22 per head ... and £66 per head in the rich Netherlands',[13] Italians, no doubt, immediately thought of the Mezzogiorno.

It is clear that in its original proposals the European Commission had intended regional policy to reflect a genuinely Community initiative but had been largely frustrated by the timidity and nationalism of member states. With successive reforms, in 1979, 1984 and, most recently in 1988, regional policy has been strengthened, funding has been increased, co-ordination (both different regional policy actions and regional policy with other Community policies) has been improved and, perhaps most significantly, a measure of responsibility for the initiation of action and for the formulation of policy objectives has moved from a national to a Community level.

The first reform came into effect on 6 February 1979. It established a 'non-quota' section of the ERDF under which specific Community measures could be financed. One purpose of such measures was to offset the regional effects of other Community policies, such as those producing industrial restructuring. Compared to quota measures they were to have three key distinguishing features. They were to be implemented as multi-annual programmes; aid was no longer to be confined to physical investments but could extend to assist small and medium-sized enterprises; and the geographic coverage of aid could, in principle, differ from areas designated by national governments.

Further symbolically significant steps were the commitment to publish a periodic report on the social and economic situation of the regions of Europe and measures to ensure that the regional impact of Community policies should be systematically assessed (via a Regional Impact Assessment or RIA) and that non-regional policies that had an adverse impact on regional development should either be modified or counter-measures should be introduced. In addition, further moves were made to ensure the co-ordination of national regional policies.

The 1984 reform, which was initiated in 1983 and actually came into effect on 1 January 1985, was specifically designed to accommodate the increasing scale of regional problems associated with industrial restructuring. Quotas were replaced by a so-called 'indicative range'[14] of aid each member state could expect over a three-year period. Allocation of aid to a particular country over the lower limits (which for all countries added to about 88 per cent) would depend on the extent to which grant applications satisfied the priorities and criteria established by the Commission. Programme financing, which was seen as making for greater coherence, was to be possible for any ERDF spending and was set to reach at least 20 per cent within three years.

Regional policy had necessarily started as simply a Community contribution toward the cost of nationally determined projects. The Community had little or no influence on what projects came forward or which would receive aid. The eligible regions for aid were nationally defined and, although in principle Community contributions were intended to be additional to regional aid member states would otherwise have provided, the reality was that Community contributions were largely treated as contributions to national exchequers. With each successive reform the movement was steadily towards a more significant role for the Community compared to member states.

The political and statutory framework of industrial restructuring

The policy 'tools' of the Community are essentially a combination of competition regulations, co-ordination and aid programmes. However, in recent years these tools have sometimes appeared to be operating in opposition to each other. The latter have come to the rescue of industries (or regions affected by the decline of these sectors of activity) that industrial and competition policies were attempting to restructure or contract.

Actions taken to restructure declining sectors can be illustrated by the examples of the steel, textile and shipbuilding industries. The Treaty of Paris of 1951 gave the Commission numerous statutory powers with respect to the steel industry when it created the European Coal and Steel Community. The Commission is responsible for periodically defining modernization objectives, orienting production and increasing production capacities, judging investment programmes, authorizing industrial concentration, and granting investment loans. The Treaty of Paris also empowers the Commission to declare a crisis situation that enables it to regulate production and limit imports from non-member countries.

By the late 1970s the steel industry faced a worldwide crisis. Much of the period since 1951 had been a period of euphoria during which policies had been geared to ensuring adequate capacity. This gave way to a severe recession, which led the Commission to decree a situation of 'manifest crisis' in October 1980. As a result it set up a production quota system, covering approximately 80 per cent of production, that is periodically renewed.

The decision of the Community to oblige national governments to restructure their steel industries, even when many were politically reluctant to do so, was implemented in two ways. On the one hand, the Community set minimum requirements for shutting down capacity. On the other hand, it demanded notification of all measures of general or specific aid granted by governments, reserving the right to review their legality. National policy-makers were thus caught in a stranglehold that obliged them, whether they wished to or not, genuinely to restructure. Up to 1980, the need for reconversion and the local adjustment problems it would cause had been concealed by national measures of protection. From 1980 onwards all countries, with greater or lesser willingness and speed, restructured their steel industries.

The Community, of course, set up financial aid for reconversions of activity and workers but its actions were directed to produce a more (sometime less) voluntary deindustrialization in order to make structural adjustments in the steel industry. By 1989, the process was still not complete, but it had produced its principal effects. The European steel industry has gained in productivity and competitiveness but in the process many jobs have obviously been lost. In many of the affected regions this job loss has not yet been offset by the emergence of new activities.

The chain of events in the textile industry was much the same. Stiff foreign competition from low-wage countries, in a context of only moderate growth of domestic demand, generated a need for a structural adjustment, aimed at reducing capacity and modernizing what remained.

The Community acted in two ways. It moved to control national aid as early as 1971 and it provided for its own restructuring measures. These latter assisted commercial co-operation and provided specific aid for modernization by means of loans from the European Investment Bank (EIB), and intervention by the ERDF and the Social Fund. The Multi-Fibre Arrangement, signed in 1973 within the framework of the GATT, is an additional part of a wider strategy. The territorial impacts of this deindustrialization were also marked, but not as obviously as those of the steel industry. The textile industry is in general more widely distributed and the process was spread out over a longer period of time.

In the shipbuilding industry, the necessary restructuring that took place after 1975 led to drastic cuts in employment (more than 50 per cent from

1975 to 1980, more than 30 per cent from 1980 to 1985, and more than 40 per cent forecast for 1990). These cuts should lead this sector to a situation in which jobs in 1990 will only represent about 20 per cent of the number that existed at the end of the 1960s.

This restructuring was organized by several directives of the Commission that fixed the ceiling rate of national aid in relation to the price of ships. The most recent directive, dated 28 April 1981, forbids both national aid, which distorts competition between member countries (thus imposing a common ceiling rate) and all investment in capacity. In exchange, it authorizes aid assigned to dealing with the social and regional consequences of restructuring (see below).

Aid to technological development

In contrast to the considerable concern shown about the necessary restructuring of declining sectors, it is noteworthy that the Community has concerned itself in only a limited way, and belatedly, with sectors of advanced technology. A few examples will illustrate this point: a European nuclear programme was cut short despite the existence of EURATOM; only one pre-competitive research programme for electronics and telecommunications (ESPRIT) was set up. The success of Airbus and the European Space Agency appear to be the only exceptions.

There are several reasons for this. The first is that advanced technology programmes have remained for the most part under the control of individual nations that are jealous of their prerogatives in this domain. When a particular area of research is judged to have both civil and military applications, as in the case of telecommunications and nuclear research, different national defence policies of member states have made a common approach more difficult. At the same time, as has already been noted, the Community has remained the prisoner of a structure of expenditure that is essentially determined by the agricultural policy and has been unable adequately to fund programmes to give worthwhile results.

A further factor is that technological development policy has never been an explicit aim. It is not mentioned in the Treaty of Rome and is still overlooked today. This is difficult to understand, especially at a time when it is widely considered to be an obvious priority for structural adjustment and for the development of new activities.

It is, finally, within the scope of regional policy programmes and aid to reconversion that the Community has become aware of the need to implement this priority through limited aid to technology transfer operations for small and medium-sized regional businesses. With regard to possible major technological programmes, the only solution seems to be extensive co-operation between big firms and the individual nations. To this the Community budget would make a marginal financial contribution that could, however, in some cases, be politically decisive.

The Community attitude toward other industrial sectors

Without going into the details of Community policy on competition, one must bear in mind the importance that controlling national aid has assumed in the last 15 years. This control is provided for in Article 92 and organized by Article 93 of the Treaty of Rome. State intervention in several member states was on an upward trend during this 15-year period. Because of the severe recession of the late 1970s, several governments were tempted to help their firms, especially when the difficulties of those firms or industries had a strong local impact. The Community controls, considered as a burden by some national authorities, have not, however, had only negative results.

They have, first, encouraged the states to question the principle of habitually bailing out firms in difficulty. Many countries have, partly as a result, begun to make radical changes to their regional or national aid systems. These modifications relate to the different types of aid and their means of application; automatic aid, for example, is being replaced by more discretionary aid. These controls have helped avoid overbidding between national authorities and have also discouraged possible overbidding by regional and local authorities. The Community has shown flexibility on occasion, however, and made use of the dispensations permitted in Article 92. This was the case when the Commission accepted increased rates of aid for particularly distressed regions, such as Lorraine, during the severe recession in the French steel industry.

The regional impact of co-ordination: the special case of transport

Transport and communications policy is another good example of Community policies pursued for other objectives that have had important regional impacts, and promise in the future to have even more profound ones. The harmonization of transport systems is being sought for a variety of reasons: to encourage development of new technologies; to help reduce overall costs; to improve the quality and timing of services; and to increase competition.

Policy for both air and road transport is dominated by the goal of deregulation, as is that for sea transport. Major improvements in competition are looked for by the end of 1992. Significant improvements have already been achieved. Recent agreements remove the veto power of individual countries on lower air fares to destinations in other member states, and on allowing lorry owners to compete for loads (on an almost equal footing with local firms) in external markets. Long-distance bus transport will soon be substantially deregulated.

For railways the most important actions have related to co-ordination of national policies and standards, although some relevant spending has been directly grant-aided from Community resources or loan finance has been provided by the EIB. These apparently quite minor actions have, perhaps in combination with the catalytic effects of 1992, had dramatic results. With support from the Commissioner for Transport the Community of European Railways announced in January 1989 an integrated plan for a high-speed rail network linking most of the major urban regions of Europe.[15] The system will run at speeds varying from 200 kph to 350 kph, from Scotland to Sevilla in southwestern Spain or Napoli in the south of Italy, and from Bretagne in western France through Bruxelles to Amsterdam and Wien. The network is scheduled to be completed by the end of the century and is integrated with the construction of the Channel Tunnel between England and France. Europe provides ideal corridors for high-speed train traffic and there can be little doubt that not only will most short-distance and medium-distance air traffic between centres served by the system switch to rail but that much additional movement will be induced. City-centre to city-centre times between London and Paris, for example, will fall from their present four and a quarter hours to less than three.

The regional impact of these transport changes will be profound. No regional impact study has yet been done because they have been treated as mere co-ordination of national policies and as sectoral policies. The deregulation of transport is likely considerably to strengthen more developed and core regions and within those core regions strengthen some parts and weaken others. High-speed rail will confer great competitive advantages on those regions that are well served (in particular the London–Lille–Paris–Bruxelles–Amsterdam, and the Barcelona–Marseille–Nice–Genova–Milano corridors) and penalize those that are not, such as southern Italy, Portugal and Greece. It will benefit particular urban regions that integrate the rail system with other transport systems, especially air transport. Here, Paris and Amsterdam have the most advanced plans. In general it will reinforce existing tendencies to recentralization in the larger metropolitan regions served by the system.

Community policy appears to be defensive, concerned first with restructuring and regulating competition and thus indirectly encouraging deindustrialization and, more recently, regional transformation. The interventions of the Commission are not limited to the above, however. It is also concerned with finding solutions to the social and territorial problems of this voluntarily accelerated deindustrialization. The tools used to help achieve this are a compensation brought into being by the Community in the name of intra-community solidarity, a type of social compensation for the most painful aspects of this structural adjustment. An examination of the tools that have been used helps us assess the efficiency with which the wounds caused by induced restructuring have been healed and new development promoted.

Community programmes for regional restructuring

The 1979 reform of the ERDF provided the first opportunity for the Community to begin to determine its own priorities for regional restructuring. The actions that were taken using the first 5 per cent of non-quota funds and then, after 1984, the intentionally vague, but potentially larger proportion of resources available, reflect these priorities. The 1984 reforms gave the Commission the means to develop a proper strategy of assistance and tied the ERDF grants more closely to Community objectives.[16] These objectives were to encourage the development of new activities in the areas affected by the restructuring of certain industrial sectors. The 1984 reforms replaced the non-quota actions with the Community Interest Programmes (CIP) and the National Programmes of Community Interest (NPCI). However, it was agreed that non-quota programmes started before this reform would continue to be implemented until 1989.

Essentially three programmes were launched:

The steel industry Between 1979 and 1983 the non-quota programme focused on the badly affected areas of Belgium, Italy and the United Kingdom. Then, in 1984, it was applied to additional areas in the United Kingdom, and to areas in the Federal Republic of Germany, Luxembourg and France.

The textile and clothing industries A programme was launched in 1984. The programme related to zones in Belgium, Ireland, Italy, the United Kingdom, the Netherlands, France and, since, 1985, the Federal Republic of Germany.

The ship-building and fishing industries This programme was agreed in 1980 because of the very severe situation of many European shipyards. It was first applied to British zones (from 1981 to 1983). In 1984, it was extended to the Federal Republic of Germany, and in 1985 to certain areas in France and Italy. The specific action for the fishing industries was only set up in 1985.

All these programmes were defined and set up along similar lines and generally resulted in support from the ERDF for the following operations:

(a) renovating or redeveloping run down urban or industrial sites;
(b) constructing new infrastructure needed to attract new activities;
(c) creating or developing advisory or economic co-ordination organizations;
(d) promoting innovation in industries and services by means of information exchange, feasibility studies, assistance to business services and the provision of suitable premises;
(e) market studies;
(f) improving access to venture capital for small and medium-sized businesses; and
(g) aid to investment, low-interest loans and capital subsidies.

With regard to shipbuilding and fishing, the programme could also intervene to help protect the coastal environment and to develop tourism in the affected maritime zones.

The ERDF devoted approximately ECU 100 million a year from 1985 to these different programmes. As aid from the Community cannot exceed 50 per cent of the cost of the accepted operations, the corresponding global investments have been two to four times higher, or about ECU 200 million to ECU 400 million per year. These sums are not negligible, but in comparison to the scale of the problems they were wholly inadequate. They must, in fact, be considered as only an element intended to trigger other initiatives.

Integrated actions

These were introduced on an experimental basis following the 1979 reform. The intention was to tackle sectoral problems effectively and at the same time provide a co-ordinated approach to the several related problems of affected regions. They went beyond tackling the sectoral problems in that they permitted a co-ordinated and coherent intervention of the three structural funds (not only of the ERDF). These actions are the result of a global conception of economic restructuring that came closer to being the sort of strategies necessary to deal with deindustrialization and regional transformation. Their methodology was built on the experience gained from pilot projects launched in 1980 to tackle the problems in Napoli and Belfast.

These projects laid the basis for the development of Integrated Operations of Development (IODs) which, in contrast to non-quota programmes, did not have a limited field of operation although the area covered had to be eligible for the structural funds. This device therefore permitted actions that were not in the form of one-off projects or for a limited time period and favoured global programmes of redevelopment. It is probably too soon to draw conclusions from operations that only really started up in 1987 but the fact that they show a change in the method of Community intervention must be stressed. Instead of only supporting a project, the Community now supports the elaboration, implementation and monitoring of a structural adjustment programme.

This trend was confirmed by the adoption in 1985 of the Integrated Mediterranean Programmes. These were intended to help the Mediterranean regions of existing Community members (Greece, Italy and parts of France) adapt to the entry of Spain and Portugal. These programmes, scheduled to be set up from 1986 to 1993, essentially concern regions that are only slightly industrialized, but they must be included in this discussion because they are likely to make a notable contribution to structural change there. They could play a positive role in the birth of a new industrialization that would make the Mediterranean part of Europe their privileged domain. In any case they are significant because they consumed resources on a large scale; approximately ECU 7 billion for seven years.

Community interest programmes

These programmes are a logical follow-up to the previous actions and were introduced following the 1984 reform. They share the same philosophy, that is the definition of coherent sets of actions lasting over a period of years that are meant to allow the development of new activities. Whereas non-quota programmes aimed at healing wounds, the aim of the Community Interest Programmes (CIPs) was to create something new, either by aid to small-scale industrial or service firms, by investments in infrastructures, or by helping create and build an 'indigenous potential'. These programmes do not directly intervene in the restructuring of declining sectors but aim to create the conditions for a fresh start.

The first two CIPs – STAR and VALORAN – were adopted on 27 October 1986. These covered a five-year period and were to receive funding totalling ECU 1,200 million. They were addressed, respectively, to improving access to advanced telecommunications and exploiting indigenous energy potential in backward regions, broadly corresponding to those identified with Objective 1 following the 1988 reform (see below).

The National Programmes of Community Interest (NCPI) are similar in conception but correspond to national objectives, if the latter are in line with the objectives and policies of the Community. A good example is that of the NCPI of Longwy. This programme is unusual in that it affects three border zones belonging to three different countries: Longwy (France), Aubange (Belgium) and Pétange (Luxembourg). The situation in these three adjoining areas is similar and results from the restructuring of the steel industry. The NCPI, or rather the three NCPIs, adopted in 1986 cover the 1986 to 1990 period and have been allotted ECU 100 million by the Commission. Given the relatively small population of the area (300,000), this sum is not insignificant.

The 1988 reform

The most recent reform of 1988,[17] although the most comprehensive to date, can be seen as the continuation of a process. As discussed in the next section, it was seen as necessary and complementary to the creation of a single European market. The intention was to consolidate and co-ordinate all structural spending and, in addition, the financial activities of other Community institutions, such as the lending policies of the European Investment Bank. The following five objectives were formulated for the Structural Funds:

(1) Promotion of the development and structural adjustment of the regions where development is lagging behind;
(2) conversion of the regions, frontier regions, or parts of regions (including employment areas and urban communities) affected by industrial decline;
(3) the combatting of long-term unemployment;
(4) facilitation of the occupational integration of young people; and
(5) with a view to the reform of the Common Agricultural Policy through
    (i) speeding up the adjustment of agricultural structures, and
    (ii) promoting the development of rural areas.

The funds available for these objectives were to be doubled in real terms by 1993 when they would reach ECU 14.2 billion to which would be added additional agricultural aid. The ERDF, in combination with other Directorates of the Commission, would be responsible for implementing Objectives 1, 2 and 5(ii). The proportion of spending pre-allocated to member states (more or less the old quota system) fell further to 85 per cent of the total available and the strong implication was that after five years it would fall still more. The pattern of spending moved further towards programme funding, with provision for both Community Programmes initiated by the Commission, and National Programmes initiated by member states.

Within this new framework, problems of deindustrialization were mainly the concern of Objective 2 although Objectives 3 and 4, for which the Social Fund was the lead Directorate,[18] were also relevant. Objective 1 related mainly to the problems of regional development in the poorer rural regions of Southern Europe although, of course, their structural adjustment involved the restructuring and capitalization of agriculture, itself a form of industrial transformation causing significant problems of adjustment. Objective 5 appeared to be mainly a political stratagem to placate Europe's strong agricultural lobby as it faced a falling real value of price support. Since we are interested in the role of Community regional policy, in the context of industrial transformation, it is worth noting that 80 per cent of spending from the Regional Fund was earmarked for Objective 1.[19] Within this total was the Integrated Programme for the Mediterranean.

Although the 1988 reform increased the Community input into European regional policy further, the outcome still reflected a process of negotiation. Although the rules for designating regions qualifying for support appear objective, producing a formula-determined outcome,[20] the reality was otherwise. The technical criteria used to designate regions for assistance under the various objectives that were identified themselves reflected national horse-trading; and the actual selection of regions deemed to meet the technical criteria for assistance reflected further national horse-trading. Northern Ireland, for example, was specifically identified as qualifying for aid under the more generously funded Objective 1. Objective 1 regions[21] were supposed to have per capita GDP below 75 per cent of Community average. GDP per capita in Northern Ireland in 1985 (the most recent year then available) was 89.7 per cent of the Community mean. Building a unified Europe is a long-term political process.

The original Council regulation of June 1988 identified both the criteria to be used and all regions qualifying for assistance under Objective 1, but only identified criteria for the selection of Objective 2 regions. Eurostat was charged with assembling the requisite data, which was published in October 1988.[22] The national horse-trading was not complete, however, until March 1989 when the list of Objective 2 regions was published. These are shown in Figure 9.4.

The criteria that the Council regulation defined for identifying regions seriously affected by industrial decline were that they should be Level 3 regions or parts of Level 3 regions in which:

(a) the average unemployment rate over the previous three years had been above the Community average;
(b) in some years since 1975 the percentage of industrial employment had exceeded the average share of the Community as a whole; and

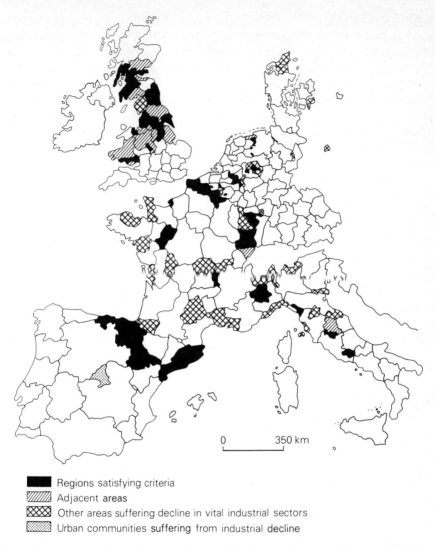

■ Regions satisfying criteria
▨ Adjacent areas
▩ Other areas suffering decline in vital industrial sectors
▧ Urban communities suffering from industrial decline

Figure 9.4 Areas qualifying for support under Objective 2 of the revised Structural Funds. *Source: Information* March 1989.

(c) there had been an observable fall in industrial employment compared to the year selected for (b).

In addition, adjacent areas could, on a discretionary basis, be added: 'urban communities' (not defined), with an unemployment rate 50 per cent above the Community mean and with a 'substantial' fall in industrial employment could be included and so could 'other areas which have recorded substantial job losses over the last three years or are experiencing or are threatened with such losses in industrial sectors which are vital to their economic development, with a consequent serious worsening of unemployment in those areas'.

These criteria were, self-evidently, not precise. As the incoming Commissioner for regional policy said, 'the division of the criteria into two categories – a set of basic statistical criteria and a separate group of optional criteria which were less precisely defined – had made for a complex process of selection'.[23] This process was not made simpler by the inability of Italy to provide any figures at all on industrial employment for Level 3 regions and the British insistence that such information, although it would be submitted to Eurostat as legally required, should be totally confidential and revealed by Eurostat to no one (despite being available in the United Kingdom on payment of a fee to the Department of Employment). In total the spread of the regions, in terms of the population covered, was as shown in Table 9.2.

In one way, however, these designations did represent a significant further development. Not all the areas designated as eligible to receive aid as Objective 2 regions were currently designated for aid by national regional policy. As the Commission explained: 'This was because "Community criteria and priorities" had been taken into account'.[24] Since Community aid still required 'additionality' and co-financing (with the rare exception of some Community programmes) the implication was that member states hoping to benefit would have to initiate spending in the relevant regions if they were to attract Community funds. Thus, for the first time, national regional policy was made to some extent subservient to Community regional policy.

Table 9.2 The significance of Objective 2 regions

	Total population in Objective 2 regions (thousands)	Percentage of national population
Belgium	2,178	22.1
Denmark	252	4.9
Federal Republic of Germany	6,992	11.4
Spain	8,577	22.2
France	10,084	18.2
Italy	3,764	6.6
Luxembourg	140	38.1
Netherlands	1,440	9.9
United Kingdom	19,868	35.0
Total	53,295	16.5

Table 9.3 Balance sheet of European Community policies (budgetary balance in millions of ECUs, 1986)

	FEOGA 6		FEOGA 8	Regional Fund	Social Fund	Balance on structural funds
Belgium	16	−1	−16	−79	−28	−123
Denmark	538	9	−3	−40	26	−18
Federal Republic of Germany	−1,404	−24	−86	−559	−475	−1,119
Greece	966	1	72	262	63	397
Spain	−1,271	−8	−51	141	13	103
France	863	−4	26	−295	−152	−421
Ireland	984	1	57	544	179	289
Italy	−69	−10	64	360	133	557
Luxembourg	−42	0	1	−5	−3	−7
Netherlands	726	−5	−34	−161	−112	−307
Portugal	−155	−1	−6	168	90	252
United Kingdom	−1,225	5	−26	146	260	380

It is still far too early to judge the efforts of this 1988 reform. First indications are that the commitment of 80 per cent of available funds to Objective 1 regions is imposing constraints. Actual projects for these areas acceptable to the Commission are not coming forward fast enough; commitments made under Objective 2 threaten to exceed the limits it was intended to impose. Nevertheless, although too early to judge, the efforts the Community has made since 1973 to establish European, as distinct from national, regional policy objectives, and to materially assist structural adjustment have been significant.

Community efforts in this respect still fall a long way behind the most energetic national programmes, such as the programme in the Federal Republic of Germany for the reconversion of the Ruhr, but they now go further than the mere display of political intent. There are real resources involved and the move to Integrated and Community programmes has also had a demonstration effect. National and regional authorities have been compelled to co-ordinate their efforts and to consider the total effects of sectoral deindustrialization and transformation on a region. These programmes may have positive results and contribute to redevelopment in ways that exceed the value of their resources.

One must, however, always bear in mind that such programmes do not carry much weight in the complex balance of the Community's structural policies. Even though they are not negligible in absolute terms, they remain marginal in comparison to the Common Agricultural Policy, for example. More generally still, the evaluation of structural policies, outlined in the 1988 reform, will depend on the attitude of member states faced with increased net contributions to the budget for structural action. Table 9.3 shows contributions for 1986, and demonstrates the net contributions made by certain countries, notably the Federal Republic of Germany. The 1988 reform will certainly not improve the net situation of the richer countries of the Community and political reactions could one day split the solidarity of the Community. The recent events in the German Democratic Republic, and Eastern Europe generally, could prompt the Federal Republic to set its sights elsewhere and to devote its energy to dealing with the German issue. This might well cause it to reconsider the extent to which it foots the bill for structural adjustment in the present European Community.

The regional impact of economic and monetary union

The big steps and the great dates in the construction of the European Community have always provided an opportunity for rethinking and revising the purpose and practice of regional policy, including both its relevance to achieving the general policy goals of the Community and its potential effectiveness. As was shown above (pp. 277–80) this happened on the occasion of the first enlargement of the Community in the early 1970s with the addition of countries such as Ireland and the United Kingdom with their regional preoccupations. It happened again in 1978–9, when a first agreement on the European Monetary System was reached, forcing each country (with the exception of the United Kingdom) to harmonize its general macroeconomic policy with that of the other partners and to keep its currency within defined limits of fluctuation with respect to a weighted average of European currencies, the ECU (European Currency Unit). It happened once again, with the approval of the Single European Act by the heads of state at the Luxembourg European Council held in December 1985. This agreement was designed to achieve a single European market. The need to revise the operating rules of the three Structural Funds was stipulated on the final signing of the Single European Act.

The economic and monetary union

In the second and third case explicitly, but implicitly also in the first case, the important theoretical idea underlying the revision of regional policy and its mechanisms of intervention was that a movement toward greater integration between the different regional economies, although beneficial in aggregate terms, would inevitably lead to a widening of inter-regional disparities within the Community. In fact, integration between previously separated local markets – through abolition of customs formalities, duties and other non-tariff barriers to trade, better infrastructures, and common fiscal and legal regulations – put different production systems, characterized by highly dissimilar productivity levels, in direct competition. While stronger economies would benefit from increasing economies of scale and favourable learning processes, weaker economies could suffer as local markets were served from distant regions and entire local sectors of production perhaps disappeared.

The case of Italy's Mezzogiorno in the early 1960s was extremely telling in this respect. The coupling of the initial effects of the Treaty of Rome creating the European Common Market with the integrating effects of the completion of the north–south transportation system (the Autostrada del Sole) led to the bankruptcy of the traditional light industry system in the south, especially in the food, textiles, and clothing sectors. A serious external deficit in the Mezzogiorno's external trade balance was opened and this was inadequately counterbalanced by the exports of the new, state-owned, state-subsidized heavy industries (petrochemicals, iron and steel) that Italian regional policies established there.[25]

The effects of European monetary union should be seen in a similar light. Monetary union means that countries that are unable to catch up with the average rate of productivity growth of the Community at large, owing to a generalized weakness of their industrial structure or to the presence of a 'two speed' internal regional economy, lose a radical policy instrument represented by 'competitive' exchange rate devaluation. In a sense, they are obliged to utilize a currency whose international value is fixed on the basis of economic indicators that have little to do with their own international competitiveness.

This is highly significant since exchange rate movements are the conventional processes by which external equilibrium is restored in the presence of a fall in international competitiveness, and by which the Ricardian law of comparative advantage may continue to hold while assuring weaker countries some specialization in the international division of labour. Short of this, nations, like regions, are left to compete under a regime of absolute advantage, which means that no specialization may be left to them unless a classical (but perhaps improbable) fall in local wages and prices takes up the task of restoring some form of external equilibrium.

These arguments were explicitly put forward in official discussions on the European Monetary System in 1978 by the Italian representatives, assisted by officials of the Bank of Italy such as Rainer Masera.[26] As a result, weaker countries were allowed to maintain a wider margin of fluctuation for their currencies. In addition, and of most relevance for the present discussion, they obtained an acknowledgement of the principle that a more important involvement in structural policies should parallel the path towards monetary union. The Italian Commissioner for regional policy, Antonio Giolitti, developed this principle in the 1979 revision of the structural funds mentioned above, with the creation of the 'non-quota' section of the ERDF.

The single European market of 1992

From 1985, under the chairmanship of Jacques Delors, the process of economic integration was accelerated, even relaunched. The plan for what was subsequently called 'Objective 1992' was presented to the European Parliament as early as January 1985, and the heads of state discussed it at the Councils of Bruxelles (in March) and Milano (in June). All this led to a real reform of the Rome Treaty, the so-called Single European Act, which was subsequently ratified by the national parliaments. Through it, new institutional laws were adopted for the achievement of the single market. In particular these included a strengthening of the power of the European Parliament, and the decision that two-thirds of the nearly 300 regulations necessary for the harmonization of national economic and administrative legislation, and for the achievement of the free movement of people, capital, goods and services could be adopted by the Council by a majority instead of requiring unanimous approval.

The most important objectives of the Single European Act are the opening of procurement markets, mutual acknowledgement of administrative regulations, harmonization of indirect taxation systems, financial integration and deregulation of capital flows and the removal of all non-tariff barriers to internal trade. Even as abstract objectives, these are likely to have a profound psychological effect and to represent an important challenge for the whole Community.

However, these objectives could not be launched in a vacuum. There needed to be a public commitment in the direction of those areas that might be threatened by the new integration process and towards other complementary policies including transport. In the Communication of the Commission to the Council of 15 February, 1987, the so-called 'Delor's package' or 'plan' was launched. This defined the financial resources that could allow the achievement of the objectives of 'social and economic cohesion' present in the Single European Act. The doubling of the 'structural funds' and the definition of the five major 'Objectives' analysed above (p. 291) were proposed, and subsequently adopted through a difficult decision process in Bruxelles in February 1988. It was also pointed out that a genuinely single enonomic area could not be achieved without major progress in policies for transport, infrastructure and the environment.

Some countries, principally France, thought that the effectiveness of policy rather than just the quantity of financial resources should be the target of the revised policy approach. However, an important element of the Delors Plan should be emphasized as a general methodological point. It points out that 'les instruments communautaires doivent cesser d'être considérés comme les éléments d'un système de *compensation financière*. Ils sont appelés à jouer ... un rôle important pour la *convergence des économies*'. (Community instruments must cease to be considered as merely the elements in a system of *financial compensation*. They are called on to play ... an important role in achieving *economic convergence*.)

The element of the total package introduced following the 1985 agreements highlights the limits of the traditional approach to regional imbalances (and related policies), which treats them as mere 'disparities' to be counterbalanced through appropriate compensating flows. The agreements reflect a revised diagnosis of inter-regional disequilibria based on a dynamic analysis. Such a view sees a need for regional policy to influence the paths of regional growth themselves.

The effects of 1992 on regional economies

The general view that we are arguing is that the effects of 1992 on regional economies have to be understood in intrinsically dynamic terms. The achievement of the single European market is in fact likely to provide a powerful stimulus both to competition and to the rate of innovation within the borders of the enlarged Community, and so provide great opportunities which are, in principle, open to each regional economy.

However, to date, these effects have been officially analysed by the Commission in purely static terms, linking them to the actual specialization and comparative advantage of the single regions and to the possible emergence of strong economies of scale (see for example the well-known Cecchini Report on the costs of Non-Europe). The role of these elements will not be negligible, but the aggregate forecast benefits from the single market will probably be reached more as a result of the dynamic elements mentioned above than of these latter, more traditional ones.

The evidence presented in this chapter suggests the need for and perhaps the likelihood of a sharp further reduction in agricultural spending, balanced by increased structural spending (and spending for other purposes, perhaps most importantly on R&D) and a redirection of structural spending from rural to urban areas. There are both equity and efficiency arguments to support such a switch. In both the old industrial areas and in the poor agricultural regions, problems are becoming more concentrated in the urban areas of the regions and this tendency is likely to continue. It is the urban areas of the Community that are emerging as the nodes of disadvantage. Although spending 80 per cent of structural funds in the backward regions does not, of itself, mean that the direction of spending has to be to the rural parts of them, the impression is that, at least historically, that has been the case. Much of regional spending has been on new infrastructure in greenfield sites and agricultural structural spending has been in rural areas. Since problems are increasingly concentrated in the cities of those backward regions as agricultural intensification produces rural–urban migration, and increases the incomes and asset values of remaining farmers, equity demands that structural spending be redirected. In the declining industrial regions the problems are concentrated in urban areas although in parts of the Community, for example, in eastern France or South Wales, these cities can be quite small. At the same time efficiency arguments favour directing intervention increasingly to urban areas. The whole drift of the argument elaborated above – including the reasons for increasing concern with the dynamic aspects of regional growth and improving growth potential, and the role of agglomeration economies and services – suggests that the potential for regional response is primarily an urban one. This potential for changing the growth path of regions is not necessarily confined to the old built-up urban areas but it is concentrated within the functional sphere of influence of cities.

10

Europe's regional–urban futures: conclusions, inferences and surmises

ROBERTO CAMAGNI, PAUL CHESHIRE,
JEAN PAUL de GAUDEMAR, PETER HALL,
LLOYD RODWIN & FOLKE SNICKARS

OUR MISSION is almost completed. In the previous chapters we have analysed the industrial and regional transformations now taking place in six countries of Western Europe. We have also provided an overview both of these findings and of the changes from the perspective of the European Community. In the course of preparing this volume, we have arrived at some conclusions and drawn some inferences and surmises that ought to be shared with the reader. They focus on two speculative issues: the broad implications for regional and urban areas of the development of the European Community in the context of an internationalizing economy and an informational revolution; and our expectations about some of the more specific spatial effects as far as we can discern them.

Anticipated effects of 1992

The building of a more integrated market for goods, people, capital and services in the European Community in 1992 will reinforce several general trends: these include increasing internationalism and integration of the national and regional economies; continuing innovation and increasing dynamic competition between firms; emerging new organizational models based on co-operation agreements and new networks; and increasing integration of information and communication between the single economies. These trends will be strengthened by 1992, in part because it will destroy surviving barriers to competitive forces, but also because it will change the way in which Europeans and non-Europeans alike perceive Europe. It will therefore reinforce some general spatial tendencies that were already apparent.

The effects of 1992 on national and regional economies must be interpreted in the light of such phenomena as diffusion of know-how, innovativeness, technology 'creation', psychological effects on firms' expectations, synergy effects and dynamic comparative advantages. These effects are much harder to model, to quantify and forecast, and are perhaps less well defined, than the traditional effects of custom unions presented by conventional economic theory (gains from trade creation and increasing specialization, or economies of scale). Nevertheless in our opinion these effects are much more important on both an analytical and an empirical level, in that they impinge on the nature and driving force of economic development: innovation and industrial transformation.

An interlocking set of influences is likely to shape national and regional economies as a consequence of the completion of the single European market.

Macroeconomic influences

The symbolic effect of 1992 will generate a strong expansionary impact for mainly psychological reasons. Individual firms are already anticipating the competitive atmosphere of the single market, and rushing to reach a stronger technological, organizational and commercial structure. The single market is virtually here: the explosive increase of fixed investment in all countries of the European Community since 1984 has produced the most continuous and durable period of expansion for many years. This can be explained largely by the anticipation of the competitive effects of 1992. The average overall annual rate of growth of investment in the Community was −0.4 per cent between 1974 and 1981; 1.1 per cent from 1982 to 1986 but 4.8 per cent in 1987 and 7.3 per cent in 1988. The biggest increases were in Belgium, Spain, Italy, Portugal and the United Kingdom.[1] The pattern is similar in Western European countries outside the Community.

This element alone is likely to influence development strongly in all regions, inside or outside of the Community. In fact, the dispersion of regional rates of output growth is not wide enough to offset the national aggregate growth effect, which remains the major component of each regional growth rate.

Mesoeconomic influences

The major part of the new trade flows created by the expanded internal market will be intra-industry trade rather than increased inter-industry trade (characterized by specialization in single types of commodities in the single regions). This is because integration is taking place across a set of regions where factor prices and resource endowments are, by the standard of their world variation, comparatively uniform.[2]

As a consequence, the gains from trade will be much more difficult to assess through traditional analytical tools of comparative advantage. They will relate as much to the level of total utility of the consumers as to a reduction of production costs and of prices. These influences are likely to create wider opportunities for regions and firms, in the form of a substantial expansion of market niches of differentiated products. Each firm will be able to create its particular competitive advantage on single segments or sub-markets through product innovation, image and marketing campaigns, product differentiation, or technological innovation, almost independently of its regional environment. The static conditions of relative efficiency and comparative advantage will be less important than the capability of the firm for innovation, which may be strongly related to its regional milieu.[3]

Thanks to this mechanism, small and flexible firms may compete effectively with big firms, if they can find in their local milieu those elements that speed up the circulation of relevant information, reduce transactions costs and dynamic uncertainty, and enhance local synergies and creativity. Big firms may take advantage of the same phenomena, which go beyond the exploitation of economies of scale, in the form of 'scope economies' and economies of joint production. New forms of economies of agglomeration are likely to become increasingly important.

Microeconomic and locational influences

At this level, in so far as it affects industry, the past can be some guide to the future. In the early stages of the Common Market, large firms appeared to be the main beneficiaries, as did more central regions. More recently – in the late 1970s and early 1980s – there were some significant diffusion effects, both down the firm hierarchy and toward outlying regions, and to regions having particular characteristics that might have identified them as peripheral. These include, for example, southwestern France, southeastern regions of the Federal Republic of Germany, central Italy, or southwestern England. These fortunate regions, however, have tended to be in the European mainstream and to have had adequate communications with central metropolitan regions. Factors in regional (and firm) success appear to have been flexibility, marketing effectiveness, and capacity to innovate. It seems likely that these factors will persist and that the positive diffusion effects will be strongest in regions that have more highly skilled labour, good communications, a greater capacity to generate new firms that develop markets external to their region and a superior adaptive capacity.

At the microeconomic and locational level of service activities, there will be strong and direct effects on the pattern of regional economic growth. Service activities throughout Europe were mainly untouched in the previous phases of development of the European Community. Banking, insurance and professional service activities, together with headquarter functions and commercial activity, will be directly influenced by the integration of national markets. In contrast to most of the preceding effects, the transformation and integration of service activities will mainly benefit urban areas and, in particular, the largest cities and metropolitan areas.

The impact among metropolitan and urban areas

For non-metropolitan and non-urban regions, the impact of 1992 will be mainly indirect, coming from, for example, increased competition and production rationalization. In the case of metropolitan and urban areas the impact will be a direct one. A unified European market already exists for the majority of industrial products, even though non-tariff barriers persist. For several growing sectors, however, which show a clear-cut metropolitan bias, e.g. the financial–banking sector, the insurance sector, or professional services, the integration process is only now beginning. National oligopolistic structures in these sectors, sheltered up to now from foreign competition by strong institutional barriers and national regulation, will undergo a rapid process of restructuring and concentration. The weaker financial structures of southern countries will suffer from increasing competition from the stronger northern European firms and institutions. An example of a vulnerable sector is Italian banking. Internationalism and cross-location of subsidiaries of European banks and insurance firms in the different national markets will expand, and there is likely to be a stronger presence of non-Community institutions (Swiss, Nordic, Japanese and US). Their aim will be to take advantage of a continental market in which the psychological effect of the integration of the capital market is likely to become of paramount importance.

The creation of a truly unified market for industrial goods, the push towards truly international markets for firms of all sizes, and the need to maintain a presence in the different national markets will orient industrial firms towards mainly 'central' locations for their subsidiaries, namely, metropolitan ones.

The continuing heterogeneity of the national markets, due to their different languages, makes it impossible for centralized, foreign remote control. It will be essential for firms to have local, subordinate headquarters at central locations in each national market. Even the markets for professional services will, in due time, become progressively integrated. The reorganization and internationalization of this market, with the consequent rationalization and concentration process (which is already apparent among software, accounting and consulting firms) will probably reinforce the quest for central locations within big metropolitan areas.

Big metropolitan areas are likely to act as 'gateways' in the internationalization process of the national economies, supplying their respective hinterlands with all the service functions needed for the purpose. Finally they will attract investment (in headquarters, commercial and service functions) of major non-European companies.

The consequence of these processes will be that the largest metropolitan regions will compete directly with each other in the attraction of these high-level functions. As far as these strategic and control functions are concerned, competition in the 1992 market will take place mainly through the highest levels of the worldwide urban hierarchy. Each metropolitan area has therefore to find a new strategy for competing in this new context, enhancing the efficiency of the territory it controls.

By specializing in some high-rank functions, but not in all of them, lower order metropolitan areas such as Manchester, Stuttgart, Lyon, Rotterdam, Zürich, Torino, or Edinburgh may acquire in these functions the status (and the consequent level of power, control and wealth) of international cities.

The same process is visible at a lower hierarchical level within some national urban systems. By specializing not directly in general high-order services, as previous centres did, but on a specific industrial production or sector *filière* (going from manufacturing to specialized technical and marketing services for that specific product) certain cities may develop a worldwide competitiveness and standing in their fields. Examples can be found in textile and fashion creation in Firenze, the silk *filière* in Como, the machine tool *filière* in Hannover, or in the cultural centres of Aix-en-Provence, Salzburg, or Oxford.

The possibility of exploiting these new processes and opportunities is, however, unevenly distributed and related to a number of factors that are difficult to quantify or even precisely define. It depends on the particular centre being able to focus strategically on some local vocation that its resource endowments, location and creative potential equip it to pursue successfully. A further development is that not only are industrial and other activities that handle goods in bulk (wholesaling and bulk retailing) continuing to diffuse out of urban areas altogether, but other activities are becoming increasingly more mobile. A given service activity may be strongly constrained to locate in an urban region of given characteristics (a continental headquarters in London, Bruxelles, or Paris; a national subsidiary in Milano or Frankfurt; a specialized advanced instrument engineering firm in Cambridge or Stuttgart) but it is increasingly free to choose in which city of those characteristics it will locate. Some cities, left to themselves, while starting from an adequate position, are likely to lose in the competitive process. Other cities, notably the smaller declining industrial cities of northern Europe, such as Sunderland or Charleroi, or the small growing cities of transforming agricultural regions of southern Europe, such as Córdoba or Messina, are inadequately endowed to compete in the first place.

The urban dimension of structural adjustment policies

Backward areas and old industrial regions, and notably the urban areas, have particular dangers and risks in prospect from the single European market. This may be even more so than for European regions that are outside the European Community competitive environment. They face a colonization of their internal markets by external, mainly foreign firms, and the consequent collapse of the local industrial fabric as a result of rationalization processes at the European level or the dismantlement of institutional barriers that have sheltered the local internal market up to now. Most industrial markets are already integrated, but if one thinks of public procurement, one can see that there are still significant markets in which the lack of external 'transparency' is high. This may be especially true in the case of infrastructure provision and building construction, two sectors that local firms and employment have traditionally regarded as their own in peripheral regions. These will be newly exposed.

In addition, local firms in such areas, depleted of human capital and entrepreneurial skills by generations of outward migration, and typically with poorly developed educational systems and research facilities, are less able to exploit the new opportunities offered by the wider European market. They have insufficient innovation capability in the spheres of technology, organization and marketing.

Old industrial areas suffer the additional disadvantage of continuing losses of employment and income from their declining traditional sectors and the negative local multiplier and accelerator effects that these entail. Urban areas in wider backward agricultural regions suffer from an exogenous increase in the supply of largely unskilled labour, both from continuing rural–urban migration and from the high rates of natural population increase that are found in such circumstances for both demographic and sociological reasons. In these backward agricultural regions of the Community and elsewhere in Western Europe, the problems of adjustment are becoming increasingly concentrated in urban rather than rural zones, even though a part of the urban problem arises from agricultural transformation. The restructuring of agriculture means that those farmers who remain have increasing asset values and incomes. The largely unskilled rural–urban migrants produced by this process of agricultural intensification are increasingly moving into urban underemployment and unemployment rather than higher-productivity advanced sectors.

Adjustment problems, especially in the European Community, are therefore likely to be increasingly concentrated within urban regions and, more specifically, within the cores of those urban regions. There will be increasing polarization between successful metropolitan areas and smaller cities where the local economy has successfully specialized, and smaller cities that have failed successfully to specialize in the competitive process. The ranks of these unsuccessful cities will be swollen by a number of declining industrial cities – particularly smaller and more remote ones and those that grew on the basis of natural resources – and by a range of cities of varying size in poor, backward and peripheral agricultural regions. Even quite large cities, such as Sevilla or Catania, can be losers in this process.

In short, although the main beneficiaries of European integration are likely to be in urban regions, particularly in more central urban regions, the nodes of disadvantage created by this process are also likely to be increasingly urban, both within and outside the European Community. We anticipate, therefore, a further significant shift in the direction of Community policy.

Some specific spatial effects

New spatial hierarchies

The changes noted have also altered the balance between nations and regions. In the past, regions were parts of national economic systems. Now some – such as the South East of England, Ile-de-France, Frankfurt, Bruxelles and Lombardia – belong to the international economy, while others – such as the South West of England, Lorraine, or the Pays Vasco – remain firmly within national economic spheres. We are familiar with the notion of a multinational company. Now we must get accustomed to multinational regions, within which both individuals and firms are internationally oriented, and that will increasingly make decisions that will impinge on the rest of Europe and the world. The agricultural, industrial and informational Europes co-exist, even within individual nation-states; some are still traditionally involved with the transformation of matter, others have largely completed the evolution to an information-handling economy.

As regions have evolved, so have the cities that form their cores. Until now the cities of Europe have formed a series of national hierarchies rather than a single urban hierarchy. These hierarchies differ according to their diverse national histories. We only need to compare the pyramidal hierarchy of centralized France with the flat hierarchy of decentralized Germany and Italy. These hierarchies are just now almost certainly beginning to merge into one common urban system. The top levels in this system are the multinational cities and their surrounding regions, the contemporary version of the world cities.

One critical question for Europe after 1992 will be the prospects for cities in this group, and of the regions around them. The two are intimately bound up together. It is because Europe is a heavily urbanized continent that there is a close relationship between economic and demographic trends in one and the other. In studying the forces that help shape the progress or decline of the one, we shall usually be able to guess the fate of the other.

Forces of centralization and decentralization

Concentration and deconcentration represent two polar forces. Both are powerful; both work at more than one spatial scale; and, of course, both create pressures in opposite spatial directions. Predicting their relative strength is the key to forecasting the way European regions and cities will grow.

The globalization of the economy will increase economies of scale and of agglomeration in space. Instead of numerous national capital cities providing locations for headquarters of financial and industrial corporations, one might emerge more powerful than the others in this new hierarchy. The span of control increases and intensifies. Technological change assists the process. Air traffic control allows plane services to interconnect through a few hub airports; high-speed train networks bring major centres in closer contact with each other; and high-quality telecommunications networks allow these centres to communicate instantly with wide tributary areas.

However, there are also opposite tendencies. Concentration in a few global cities brings with it increases in rents and other costs, providing an incentive to large firms to decentralize many of their more routine operations to lower-cost centres. Overcrowding in the main world cities produces a search for less-congested, higher-amenity locations that ease the stress on senior personnel and their households. Sophisticated telecommunications linkages, with reductions in message transfer costs, make it possible to operate major operations in locations far distant from headquarters. Better transportation linkages, especially high-speed trains, shrink distances between these operations, making it possible for executives to travel back and forth between the two in a single day.

Given these inter-related, countervailing forces, what will be the ultimate regional urban outcome? The evidence, from recent empirical studies of change in Europe, is that for a few strategic functions it will reinforce centralization at the larger continental scale, and for many other functions it will prompt deconcentration at the more local scale, both within and outside the macro region. Specifically, the top-level global centres are attracting the highest-level service functions but are exporting the lower-level, more routine information-processing functions. These decentralized operations, and the people who operate them, move to areas within relatively short journey times, typically at most one hour distant.

Since service activities in these top-level urban regions are growing rapidly, this process is likely to constitute loss only in a relative sense. Not only the headquarters elite, but also the constellation of smaller, specialized firms in business services, to whom they symbiotically relate, remain attracted to the centres. This attraction will be reinforced by the development of high-speed rail links in the 1990s. This pattern has spread from Northern and Western Europe to Southern Europe in the 1970s and early 1980s.

At the same time, in the largest and most complex Functional Urban Regions – the areas embracing the cities and their surrounding spheres of influence[4] – a pattern of recentralization is also evident. Growth passes from a core to a series of smaller satellites around it. London presents an extremely striking case. In London the fastest population growth is now occurring 100–160 kilometers from the centre, and with a strong decentralization of many more routine informational services to medium-sized growth centres in the population range 100,000–250,000 within this ring.[5] This is true also for Paris, Milano, Frankfurt, Madrid, Stockholm and other major European cities. Thus there is a paradox: in the top-level centres, overall population and employment levels may be static, even while they are increasing their hegemony over the rest of the urban system. The fastest rates of growth tend to be recorded in axial bands, sometimes, as in the Federal Republic of Germany, following major transportation corridors or axes between the top level centres, sometimes, as in the United Kingdom, arrayed at right angles across them, at one hour or more travel time from the global centres themselves.

The battle between the world cities

In this process, we detect shifts occurring at the top of the European urban hierarchy. London is still the top European world city in terms of the controlling nature of its economy, dominated by high-level service functions. Its pre-eminence reflects the enormous accumulated commercial power of the United Kingdom and its unique position at the intersection of three major trading economies, the Anglo-American, the Commonwealth and the European. However, there is persuasive evidence of slippage. The British economy has lost traditional manufacturing and goods-handling functions, and has so far failed to replace them sufficiently with information-handling and knowledge-creation functions. The decline in growth potential partly reflects a process of local decentralization within the United Kingdom. Paris is London's major rival as a comprehensive high-level service city, though it has a long way to go to catch up. Frankfurt is also a rival in commercial and communications functions, and is rapidly improving its position.

A vital role will be played by the rapid integration of the European economy after 1992. Bruxelles' already strong performance reflects its attraction as the political capital of Europe, a role that is likely to be enhanced as remaining European Community functions, especially the parliament, settle there. An equally important role on the commercial side will be played by the location of the European central bank. These pan-European forces could well reshape the historic hierarchy of Western Europe, as Bruxelles could become the Washington of this integrated Europe and Frankfurt its New York City.

A crucial role will also be played by inter-regional transport and communications. In the future international economy, the movement of people and electronic communications will be more important as location factors than the movement of goods. The impacts of new technology will make themselves felt both intra-regionally and inter-regionally. The first will be the more important; but dramatic changes are likely to occur in the inter-urban pattern also.

Having the world's top international airports, London presently leaves both Paris and Frankfurt far behind in terms of intercontinental transportation. However, the next two decades are certain to see the growth of a European high-speed train system that will drain off most of the short-distance European air traffic. Key locations are likely to be Lille, the hub of the new European trunk line (London or Paris–Lille–Bruxelles–Köln–Frankfurt–München) with Bruxelles close by. Locations that are developed as interchanges between air and rail will also have an advantage. Both Paris and Frankfurt are taking an early lead through intelligent planning of their high-speed rail networks, which will be routed to give direct airport access. Bruxelles will go the same way, but in London in early 1990 all planning was bedevilled by uncertainty over the routing and funding of the high-speed link from the Channel Tunnel and there is as yet no proposal to link London's two airports directly to high-speed rail.[6]

Advances in telecommunications are likely to reinforce the dominance of the present information-rich metropolitan regions. Such regions enjoy a kind of benign circle: their concentrations of information industries create a high level of demand for advanced telecommunications. These in turn bring forth the supply, and this proves attractive to the kinds of activities that depend on these services. Peripheral regions, in contrast, suffer from low take-up of advanced services, which inhibits their development. The specific outcomes are still somewhat uncertain because of the lack of an international telecommunications entity in Europe. Much will depend on the speed and pattern of the integrated telecommunications system in the 1990s as it emerges.

For these and other reasons, one must exercise care in interpreting, prognosticating, or comparing the rates of growth in the vicinity of large urban cores. Cities and regions that have good communications channels, situated outside the established core regions but fairly close to them, are often the winners in the urban system level immediately below the leading European cities. This is especially true if these cities and regions have efficient systems of transport and communications. Tertiarization assists this process of recentralization by strengthening not only the great metropolitan centres but also the lower-order and environmentally attractive service cities and towns close by. In contrast, old industrial cities are declining or being transformed, and goods-handling activities are becoming more and more diffused.

One feature of tertiarization is the enormous freedom of choice, compared with any previous historical period, about where to live and where to establish activities. Given the agglomeration economies of new service activities and the benign circle that the immigration of richer inhabitants is likely to bring, many growing cities may become nicer, not nastier, places to live in than they were in the past. Such regions may be found in most countries of Europe. This phenomenon is even more likely if the delineation is extended beyond individual metropolitan regions to embrace complexes of such regions. South East England, as noted, has recently been growing most rapidly at its very fringe and beyond, 100 kilometers and more from London, in a rural belt dominated by small and medium-sized country towns. The Toscana region in Italy has some of the same properties, as does the Toulouse region in France and the Cataluña region in northeastern Spain. In the Federal Republic of Germany the region between Frankfurt and München is another such decentralized growth pole. Farther north, both the København–south Sweden region, and the central Swedish region oriented toward Stockholm have some of the same features that bode well in the tertiarized European economy.

Not all regions, however, will be winners in the economic and political transformation of Europe. Some of the worst adjustment problems will occur where regional economies are inflexible. This inflexibility is particularly associated with old resource-based industrial regions, but adaptive capacity varies for other reasons as well. It varies as a result of institutional and cultural differences, and because educational levels, human capital and access to information are not evenly distributed. Regions with a larger stock of recent migrants may also have a greater adaptive capacity. This has important policy implications.

The transformation is easier for the old industrial cities than for the even older agricultural service cities in Southern Europe's rural periphery. If the inflexibility of the large industrial city can be corrected, and if its environment has not been altogether degraded through pollution from industrial activities, such a city can survive and prosper. These are big ifs, and we do not know how to ensure their realization. We do know that the costs of such a revitalization are not insurmountable if local entrepreneurship can be encouraged before the decline of the economy has proceeded beyond repair. Some heavy industrial cities, and regions like the Ruhrgebiet, have inherited a highly competitive social, political and communications infrastructure. They can become cornerstones in the European informational revolution.

Europe and Japan: a comparison

Some parallels may be drawn between the European and Japanese urban systems. Both are very old, although for historic reasons the European system is both larger and more diverse. In both, there is a tendency for urban growth to cluster along megapolitan corridors (the Tokaido megapolis, and the Mediterranean corridor). New transportation and communications systems in the form of high-speed trains and fibre-optic links have reinforced this axial tendency in Japan and may do so also in Europe. Within these corridors, improved communications may strengthen the position of the top-level centres, which may extend their sphere of control. This clearly happened with Tokyo after the opening of the Tokaido Shinkansen rail system.[7] In France, however, the impact of TGV-SE (Train à Grande Vitesse – Sud Est) seems to have strengthened

Lyon rather than Paris, at least for the time being.[8] The precise parallels may be misleading, and firm conclusions cannot be drawn at this stage. One certain effect of high-speed rail will be to centralize activity at those points, mainly the centres of the larger cities, that have direct access to it.

Japanese national infrastructure programmes rest on a much more deliberate public policy than in the United States. Investments in high-speed trains were started in the early 1960s. The urban mass transit problem has been solved through public initiatives, offsetting any pressing current need to spend economic resources on transport and communications. These policies, despite minimal costs for military defence, have kept the investment share of GNP high. This has supported major technological breakthroughs and a strong position in global financial markets. However, private consumption is still low and the Japanese are far behind the United States and Europe in housing, community facilities and space standards, and in the levels and quality of certain public services.

Another interesting feature of the Japanese experience of special interest to European scholars is the issue of density and dispersal. The Japanese have been forced to concentrate their population in view of limited land resources. Contemporary Europe has both a dispersed settlement system, US style, and a Japanese-type multi-centred settlement system. The odds are that rising income in Japan, and rising costs in Europe, may impel greater convergence between the different systems.

The Greater Europe of the 1990s

The astonishing, and probably least predictable, element for the 1990s will be the totally unexpected transformation of the planned economies of Eastern Europe and the speed and character of their incorporation into the European and global economy. These changes and their potential impact overshadow even the economic transformation of Western Europe, including the arrival of the single European market. Some analysts believe that the prospects of an economic integration between Western and Eastern Europe are clearly several orders of magnitude more important to the global economy than the future links between Europe and the developing countries.

There will certainly be major regional effects as the European centre of gravity shifts eastward. Such an opening of paths to the east may have huge consequences for the development of the whole urban system of Europe. The historical period of the Iron Curtain is short in comparison to the centuries it took to form the urban hierarchies in the pan-European scene.

In particular, Berlin, and to some extent Wien, might well recapture much of their former roles as key junction points for international traffic, making them much more attractive for high-level service activities. Up to 1914 and even to 1939, these cities were major transportation and transshipment points, both for people and goods. On that basis, they each assumed a considerable industrial power. As capital cities of vast continental empires, they acted as sub-continental centres of finance, culture, education, publishing and conspicuous consumption. They competed as cultural centres of German-speaking Europe, a region that, until the 1930s, still constituted a very distinctive area of European cultural and scientific leadership. Both were, in a strict sense, world cities.

Since the end of the Second World War and the division of Europe, there has been no such city in this part of Europe; there is a vacuum waiting to be filled. Both Wien and Berlin have been reduced to shadows of their former selves, the first because of the dissolution of the Habsburg Empire in 1919, the second because of the division of Germany and of the city itself, coupled with the transfer of the Federal Republic's capital functions to Bonn. The lost role is symbolized by the vast office structures of the German Reich on the Wilhelmstrasse and the nearby Anhalter Bahnhof, once one of the city's principal entry points: the first occupied by the decaying ministries of the German Democratic Republic, the second a ruin, the two still separated in early 1990 by the soon-to-be-dismantled wall.

Whatever the precise political outcome of the momentous changes unfolding in Eastern Europe in the autumn of 1989, Berlin's role in the European hierarchy seems certain to be enhanced. Even before the two Germanies were reunited, the breaching of the wall and the progressive reintegration of the two halves of the city were bound to enhance Berlin's global-level service functions, symbolized most dramatically in December 1989 by the joint decision of the mayors of East and West Berlin to rebuild the Potsdamer Platz, the city's heart, divided in half by the wall for nearly 30 years.

Berlin's new role, in turn, is likely to reorient the relationships between the first-order and second-order European centres, which are presently dominated by the strong cities of northwestern Europe that constitute the core of the present European Community. So entrenched is the position of these cities that it is difficult to see it being seriously challenged. However, they will surely face an important counterweight, at last, in the centre of the continent.

This shift will still follow the ground rules of the information revolution: tertiary activities, linked by the transmission of information, will determine the fortunes of Europe's cities. Berlin was once the high-tech manufacturing centre of the world, the late nineteenth-century Silicon

Valley, in addition to its other roles as a social, economic and political capital. It seems unlikely to recapture that high-tech pre-eminence, but this does not matter greatly. Berlin, like London, Bruxelles and Paris – and like New York, Tokyo, Hong Kong and Singapore with their extraordinary infrastructures, scale and dynamism – will base its fortunes on a changing economic foundation: the linkage of its resources and capabilities to the informational revolution.

IV
Postscript

Structural transformation in Japan: issues and prospects for regional development in the coming years

HIDEHIKO SAZANAMI

Backdrop of global economic change

CHANGES OVER the past two decades have ushered in an era in which the United States is no longer seen as the principal engine of global economic growth. The world economy is now recognized as multipolar. The nations of the European Community have promulgated the Single European Act to unify the Western European market and, hopefully, its currency, starting on 1 January 1993. The momentous events at the close of 1989, with the dissolution of the communist bloc in Eastern Europe, have added a fresh dimension to this. But, whatever the hardships in the short term, the further strengthening of Europe is bound to be the long-term outcome.

Japan, the third member of the emerging triad, has embarked on a series of structural changes to strengthen its economy. Efforts from the early 1970s to the middle of the 1980s largely concentrated on 'inward-looking' structural change, directed at improving the efficiency of Japanese industry at home and its competitiveness abroad. The phenomenal success of this endeavour made the nation an exporter of surplus capital and production expertise to both the developed and developing world. But with such success came the call for Japan to take on new roles and new responsibilities in the international arena. By the mid-1980s, the recommendations of the Okita and Maekawa reports, and the Plaza Accord decisions, had set in motion a process of 'outward-looking' structural change, that located Japan at the centre of the global economy.

The recommendations contained in the reports of the Advisory Committee for External Economic Issues (Okita Report) of April 1985 and the Study Group on Economic Structural Adjustment for Economic Cooperation (Maekawa Report) of April 1986, also an advisory committee of the Prime Minister, exemplify these 'outward-looking' changes. The reports urged the transformation of Japan's industrial structure in order to achieve: (a) sustained domestic demand-led growth, (b) increased imports of manufactured goods, and (c) increased Japanese direct foreign investment abroad. The rapid appreciation of the Japanese yen, which virtually doubled its value *vis-à-vis* the dollar following the Plaza Accord of September 1985, left little choice but to take these prescriptions seriously.

At the same time, domestic demand has greatly expanded in Japan in the past few years, as has the value of manufactured imports. Japan's trade surplus has also been falling and the gap between Japan and the Federal Republic of Germany, as the nation with the world's second largest trade

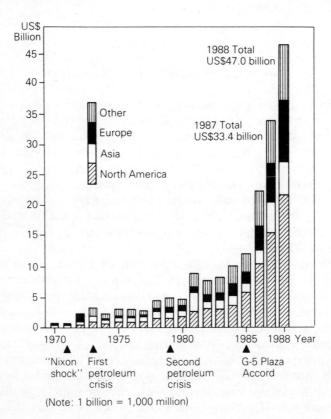

Figure 11.1 Annual trends in Japanese direct foreign investment abroad, 1970–88.
Source: Ministry of Finance, *Statistics of JDFIA Approvals and Notification* (Tokyo).

surplus, has also been shrinking rapidly. Some estimates point to the likelihood of the Federal Republic having the biggest surplus by the end of 1990. In contrast to its falling trade surplus, Japan's external assets have grown. In 1985 these stood at US$437.7 billion, approximately half the external assets of the United Kingdom and the United States. Estimates for 1989 indicate that, at US$1.47 trillion, Japan's external assets have more than tripled, to edge into second place. A parallel feature is the phenomenal growth of Japanese direct foreign investment abroad since 1985, as shown in Figure 11.1. The implications of this are discussed later; suffice it to say at this juncture that such changes will drastically alter Japan's industrial structure, with profound implications for regional economic transformation in the coming years.

Industrial restructuring trends

The transformation of Japanese industry in the post-war period has been a continuous process of response to changes in domestic and international markets. Japanese entrepreneurs have invariably sought industries with high growth potential and have redirected capital and manpower with a view towards increasing profitability, expanding market shares and diversifying product lines.[1] These resource transfers have usually been carried out gradually, thus avoiding the high costs of social dislocation that could have come about if unemployment had been generated by rapid industrial transformation. Part-time labour, sometimes considered to be employed at marginal terms, has nevertheless functioned as a buffer in effecting transformation. Sensitive corporate responses and public assistance policies have helped minimize the hardships accompanying structural changes in Japanese industry since the first petroleum crisis of 1973.

As the wages : capital goods ratio increased between 1960 and 1973, the period commonly known as that of 'high economic growth', Japanese industry began to shed labour and to shift to the production of goods with higher manufacturing value added (MVA). After the first petroleum crisis, the maintenance of competitiveness required manufacturing industry to be energy-efficient as well as labour-saving. Just before this, the 'Nixon shock', leading to upward revaluation of the yen against the dollar, required Japanese industry to learn to withstand currency value changes under freely floating exchange rates. The nation's response can be seen in the external trade structure, which was characterized by increasing import specialization in raw materials and resources, coupled with export sophistication in high-MVA, energy-efficient products.

Chemicals and heavy industries (such as basic metals) gave way to machinery, appliances and instruments, with increasing application of electronics. The transformation of Japan's industrial structure in the immediate post-petroleum crisis years strengthened the yen. Japan's export expansion exceeded the world economic growth rate and its stronger currency reduced the burden of import costs of raw materials. A yen weaker in relation to other currencies, especially those of the United States and other OECD countries, stimulated the demand for Japan's highly income-elastic, high-MVA manufacturing products. In other words, Japan's industrial structure pushed the economy towards a balance of payments surplus[2] – regardless of the strength of the yen – a situation that prevailed until the unimaginably high currency appreciation following the Plaza Accord of 1985. In this manner, Japan overcame the uncertainties of the petroleum crisis and the free floating exchange rate, but ran headlong into trade friction.

The transformation of Japan's industrial structure after the petroleum crisis was a result of 'inward-looking' efforts to increase export competitiveness. But the very success of these efforts, which led to increased market shares for Japanese products in the markets of OECD countries, was the source of trade friction among Japan and its major trading partners in the developed world. This led to what may be regarded as the beginnings of 'outward-looking' restructuring efforts. 'Voluntary' export restraints were advised by the Ministry of International Trade and Industry (MITI) and many manufacturers, including the powerful automobile industry, heeded the warning. In this period a major change in the nature of Japan's direct foreign investment abroad (JDFIA) also took place.

Japan's direct foreign investment abroad

Rapid appreciation of the Japanese currency in the years immediately following the Plaza Accord led to the adoption of drastic cost-reduction measures by Japanese manufacturing firms. In yen-terms, these were successful in reducing high wage and material costs of their products but, in dollar-terms, wages rose by 20–28 per cent and materials costs by 10–23 per cent between 1985 and 1987.[3] As currency appreciation outstripped cost reduction, exports became less competitive. The obvious alternative was the outsourcing of production abroad, with the stronger yen effectively lowering the costs of investment overseas, especially in countries whose currencies were pegged to the dollar.

The changing trends of the cumulative value of JDFIA in the manufacturing and non-manufacturing sectors in 1980, 1985 and 1988 can be seen in Figure 11.2. By the end of 1988, JDFIA in the manufacturing sector had risen to US$49.8 billion, or nearly four times its 1980 level, with the electrical, transport and general machinery sectors taking the lead. JDFIA in the non-manufacturing sectors increased by nearly six times, to reach US$131 billion over the same period. There was also a rapid increase in

Structural transformation in Japan 323

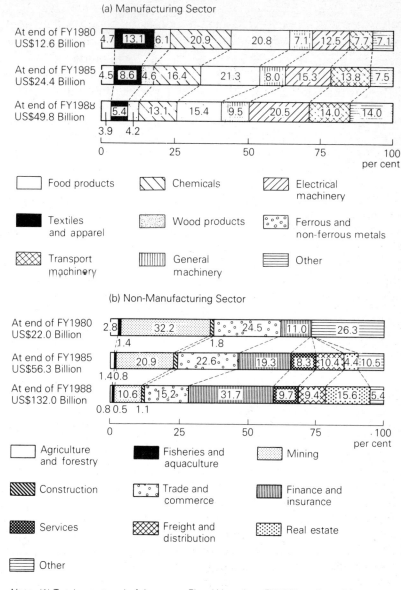

Figure 11.2 Changes in the cumulative value of JDFIA by sector.
Source: Ministry of Finance, *Statistics of JDFIA Approvals and Notifications* (Tokyo).

the share of the finance, insurance and real estate sectors, whose cumulative value in absolute terms at the end of 1988 exceeded the cumulative total of all manufacturing investments by as much as 1.25 times. This is noteworthy given the sensitive investments, particularly in real estate transactions, and an increasing number of large-scale mergers and acquisitions (M&A), carried out overseas by Japanese corporations in recent years.

There were also significant transformations in the production-related sectors. Previously concentrated in the energy, mining and logging sectors, or in labour-intensive manufacturing operations in the Third World, JDFIA has begun to move to the OECD countries. Offshore production by Japanese companies in the major OECD export markets helps circumvent barriers to Japanese exports to these countries to some extent. Profitable JDFIA operations in third world countries exporting to OECD markets, notably the newly industrializing economies of Asia (Asian NIEs), will also help to soften the blow until these countries in turn face export barriers.[4]

Although the OECD countries of North America and Western Europe are the principal destinations of JDFIA, it has also substantially increased in Asia. JDFIA in Asia rose from 685 projects in 1985 to 1,342 projects in 1987. In the manufacturing sector, it rose from 322 in 1985 to 789 in 1987.[5] This last figure was higher than the corresponding figure for North America. In terms of investment value per project the figure for Asia in 1987 was US$2.12 million, while it was considerably higher for Europe (US$5.35 million) and North America (US$9.30 million). The disparity is indicative of recent JDFIA trends in Asia, where many of the investors are small and medium-sized firms. While much of the JDFIA in North America and Europe is to secure market access in an environment that appears to be 'hostile' to Japanese imports, recent JDFIA in Asia is more closely related to the rationalization or extension of the 'home' production line in Japan.

Against the background described above, the following sections of this postscript will provide a brief overview of trends in regional economic transformation in Japan. The phenomenal growth of the Tokyo region following the petroleum crises of the 1970s will be highlighted, as will the significance of these changes for other metropolitan and non-metropolitan regions in Japan and the prospects for the development of regional economies in Japan over the coming years.

Regional economic transformation in Japan

The Pacific seaboard to the west of Tokyo, popularly known as the Tokaido Megalopolis, which links Tokyo to the other two large metropolitan regions, centring on Nagoya and Osaka, is the heartland of the Japanese economy (Figure 11.3). This spatial structure was established at the turn of the twentieth century as Japan sought to 'modernize' its industrial base and catch up with the more industrialized nations of Europe and the United States. Successive post-war efforts at regional development have worked towards minimizing the disparities between these major metropolitan regions and the outlying regions to the northeast, the north and the far west. The principal means was to strengthen the local economic base and modernize it through the dispersion of manufac-

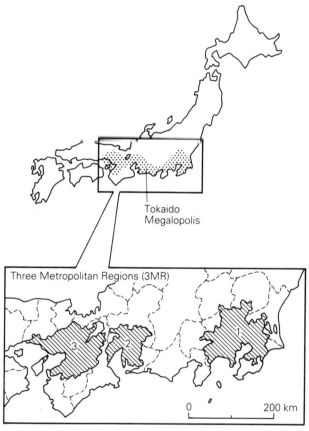

Note: 1. Tokyo M.R. (Keihin M.M.A. in source document)
2. Nagoya M.R. (Chukyo M.M.A. in source document)
3. Osaka M.R. (Keihanshin M.M.A. in source document)

(M.R. = Metropolitan Region)
(M.M.A. = Major Metropolitan Area)

Figure 11.3 The Tokaido Megalopolis and the three major metropolitan regions (3MR).
Source: Statistics Bureau, Prime Minister's Office, 'Population of Major Metropolitan Areas' in *1985 Population Census of Japan* (Tokyo).

turing industry from the three major metropolitan regions (3MR) to the non-metropolitan regions (NMR).

To strengthen the economic base of outlying regions and localities, investment in infrastructure and other economic overheads was carried out. Government subsidies through sectoral programmes of the different ministries largely finance these programmes, with prefectural governments acting as the executing agencies. The agricultural structural improvement programmes (ASIP I, II and III) of the Ministry of Agriculture, Forestry and Fisheries are indicative of investments made to strengthen local economic bases. The highway building and road construction

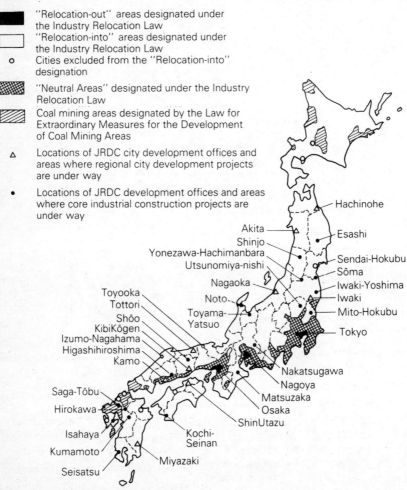

Figure 11.4 Endeavours at industrial dispersion.
Source: H. Kobayashi, 1989, Industrial Estates and Regional Development in Japan: Current issues and vision for the future, UNCRD.

Figure 11.5 Areas designated by the New Industrial Cities Act and the Special Industrial Development Areas Act.
Source: H. Kobayashi, 1989, Industrial Estates and Regional Development in Japan: Current issues and visions for the future, UNCRD.

subsidies programmes under the Ministry of Construction are examples of infrastructure investment made under these line ministries.

Another major source of revenue for public works designed to minimize regional disparities is an elaborate system of tax transfers to prefectures and municipalities (both of which are autonomous local government bodies). This favours the more backward regions. The transfers are made under the purview of the Ministry of Home Affairs, and are governed by the stipulations of the Local Autonomy Law and the Local Government Finance Law. The major guidelines for making these regional development investments are indicated in the National Comprehensive Development Plans (NCDP), under a law established in 1950.

The National Comprehensive Development Plans

The First NCDP (1962) and Second NCDP (1969), which span Japan's post-war economic recovery and rapid growth, concentrated on mitigating the over-concentration of population and industry in the three major metropolitan regions along the Pacific Coast and the rapid depopulation of outlying areas in the non-metropolitan regions. The dispersion of manufacturing industry from the 3MR to the NMR was central to these efforts (Figure 11.4) as seen in the establishment of industrial complexes in the New Industrial Cities and the Special Industrial Development Areas. The Japan Regional Development Corporation (JRDC), a government corporation with its roots in the provision of assistance to depressed coal-mining areas, worked together with local governments to establish a series of industrial estates across the nation in the NMR.[6] The infrastructure investments made underline ministry programmes and, using local government tax transfer receipts, also contributed greatly towards making industrial location viable in the NMR.

Many of the industries dispersed in the 1960s and early 1970s were large-scale, capital-intensive or energy-intensive manufacturing plants of leading firms with head offices in the 3MR. While it is true that these contributed much to local government revenues, the petroleum crisis of 1973 introduced a new element of uncertainty about their long-term viability in the outlying NMR of Japan. This period coincided with the first large wave of Japanese direct foreign investment into neighbouring Asian countries. By 1980, just after the second petroleum crisis of 1979, more than 40 per cent of JDFIA was in the chemicals, ferrous metal and non-ferrous metal industries, i.e. in sectors that were both capital-intensive and energy-intensive.

Nevertheless, regional disparities between the 3MR and NMR were narrowed towards the end of the Second NCDP. By 1975, net inter-regional migration from NMR municipalities to the 3MR had come to a halt. This is reflected in the terms 'U-turn phenomenon' and 'U-turn phenomenon', that characterize regional development documents and literature of this period. These terms were used to describe the return of youth studying or working in the 3MR to their home towns in the NMR or to local cities in the vicinity.

The Third NCDP of 1977, which was prepared in the aftermath of the first petroleum crisis, reflects the concern over energy and resources. The economic context in which it was formulated was a period of low but stable growth. The plan's central concept emphasized investment in social overheads, maximizing the quality of life in communities across the nation. The hard-earned reversal of migration into the 3MR from the NMR, which had taken 25 years of sustained investment at bettering the comparative advantage of the NMR, was central to the plan, which espoused the formation of integrated and stable settlement areas as the basis of regional development.

In contrast to previous NCDPs, the Third Plan made little mention of economic investment beyond optimistic expectations of the performance of indigenous or community-based industries. These expectations were subsequently not realized in the actual performance of these indigenous industries, indicating an inherent weakness of the plan. However, the plan did achieve a heightening of the awareness of the importance of making integrated social and welfare investments on the basis of functional areal units, which went beyond the boundaries of municipal jurisdiction. In less tangible terms, it enhanced local communities' awareness of indigenous heritage and the possibilities of stimulating development through its astute promotion, setting into motion a series of village revitalization movements across Japan.

Although depopulation continued in the more remote communities, population loss was greater in central cities in the NMR rather than in the 3MR. In the 3MR, a new phenomenon could be seen. In the period of structural adjustment following the petroleum crisis, inter-regional disparities between the Tokyo metropolitan region and the other two metropolitan regions of Osaka and Nagoya began to increase. The economic vitality exhibited by the Tokyo metropolitan region influenced considerably the outlook of the Fourth NCDP of 1987. The underlying reasons for the economic and social transformation that sustained the phenomenal growth of the Tokyo metropolitan region from the 1970s through to the 1980s are discussed later.

Because it appeared to recommend the strengthening of the functions of the Tokyo metropolitan region, the Interim Report of the Fourth NCDP, published in 1986, created a storm of protest from local authorities all across Japan. Pragmatic as it was, the Interim Report had disregarded a value judgement implicit in all previous NCDPs: the minimizing of regional disparities and the achievement of balanced development opportunities across all of Japan. Indeed, in the final document of the Fourth NCDP (1987), the proposals are for more balanced development, with the aspiration being a distribution of the nation's central functions, currently concentrated in Tokyo, across the 3MR along the Pacific seaboard. The principal targets of the New Industrial Location Plan, indicated in Table 11.1, are also indicative of these aspirations towards dispersion.[7]

Table 11.1 Principal targets of the former and new industrial location plans, and achievements as of 1985

	Former plan and objectives	Achievement (1985)	New plan and objectives
Framework of economic growth	Annual growth rate of 5.7–6.3% (1976–85)	4.3% (1976–85 annual average)	Stable growth (1986–2000)
Target figure for industrial relocation	30% reduction of industrial land areas in 'relocating-out' regions by 1976, in comparison to 1985	15% reduction compared to 1974	20% reduction of industrial land areas in 'relocating-out' regions by 2000, compared to 1985
Target figure for expansion/construction in 'relocating-into' regions	70% of all new industrial construction (in terms of land use) should be in the 'relocating-into' regions in the plan period of 1976–85	67% of all new industrial construction (in terms of land use) was in the 'relocating-into' regions in the plan period of 1976–85	75% of all new industrial construction (in terms of land use) should be in the 'relocating-into' regions in the plan period of 1986–2000
Target share of shipment of industrial products	1974 / 1985	1985	2000
'Relocating-out' regions	23% / 11%	18%	11%
'Neutral' regions	53% / 59%	55%	54%
'Relocating-into' regions	24% / 30%	27%	35%
Economic/industrial infrastructure; industrial water/land — Total area of industrial sites	151,000 ha / 220,000 ha	160,000 ha	175,000 ha
Total demand for industrial water	48 million cu.m./day (1970) / 73 million cu.m./day	38 million cu.m./day	57 million cu.m./day

Source: Industrial Location and Environmental Protection Bureau, Ministry of International Trade and Industry, Japan.

The Fourth NCDP

In the coming years, the Fourth NCDP envisions the further strengthening of the role of the Tokyo metropolitan region as a leading finance, information and business centre serving the global economy. The Osaka metropolitan region is likely to build upon its cultural heritage, academic and research institutions, consumer goods manufacture and trading tradition to become the counterpole to the Tokyo metropolitan region in its national and international operations. The Nagoya metropolitan region, Japan's manufacturing heartland, will aim to become a global centre of industrial technology. This composite megalopolitan strip along the Pacific seaboard, where the 3MR lie, is envisaged to break the unipolar concentration in the Tokyo metropolitan region; it will function, in effect, as a 'national capital belt' (Figure 11.6).

Many of the mega-infrastructure projects being planned or under way are intended to facilitate the realization of a 'national capital belt'. This includes the Linear Chuo Shinkansen, a very high speed mag-lev train running further inland and parallel to the Tokaido Shinkansen superexpress, now a venerable 26 years old. The new train will eventually cut travel times between Osaka and Tokyo (approximately 550 kilometers) to less than an hour and a half, and bring Nagoya (330 kilometers distant) to within an hour of Tokyo. Construction of the Second *Tomei* (Tokyo–Nagoya) motor expressway is also being planned, parallel to the existing expressway, and the *Meihan* (Nagoya–Osaka) motor expressway is already partly complete. These high-speed transport routes will link the 3MR even more closely. A large new offshore international airport is

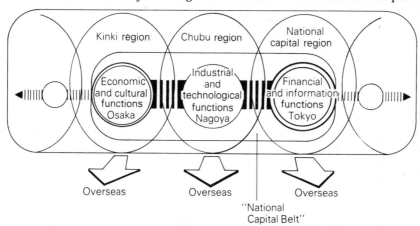

Figure 11.6 Conceptual image of Japan's 'National Capital Belt' and the division of global city functions among the 3MR.
Source: Modified version of figure in 'Outline of the Aichi Prefectural Plan for the 21st Century' (Nagoya, Aichi Prefectural Government, 1990).

under construction. Expected to be opened in 1993, it will serve the Osaka metropolitan region and its hinterland in western Japan. A similar airport is planned for central Japan, where the Nagoya metropolitan region is located. Both airports are joint ventures between the public and private sectors. A proposal to construct a third international airport to serve the Tokyo metropolitan region, to be located inland about 100 kilometers away from central Tokyo, is also being discussed.

The Fourth NCDP is a pragmatic effort to diffuse the dynamism that currently exists in the Tokyo metropolitan region to the other two MRs, while consolidating the existing strengths of each region. The meaning of this effort for the rest of Japan can only be speculated upon. One can say that the growing disparity between the Tokyo metropolitan region and the rest of Japan, particularly visible in the decade of the 1980s, is now likely to become a similar disparity between the re-established megalopolitan conurbation on the Pacific seaboard and the rest of the nation. The management of this phenomenon will be the major challenge for regional development in Japan in the coming years.

Implications of Tokyo's phenomenal growth

Following the restructuring measures undertaken in the wake of the petroleum crises of 1973 and 1979, the Tokyo metropolitan region has experienced over a decade of sustained growth, eclipsing the performance of the other two major metropolitan regions. While these trends have resulted in a heightened concentration of central functions over the decade of the 1980s, further strengthening the region's economic base, negative externalities arising from the impact of abnormally high land price increases (collectively referred to as the 'Tokyo problem') now require serious attention.[8] The gravity of the problem led to the appointing of a Special Ad Hoc Committee by the Prime Minister's Office to investigate the causes and suggest suitable prescriptions to overcome the hardships sustained, particularly by the residents of the metropolitan region. Picked up by the mass media, the 'Tokyo problem' has also entered the consciousness of the average Japanese citizen.

Restructuring measures and heightened concentration

In response to the petroleum crises of 1973 and 1979, the manufacturing industries in the Tokyo region shifted from capital-intensive heavy industries (steel, petrochemicals, etc.) to a new generation of high MVA engineering-oriented industries (including the revamped machinery and automobile industries), many of which were footloose and far less location bound (electronics, high technology materials, etc.) than the leading industries of the preceding generation. At the same time, high-level white-collar urban service industries also grew. The workforce required to man these emerging new sectors was plentiful in Tokyo compared to other metropolitan centres because more than 40 per cent of the nation's higher educational institutions were located in the region.

The pattern of industrial location also began to change in response to these developments. Inland industrial locations in the vicinity of high-speed motorway intersections began to gain favour. The price of land was already high and rising. This, coupled with pressures to improve urban amenities to at least a 'civil minimum',[9] caused many large and medium-sized industries to leave their central city locations, beginning in the 1970s. High-rise offices and residential developments were built on vacated sites. This trend extended to the large coastal industrial complexes along Tokyo Bay, which had led Japan's phenomenal economic drive until the 1970s. By the 1980s, the redevelopment of these vacated sites along the bay had become a major concern of the affected local authorities in metropolitan Tokyo as well as in the neighbouring coastal prefectures of Chiba and Kanagawa, which saw it as an opportunity to revitalize their economies and provide improved urban amenities.

With these transformations under way, Tokyo began to draw ahead of Osaka and Nagoya in terms of economic vitality. The 1980s brought about another structural change that further enhanced Tokyo's position. The liberalization of global financial markets firmly established Tokyo, together with London and New York, as the premier global centres for international finance in each time zone. Foreign financiers, insurance companies, security houses and allied enterprises sought prized central Tokyo locations for their offices. The concentration of international business and other central functions is indicated in Table 11.2. By 1987, metropolitan Tokyo was home to the head offices of around 100 of the world's top 800 manufacturing enterprises and over 20 of the world's 500 leading banks. These factors pushed up the demand for office space in a land market that has traditionally been troubled by supply constraints, and caused the price of land in metropolitan Tokyo to rise at an even more rapid pace.

Tokyo's land-price spiral

A wave of speculative real estate transactions, often financed by the major banks, also stimulated land prices in metropolitan Tokyo, especially in the city's central wards. Between 1986 and 1987, land prices in Tokyo are estimated to have increased by 24 per cent overall, while spot increases at times exceeded 100 per cent. Between 1983 and 1988, the residential land

Table 11.2 Concentration of international business and other central functions in the Tokyo Metropolitan Region

		Regional distribution of city functions Ratio to national total (%)			
Items	Year	Tokyo Metropolitan Region	Osaka Metropolitan Region	Nagoya Metropolitan Region	Non-Metropolitan Regions
Population[1]	1970	23.0	14.8	8.3	53.9
	1975	24.2	15.0	8.4	52.4
	1980	24.5	14.8	8.4	52.2
	1985	25.0	14.7	8.5	51.8
Finance					
Bill clearing[2]	1970	53.2	24.3	8.0	14.5
	1975	57.9	22.0	6.8	13.3
	1980	66.9	17.3	5.8	10.0
	1985	78.8	11.5	3.6	6.1
Balance of bank lendings, national[3]	1970	47.6	22.0	7.0	23.5
	1975	43.5	20.5	7.3	23.1
	1980	49.6	19.6	5.9	24.9
	1986	54.0	18.0	5.3	22.7
International					
Number of persons employed by foreign banks in Japan	1969	66.9	29.5	3.5	0.0
	1975	76.3	19.5	2.5	1.7
	1981	85.6	12.8	1.0	0.6
Number of foreign business establishments	1972	58.1	19.2	4.8	17.9
	1975	62.6	17.2	5.0	15.3
	1981	66.8	17.4	3.4	12.4
Information					
Number of persons engaged in information and advertisement services	1969	52.5	19.3	5.7	22.5
	1975	53.5	16.8	5.2	24.5
	1981	55.9	16.3	5.0	22.7
Business Services					
Number of persons engaged in professional services	1969	27.8	14.4	7.4	50.3
	1975	32.3	14.8	7.6	45.3
	1981	33.8	15.3	7.6	43.3
Number of persons engaged in other services	1969	37.3	16.5	7.4	38.8
	1975	34.5	18.8	6.7	40.0
	1981	35.0	17.3	7.2	40.5
Commerce					
Wholesale trade	1970	38.9	25.1	10.9	25.1
	1976	38.8	22.1	9.8	29.4
	1979	37.7	21.4	10.1	30.9
	1982	42.3	19.4	9.5	28.7
	1985	41.5	19.5	10.1	28.8
Business Management					
Number of corporate head offices with capital over ¥1 billion	1970	59.5	22.1	5.8	12.6
	1975	58.4	20.1	5.4	16.1
	1980	59.4	19.1	5.5	16.0
	1985	58.9	18.3	5.5	16.9
Research & Development					
Number of persons employed by academic institutes	1969	47.4	12.8	4.5	35.4
	1975	49.2	13.5	4.2	33.1
	1981	46.3	13.4	5.0	35.4
Education					
Number of university students[4]	1970	50.6	20.4	6.7	22.3
	1975	48.2	20.9	7.1	23.9
	1980	45.0	20.1	7.2	27.7
	1986	43.7	19.5	7.3	29.5

Table 11.2 Concentration of international business and other central functions in the Tokyo Metropolitan Region – *continued*

		Regional distribution of city functions Ratio to national total (%)			
Items	Year	Tokyo Metropolitan Region	Osaka Metropolitan Region	Nagoya Metropolitan Region	Non-Metropolitan Regions
Production					
Industrial shipment[5]	1970	29.7	20.4	12.6	37.3
	1975	27.0	18.0	12.7	42.2
	1980	26.6	16.5	13.2	43.7
	1985	25.6	15.8	14.4	44.2
Culture					
Number of employees in	1970	50.9	17.5	5.7	25.9
cultural institutions	1975	50.9	16.4	5.9	26.8
(on a workplace basis)	1980	50.7	15.8	6.1	27.4

Source: Industrial Location and Environmental Protection Bureau, Ministry of International Trade and Industry, Japan.
Notes: 1) Compiled from Population Census, Establishment Census, Census of Manufactures, Census of Commerce, Prefectural Economic Statistics, Annual Reports of Taxation Statistics and Basic Survey of Educational Institutions; 2) Calendar year total; 3) As of March 31; 4) As of May 1; 5) Tentative for 1985.

Table 11.3 Land prices and land use intensity in the 23 central wards of Metropolitan Tokyo, 1983 and 1988

	1983	1988
Land prices in residential areas*	356.2	1,361.4
	(100)	(382)
Land prices in commercial areas	1,495.9	8,238.0
	(100)	(551)
Designated floor area ratio	242%	242%
Realized floor area ratio	90.6%	99.3%

Source: Onishi, T. 1989. *Tokyo-ken: Nihon no kouzouteki saihensei no kadai.* (Tokyo Region: Issues in Industrial Restructuring in Japan). UNCRD.
Note: Land prices*: in thousand yen/sq. metre; officially announced land prices.

price rose 3.8 times and commercial land 5.5 times in the 23 central wards of Tokyo, as shown in Table 11.3. Local authorities and government agencies were caught in a bind as the cost of land made it virtually impossible to carry out public sector projects requiring pre-emption.

An additional factor aggravating the surge in land prices relates to the recent sale of public land. Following the central government's commitment to administrative reform, local government also tried to pare finances and curb the escalation of deficit financing in government bond issues. Public service efficiency was to be improved by privatizing state-run enterprises. The sale of their assets would provide the finances for administering reform and raising performance levels. The extensive land holdings in prime central urban locations, especially in Tokyo, of the deficit-ridden Japanese National Railways (JNR), were a prime target of the privatization programme. Other large tracts of publicly owned central urban land were also put up for sale under a bidding system that would transfer these holdings to the private sector. The cumulative effect of all this was to push land prices even higher. Short-term relief of the strain on the public exchequer was obvious. The public evinced concern over the sale of public land. The greater concern, however, was whether this would not compound in the long term the problem of enhancing the urban amenity levels of Tokyo, hardly high at present for a nation of Japan's economic stature.

In an attempt to slow rising land prices the Tokyo Metropolitan Government introduced an ordinance imposing high property transfer taxes within its area of jurisdiction. Departing from precedent, the rule applied even to the sale of a residential property followed by a new purchase for the same purpose. The objective was to encourage retention of holdings, or at least land price increases to central areas and prevent a similar effect in other areas of metropolitan Tokyo. This measure was quite ineffective. Those displaced from central Tokyo had to find new homes. The difficulties in metropolitan Tokyo served to make more attractive the properties on the inner borders of the neighbouring prefectures, which did not fall under the ordinance. In this manner the 'Tokyo problem' spilled beyond the boundaries of metropolitan Tokyo into the neighbouring prefectures of Chiba, Kanagawa and Saitama. High rates of land price increase, although smaller in absolute terms than those in metropolitan Tokyo, are now evident in the central cities of the Osaka and Nagoya metropolitan regions as well as in the adjacent municipalities.

In 1989 the Basic Land Law was enacted which stressed the use of land as a 'public good'. The same conceptual framework will be used in an expected revision of the National Land Use Law of 1974. Measures to increase land supply in metropolitan areas and impose high taxes on vacant urban land (especially agricultural land) in the Urbanization Promotion Areas, designated under the City Planning Act, are foreseen in the near future. Given the high price of land in the major metropolitan regions, it is likely that the basic policy stance promoting home ownership will give way to provision for increased rental housing under higher building densities as a viable alternative in central metropolitan areas.

Physical transformation and social dislocation

Much of metropolitan Tokyo's urban landscape consists of miniscule plots, many bearing hastily built two-storey mortar or wooden structures, or pencil-slim, medium-rise concrete structures of poor quality. The discrepancy between designated and actual floor area ratios (FAR) (see

Table 11.3) is indicative of this. The 'Tokyo problem' has led to consolidation of this finely textured and dense, but relatively low-rise, urban fabric creating higher building densities in the more desirable locations. Although the prices are unparalleled in other countries, and the end user is only reached after a series of transactions, many of which are blatantly speculative, the prime area developments reflect business confidence in the area. They have also contributed towards a measure of urban beautification. The problem is the fact that the only immediate beneficiaries are big business, commercial enterprise and wealthy individuals whose incomes are way above the metropolitan norm.

An average salaried worker will be lucky to find an 80 square meter apartment about an hour's commuting distance from central Tokyo at a cost of from 12 to 15 times his annual income. Low-rent apartments that have been incubators for metropolitan industries such as fashion and industrial design, which also housed the low-budget university student, and eased the uninitiated and newly married into metropolitan life, are fast disappearing from central Tokyo. They are being replaced by office buildings and luxury residences symbolizing corporate wealth. Community residents, facing harassment from pressures exerted by an unscrupulous squad of speculative real estate agents, sometimes band together to resist displacement or to seek better compensation. But the invariable outcome is the breakup of communities and the displacement of small neighbourhood stores.

There is no mechanism to ensure that those displaced by urban redevelopment will benefit from amenity improvement. Tenants will lose central locations and perhaps gain little else, even though a measure of financial compensation may be gained after protracted negotiation. Small landowners in central Tokyo obtain phenomenal prices for their land enabling them to buy luxury apartments or more spacious suburban dwellings further from the centre. The supply constraint on such high-quality residential units and areas in metropolitan Tokyo and the neighbouring prefectures functions to push up prices even in non-central locations.

The rising price of land and the accompanying spatial transformation has extended the commuter catchment of central Tokyo even farther out. Today, the average Tokyoite is forced into the outer fringes of suburbia and beyond, in his or her search for a place to call home. Meanwhile, the most desirable employment opportunities in the nation remain concentrated in the centre of metropolitan Tokyo. The Tokyo region is unrivalled in its concentration of political, administrative, and economic power. It is also the springboard of contemporary Japanese culture. The crux of the 'Tokyo problem' is that the creation of great wealth and power leads to the impoverishment of life for the majority of the region's citizens.

As Japan's economy is shifting from one in which growth is led by exports to one propelled by expanding domestic demand, increased infrastructural investments are likely to be made, with the object of stimulating domestic demand. Given Tokyo's phenomenal accumulation of wealth and power, a rapid surge of investment can be expected, especially from the private sector, which views Tokyo as a low-risk area in comparison with other areas of Japan. Whether such investments will alleviate or aggravate the 'Tokyo problem' for its residents is a major policy issue. But even more important is the significance of Tokyo's growth to the other regions of Japan.

Challenges of balanced growth

Another fear is that Tokyo's unbridled growth will drain the vitality of Japan's other metropolitan and non-metropolitan regions. While the 'national capital belt' concept, described earlier, may help restructure and strengthen the three large metropolitan regions, what will be the effect on the other regions of Japan? If this problem is not addressed seriously, the 'Tokyo problem' may well be extended along the Tokaido belt. To prevent this, it is necessary to ensure that economic development is stimulated and sustained in all the regional blocks demarcated in the Fourth National Comprehensive Development Plan. This may not be easy under the present structure of local government administration, following the prevailing principles upon which regional development has been carried out. Recent changes in the industrial structure and the labour market have compounded the problem.

Efficacy of principles underlying regional development

In the past three decades, all of Japan's National Comprehensive Development Plans have called for balanced development. Nevertheless, the 'grow first, distribute later' principle has resulted in the creation of a powerful engine of growth along the 3MR, propelling Japan's burgeoning economy. The wealth created in the 3MR has been distributed to the other regions through a series of sectoral investments, tax transfers and subsidies. The guidelines for making these investments have been fundamental to each NCDP for achieving balanced regional development. Despite the variation in their particular thrust, the regional development strategies directed at the NMR in all the NCDP have, either explicitly or implicitly rested upon the following:

(a) infrastructure development as the basis for strengthening the comparative advantage and economic base of the NMR;
(b) reliance on manufacturing industries for labour absorption and stimulation of economic growth in the NMR; and
(c) dispersion of branch plants of Japanese firms with head offices in the 3MR as the nuclei of economic growth in the NMR.

However, all of these principles of dispersing economic activity and stimulating growth in the NMR have recently come into serious question.

Infrastructure
First, the nature of infrastructure investment has taken an increasingly sophisticated turn, especially with the advent of the computer and telecommunications revolution and the development of very high-speed surface transport (VHST) technologies such as the mag-lev linear motor railway. Many high-MVA industries and high-tech R&D developments have sprung up in the vicinity of international airports. New business and convention facilities are housed in 'intelligent buildings', reflecting new centres of economic growth. The scale of investment required for such mega-infrastructure projects is immense.

The costs of investing in a mega-infrastructure network across Japan must be viewed in the context of the financial situation of most prefectural and local governments in the NMR development. Administrative reforms pushed through in recent years have favoured the privatization of such expensive projects. However, private capital has a propensity to pour into low-risk locations and sectors. Many of these sophisticated mega-infrastructure projects gravitate towards the 3MR, or the 'national capital belt', rather than towards the outlying locations in the NMR.

Manufacturing
Second, recent trends indicate that the production functions of manufacturing industry alone cannot become the major source of regional economic growth in a sophisticated market economy like Japan's. Following the petroleum crisis in the 1970s, the structural transformation of Japan's manufacturing industry led to a strengthening of the export performance of Japanese goods, with the high MVA manufacturers going to the OECD markets. These goods were targeted at product-competitive market environments, where sophisticated market information, technology development and management skills require a level of business support services and a pool of human resources that the 3MR appear best equipped to provide. Regional centres are now required to deal with a host of new factors to gain comparative advantage. These are far removed from the more simple infrastructure networks and stock that inter-

regional transfers from the 3MR to the NMR had built in the past to enhance the production functions of manufacturing in outlying areas.

A pool of talent, a creative environment for research and design and infrastructure that can accommodate global flows of information and finance appear to be critical determinants of success, even for regions that have a strong manufacturing base. Today, within manufacturing firms, labour shifts from blue-collar to white-collar occupations are not uncommon. This is understandable because high-MVA products, with segmented but sophisticated markets, need a great deal of effort ancillary to mere production. This includes marketing, servicing, monitoring and research for product development. Furthermore, the marriage of the electronics and machinery (popularly called 'mechatronics') industries places increasing demand on computer software technologies. White-collar functions are integrated with the 'grease and grime' operations of the blue-collar shop floor. Company-centred human resource development programmes, facilitating the shift from blue-collar to white-collar operations, characterize operations in many of the larger Japanese firms. Regional associations of industry and trade and local governments may have to provide these services for smaller firms which lie scattered across the manufacturing landscape of the NMR.

Branch plants
Third, many of the manufacturing activities dispersed to the NMR were branch plants of leading Japanese firms with head offices located in the 3MR, particularly in central Tokyo. The commitment of these branch plants to the regional economy was taken for granted, but with the onset of the petroleum crisis the first cracks began to appear. Many industries consuming large amounts of energy, such as metal smelting plants that had been dispersed to the NMR, scaled down or closed their operations in Japan, sometimes moving to invest abroad. With currency appreciation following the Plaza Accord a decade or so later, the commitment of Japanese firms to remain in the NMR became even more tenuous.

In the past few years, following the Plaza Accord, high-MVA manufacturing has remained in Japan, except where export barriers have necessitated offshore production in the market countries. However, many operations where the comparative advantage was in cheaper labour or land have begun to move production abroad. Subcontracting firms began to follow parent firms across the globe as the larger transnational firms launched their global production strategies. Even local firms with head offices rooted in the NMR were finding it increasingly difficult to retain production functions in Japan. As a result of these trends, the third cornerstone of Japanese regional development policy – the dispersion of manufacturing from the 3MR to the NMR – is seriously endangered. It is unclear what mechanism will absorb the labour in the NMR previously engaged in manufacturing.

Options for regional development in coming years

As mentioned earlier, the key to overcoming the 'social divide' caused by the phenomenal growth of the Tokyo metropolitan region is a strengthening of the 3MR, creating a 'national capital belt'. The nature of economic activity along this belt is likely to be global in scale, dominated by central business functions and sophisticated high-MVA manufacturing. Branch plants of Japanese manufacturers with head offices in the 3MR are less likely to be bound to production in Japan if comparative advantage can be found abroad. All of this points to increasing disparities between the 3MR and the NMR. These will be particularly severe in the regions away from the Pacific coast. Viewed in this light, what are the prospects for regional development in the years ahead if the NCDP continues to emphasize the minimization of inter-regional disparities?

The hard-earned reversal of out-migration from the NMR, achieved in the mid-1970s, is in danger of being lost if nothing is done. If the economic base of the NMR is weakened, the younger and more productive age cohorts will move away, leaving behind large clusters of aged communities. There are several options to counter these trends.

Resorts/recreational cities

The NMR might become the nation's repository of cultural and recreational activities, as structural changes in the 3MR give the workforce more time for the pursuit of leisure. This would consolidate the economic base that many of the village revitalization movements have tried to create, and perhaps provide a new stimulus to some of the indigenous industries in the NMR. The Third NCDP implicitly envisaged this plan for many localities in the NMR. Today, under the Fourth NCDP, many prefectural and municipal governments are working towards the creation of resorts and recreational cities.

Most of these settlements, however, are directed at attracting tourists or transitional residents from the 3MR. Over-reliance on an economic base that provides such personal and consumer services to the 3MR may be even more precarious than relying on a 'branch plant economy' in the manufacturing sector. Residents of the 3MR can easily board an airplane and travel to recreational resorts overseas, many of which are more economical than a holiday in the NMR at home. Many localities in the NMR, especially those not favoured with an exceptional advantage in recreational or cultural potential, have no alternative but to strengthen their production or services functions.

Technopolises and science and research parks

It is increasingly evident that for manufacturing to remain in Japan, even in the NMR, it is essential that it be high-MVA in nature. The rapid appreciation of the yen between 1985 and 1989, increasing wage rates across Japan and the rising competition in the NIEs, make this so. The need to prepare for an eventual situation in which only high-MVA, high-tech industries might be able to remain in Japan was recognized in the Technopolis concept, first proposed by the MITI in 1979. The aim was

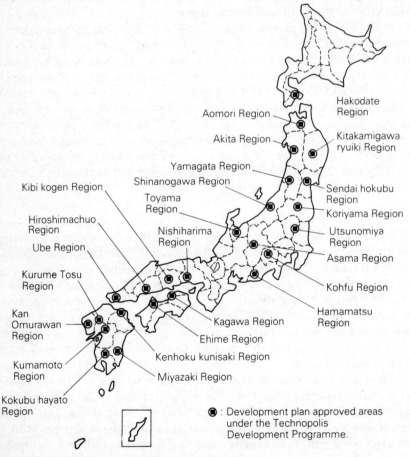

Figure 11.7 Technopolises established across Japan.
Source: H. Kobayashi, Industrial Estates and Regional Development in Japan: Current issues and vision for the future, UNCRD 1989.

to create a core of 'footloose' high-tech industries in or around key central cities in the NMR.

Of the twenty-five Technopolises established across the nation (Fig. 11.7 and Table 11.4) those along the Pacific belt have performed better than those in more remote localities. The successful ones have better access to good airports, superior infrastructure, and information and service support to manufacturing. The latter is contingent upon the quality of human resources. Recognizing this, several prefectural and municipal governments in the NMR have begun to establish or prepare for the establishment of science and research parks to be at the core of their high-tech, high-MVA industrial base. This can be seen in the increasing number of applicants for designation under MITI's concept of developing the cerebral skills of Japan's labour force.

The establishment of new universities or higher educational institutes under the jurisdiction of the Ministry of Education is a protracted process. To establish an academic and research core for a new generation of settlements, local governments turn to existing universities, usually private ones, to establish a new campus or faculty. Other alternatives include attracting the R&D arms of leading private firms and strengthening public sector research institutes. In recent years, foreign collaboration has been highly sought in establishing the academic nuclei for high-tech manufacturing, even in the NMR. For actual production, NMR operations would do well to attract 'branch plants' of foreign firms interested in securing a foothold in the lucrative and affluent Japanese domestic market.

Science and research parks in Japan have thus far reflected the needs of the leading edge in manufacturing. In the coming years, industrial and regional development policy makers will have to grapple with how to support the transformation of the more labour-intensive sectors of manufacturing industry, especially in the NMR.

As the levels of wages and the value of the yen remain high, most of Japan's labour intensive industries, which cannot make a sudden transition to automation, will have to find cheap labour, perhaps foreign, or move their production functions abroad. Small independent firms need the greatest support to survive this shift. Small companies linked under subcontracting arrangements to bigger firms with poor prospects for increasing MVA are faced with the same problem. They must find a way to endure the labour shortage in the short term and to diversify or move their production functions abroad in the medium to long term. Service support systems will have to be devised that will make these small and even micro-scale transnational enterprises prosper at home and abroad. Such support includes strengthening head office functions and access to information on markets and investment opportunities abroad. What is evident is that the future of manufacturing industry in the NMR and

Table 11.4 Basic characteristics and conceptual features of designated Technopolises

Area	Prefecture	Main city	Number of cities/towns/villages in Technopolis (not including main city)	Population (Technopolis area) (main city)	Features of industrial complex	Features of R&D concept
Hakodate	Hokkaido	Hakodate	3 towns	380,517 320,152	Marine industries, resources utilization industries, frigid area community development	Integrated regional marine research centre, urban development centre, resource utilization research
Aomori	Aomori	Aomori	4 cities 2 towns 2 villages	604,325 287,597	Mechatronics-biotechnology industries	Local industry research, modern technology research laboratory, institutes of industry and technology, etc.
Akita	Akita	Akita	2 towns	304,823 284,863	New materials, resource energy development, electronics, mechatronics related industries, biotechnology related industries, etc.	Metal frontier centre, local technology centre, medical centre for the elderly, etc.
Utsunomiya	Tochigi	Utsunomiya	1 city 2 towns	469,944 377,746	Mechatronics, electronics, fine chemicals, new materials, etc.	Mechatronics laboratory, regional industrial institutes, institutes of physics/technology, etc.
Shinanogawa	Niigata	Nagaoka	7 cities 6 towns 1 village	638,509 183,756	High-dimension systems, new materials processing, urban business, fine industry, fashion	Development Education Research Promotion Centre of Nagaoka Technical Univ. (established), Technopolis Development Centre, Kashiwazaki Softpark
Toyama	Toyama	Toyama Takaoka	3 towns	568,291 480,110	Biotechnology, mechatronics, new materials, etc.	Toyama Tech. Development Corp., Bioscience Research Centre, Modern Technology Interchange Centre
Hamamatsu	Shizuoka	Hamamatsu	2 cities 2 towns	619,621 490,824	Photo-industry, musical instruments, sophisticated mechatronics, information communication system, etc.	Photo-information Technology Integrated Research Centre, Electronics Centre, life behaviour research organs, etc.
Nishiharima	Hyōgo	Himeji	3 cities 10 towns	716,679 446,256	High-tech industries, medical and social welfare industries, etc.	Life science laboratories, etc.
Kibikōgen	Okayama	Okayama	2 cities 5 towns	660,183 545,765	Medical, pharmaceutical and chemical industries, agroindustry, etc.	Enzyme/bacteria bank, experimental organism centre, graduate schools, etc.

Table 11.4 Basic characteristics and conceptual features of designated Technopolises – continued

Area	Prefecture	Main city	Number of cities/town villages in Technopolis (not including main city)	Population (Technopolis area) (main city)	Features of industrial complex	Features of R&D concept
Hiroshima-chuo	Hiroshima	Kure	2 cities 2 towns	375,855 234,549	Mechatronics, ship and ocean electronic, home electronics, regional community systems, etc.	R&D organs, modern technology development centre, international materials science research centre, etc.
Ube	Yamaguchi	Ube	3 cities 4 towns	408,774 168,958	Fine chemicals, biotechnology, new mechatronics, electronics, mechatronics, etc.	Technology promotion corporations, new materials development organs, ocean development organs, etc.
Kagawa	Kagawa	Takamatsu	4 cities 7 towns	635,705 316,661	Ultra-precision measuring instruments, living systems, ocean development, etc.	Experiments/research, life technology experimental cities.
Kurume-Tosu	Fukuoka, Saga	Kurume	1 city 5 towns	332,487 216,974	High-system industry (information-associated industry, community development, mechatronics), new materials biotechnology, etc.	R&D park, integrated information centre, etc.
Kumamoto	Kumamoto	Kumamoto	1 city 12 towns 2 villages	738,558 525,662	Applied machine industry, biotechnology, computers, information systems, etc.	Research park (Bio wood), electronics applications machine technology research laboratories, etc.
Kan-Omurawan	Nagasaki	Sasebo	2 cities 1 town	440,778 251,188	Ocean development associated instruments, resources and energy development based on mechatronics, etc.	Research Park (Bio wood), laboratories for research in electronics applications for machine technology and for semi-conductor applications.
Kenhoko-Kunisaki	Ōita	Ōita, Beppu	4 cities 13 towns 2 villages	281,513 496,963	IC, LSI, new materials, soft engineering, techno-green industry, regional resources utilization, etc.	Regional economy information centre, industry–university–government co-operation system, etc.
Miyazaki	Miyazaki	Miyazaki	6 towns	356,876 264,855	Electronics, mechatronics, new materials, biotechnology (fine chemicals, biomass, etc.)	Co-operative Research Development Centre, IC laboratory, etc.
Kokubu-Hayato	Kagoshima	Kagoshima	1 city 12 towns	691,909 505,077	Advanced equipment (electronics, mechatronics), new materials (fine ceramics), regional industry (modern fishing and agroindustry), biotechnology, etc.	Technology promotion organs, material resources research centre, regional industry promotion associations, etc.

Source: Industrial Location and Environmental Protection Bureau, Ministry of International Trade and Industry, Japan.

indeed in all of Japan, depends on the ability of firms to remain strong both at home and abroad. The creation of a human resource base in the NMR, utilizing the attractions and amenities of such less congested regions as well as new academic and information-related stimuli is critical. A vital issue is the extent to which the physical dimensions and settlement structure of Japan's national land can be transformed, particularly under the current system of local government administration.

Institutional adequacy of local government

Japan's two-tiered local government structure, consisting of prefectures and municipalities, is complicated by the existence of 11 specially designated cities. These cities are large metropolitan centres with prefectural level responsibilities. They are sometimes more powerful than the prefectures which execute central government mandates over them. Such a structure does not foster the growth of municipalities into autonomous metropolitan governments with the strong support of a regional authority, whose jurisdiction covers the metropolitan hinterland. In many instances, the area of jurisdiction of prefectural governments covers only one part of the metropolitan area of influence. Overlapping functions and mandates between prefectural governments and large metropolitan municipalities often create competition, rather than co-operation.

If metropolitan Tokyo and the other metropolises along the proposed 'national capital belt' are to act as powerful counter-magnets across Japan's national space, the growth of strong regional metropolises must be fostered in the NMR. Amalgamating several prefectures into larger regional blocks equivalent to provincial or state governments is likely to further this process. Such fundamental reforms are perhaps now necessary in view of the powerful role that regional metropolises can play as engines of sub-national and national growth. Their close and interdependent linkages with a network of cities, constituting a virtual regional urban system, should break the polarization towards metropolitan Tokyo and the 3MR. Only radical reform in the institutional background of local government will ensure that true decentralization of functions can take place. The success of this reorganisation will ensure the future for Japan's non-metropolitan regions.

Appendices

Appendix to Chapter 3

FRANZ JOSEF BADE & KLAUS R. KUNZMANN

Table 3A.1 The sectoral development of production (actual prices)[1]

	Billions of DM				Percentage share of all sectors				Percentage change to 1986		
	1960	1973	1980	1986	1960	1973	1980	1986	1960	1973	1980
Agriculture	27.9	48.0	63.8	68.3	4.4	2.7	2.5	1.8	144.9	42.2	7.1
Goods-producing industries	388.9	1,022.2	1,651.8	1,994.8	61.8	57.7	63.9	52.3	412.9	95.2	20.8
Energy, water	15.8	46.2	105.4	166.0	2.5	2.6	4.1	4.4	952.6	259.7	57.5
Mining	13.5	16.5	30.6	33.2	2.1	0.9	1.2	0.9	145.9	100.8	8.4
Manufacturing industry	316.8	825.5	1,319.0	1,594.8	50.4	46.6	51.0	41.8	403.4	93.2	20.9
Construction	42.9	134.0	196.7	200.9	6.8	7.6	7.6	5.3	368.7	49.9	2.1
Total service industries	212.2	699.9	1,165.3	1,748.5	33.7	39.5	45.1	45.9	723.9	149.8	50.1
Trade[1]	59.1	147.2	234.0	285.8	9.4	8.3	9.1	7.5	383.6	94.1	22.1
Transport and communications	33.4	91.9	158.4	200.0	5.3	5.2	6.1	5.2	499.2	117.6	26.2
Financial and insurance institutions	11.7	51.8	100.4	154.8	1.9	2.9	3.9	4.1	1,220.5	199.1	54.2
Other service enterprises	56.3	204.9	295.1	612.0	8.9	11.6	11.4	16.1	987.7	198.7	107.4
Hotels and lodging places	14.0	29.9	47.4	60.0	2.2	1.7	1.8	1.6	329.6	101.1	26.8
Education, publishing, and media	7.3	20.6	40.7	62.4	1.2	1.2	1.6	1.6	760.4	202.4	53.4
Health	4.7	21.9	46.8	59.6	0.7	1.2	1.8	1.6	1,166.0	172.4	27.3
Remaining services, n.e.c.	30.3	132.5	160.3	430.0	4.8	7.5	6.2	11.3	1,317.3	224.5	168.3
Government and non-commercial organizations	51.8	204.2	377.3	495.9	8.2	11.5	14.6	13.0	857.9	142.9	31.4
All sectors	629.0	1,770.1	2,585.7	3,811.6	100.0	100.0	100.0	100.0	506.0	115.3	47.4

Source: Authors' calculations on the basis of National Account Tables.
Note: 1) Production values are calculated by reduction of intermediate input.

Table 3A.2 The sectoral development of gross value added (actual prices)

	Billions of DM				Percentage share of all sectors				Percentage change to 1986		
	1960	1973	1980	1986	1960	1973	1980	1986	1960	1973	1980
Agriculture	17.7	26.6	30.4	34.0	5.8	3.0	2.1	1.8	92.5	27.8	12.0
Goods-producing industries	160.8	430.9	632.2	790.2	53.2	48.8	44.4	42.0	391.4	83.4	25.0
Energy, water	7.3	20.6	36.3	52.7	2.4	2.3	2.6	2.8	622.6	155.2	45.0
Mining	8.4	8.8	13.9	15.1	2.8	1.0	1.0	0.8	81.1	72.4	9.2
Manufacturing industry	121.9	333.3	482.8	622.7	40.3	37.7	33.9	33.1	411.0	86.9	29.0
Construction	23.3	68.2	99.2	99.7	7.7	7.7	7.0	5.3	327.9	46.0	0.5
Total service industries	123.7	425.4	673.8	1,055.9	40.9	48.2	47.4	56.2	753.3	148.2	56.7
Trade	36.3	89.4	140.0	174.3	12.0	10.1	9.8	9.3	380.7	95.0	24.5
Transport and communications	19.7	51.9	85.7	110.5	6.5	5.9	6.0	5.9	460.7	112.8	28.9
Financial and insurance institutions	7.3	34.4	66.4	101.6	2.4	3.9	4.7	5.4	1,299.3	195.0	53.0
Other service enterprises	33.9	136.9	182.2	412.4	11.2	15.5	12.8	21.9	1,117.7	201.2	126.4
Hotels and lodging places	4.4	11.0	18.9	26.8	1.4	1.2	1.3	1.4	514.0	143.5	42.0
Education, publishing, and media	3.9	9.9	20.2	31.8	1.3	1.1	1.4	1.7	718.0	222.1	57.2
Health	3.5	16.2	32.6	41.8	1.2	1.8	2.3	2.2	1,086.4	157.9	28.2
Remaining services, n.e.c.	22.1	99.9	110.5	312.0	7.3	11.3	7.8	16.6	1,312.5	212.5	182.5
Government and non-commercial organizations	26.7	112.8	199.6	257.1	8.8	12.8	14.0	13.7	864.7	128.0	28.8
All sectors	302.2	882.9	1,422.9	1,880.0	100.0	100.0	100.0	100.0	522.1	112.9	32.1

Source: Authors' calculations on the basis of National Account Tables.

Table 3A.3 The sectoral development of employment

	Thousands of persons				Percentage share of all sectors				Percentage change to 1986		
	1960	1973	1980	1986	1960	1973	1980	1986	1960	1973	1980
Agriculture	3,581	1,924	1,437	1,338	13.7	7.2	6.1	5.2	−62.6	−30.5	−30.5
Goods-producing industries	12,497	12,723	11,586	10,506	47.9	47.4	49.4	40.9	−15.9	−17.4	−17.4
Energy, water	195	256	267	277	0.7	1.0	1.1	1.1	42.1	8.2	8.2
Mining	552	259	234	212	2.1	1.0	1.0	0.8	−61.6	−18.1	−18.1
Manufacturing industry	9,624	9,861	8,995	8,263	36.9	36.7	38.4	32.1	−14.1	−16.2	−16.2
Construction	2,126	2,347	2,090	1,754	8.2	8.7	8.9	6.8	−17.5	−25.3	−25.3
Total service industries	9,985	12,202	13,255	13,858	38.3	45.4	56.6	53.9	38.8	13.6	13.6
Trade	3,299	3,492	3,505	3,325	12.7	13.0	15.0	12.9	0.8	−4.8	−4.8
Transport and communications	1,460	1,523	1,469	1,449	5.6	5.7	6.3	5.6	−0.8	−4.9	−4.9
Financial and insurance institutions	383	678	740	803	1.5	2.5	3.2	3.1	109.7	18.4	18.4
Other service enterprises	1,981	2,445	2,842	3,188	7.6	9.1	12.1	12.4	60.9	30.4	30.4
Hotels and lodging places	623	690	776	827	2.4	2.6	3.3	3.2	32.7	19.9	19.9
Education, publishing, and media	243	219	241	270	0.9	0.8	1.0	1.1	11.1	23.3	23.3
Health	234	352	500	584	0.9	1.3	2.1	2.3	149.6	65.9	65.9
Remaining services, n.e.c.	881	1,184	1,325	1,507	3.4	4.4	5.7	5.9	71.1	27.3	27.3
Government and non-commercial organizations	2,862	4,064	4,699	5,093	11.0	15.1	20.1	19.8	78.0	25.3	25.3
All sectors	26,063	26,849	23,436	25,702	100.0	100.0	100.0	100.0	−1.4	−4.3	−4.3

Source: Authors' calculations on the basis of National Account Tables.

Table 3A.4 The functional development of employment (changes in occupations)

	Thousands of persons				Percentage share of all functions				Percentage change to 1986		
	1961	1973	1980	1987	1960	1973	1980	1987	1960	1973	1980
Production	13,320	11,063	9,570	8,756	51.1	41.2	36.4	33.8	−34.3	−20.9	−8.5
All service functions	11,977	15,479	16,313	16,552	46.0	57.7	62.1	63.9	38.2	6.9	1.5
Producer services	4,676	6,920	7,418	7,511	17.9	25.8	28.2	29.0	60.6	8.5	1.3
Technical services	838	1,390	1,460	1,460	3.2	5.2	5.6	5.6	74.2	5.0	0.0
Research and development	n.a.	482	520	536	ERR	1.8	2.0	2.1	n.a.	11.2	3.1
Administration services	3,838	5,088	5,395	5,428	14.7	19.0	20.5	21.0	41.4	6.7	0.6
Top executives	n.a.	824	824	969	ERR	3.1	3.1	3.7	n.a.	17.6	17.6
EDP, marketing and strategical plan	n.a.	442	563	624	ERR	1.6	2.1	2.4	n.a.	41.2	10.8
Distribution	4,468	4,354	4,140	4,010	17.1	16.2	15.8	15.5	−10.3	−7.9	−3.1
Transport, stock-keeping	2,250	2,108	1,924	1,775	8.6	7.9	7.3	6.9	−21.1	−15.8	−7.7
Trade	2,218	2,246	2,216	2,235	8.5	8.4	8.4	8.6	0.8	−0.5	0.9
Personal and consumer services	2,833	4,206	4,755	5,029	10.9	15.7	18.1	19.4	77.5	19.6	5.8
Health and education	1,118	1,758	2,303	2,547	4.5	6.5	8.8	9.8	127.8	44.9	10.6
Other personal services	1,715	2,448	2,452	2,482	6.6	9.1	9.3	9.6	44.7	1.4	1.2
Not otherwise specified	766	307	395	583	2.9	1.1	1.5	2.3	−23.9	89.9	47.6
Total employment	26,063	26,849	26,278	25,891	100.0	100.0	100.0	100.0	−0.7	−3.6	−1.5

Source: Authors' calculations on the basis of National Account Tables and of population census.

Appendix to Chapter 5

ROBERTO P. CAMAGNI

Table 5A.1 Employment changes by sector, 1970–84

	Employment (thousands)					
	1976–70		1980–76		1984–80	
	Absolute change	Percentage change	Absolute change	Percentage change	Absolute change	Percentage change
Piemonte						
Manufacturing	−5.1	−0.72	−5.6	−0.79	−131.2	−18.7
Market services	38.4	15.89	33.5	11.96	46.4	14.8
Total	83	6.573	40.3	2.995	−72	−5.19
Valle d'Aosta						
Manufacturing	0.9	8.108	1.4	11.67	−2.4	−17.9
Market services	1.8	24.32	1.8	19.57	2.5	22.73
Total	4.2	14.09	3.2	9.412	1.2	3.226
Lombardia						
Manufacturing	−4.6	−0.32	−47.2	−3.33	−151.9	−11.1
Market services	96.7	16.38	71	10.33	116.5	15.37
Total	121.9	4.598	68.7	2.478	−11.9	−0.42
Trentino-Alto Adige						
Manufacturing	9.7	16.72	0.1	0.148	−9.9	−14.6
Market services	7.8	10.7	15.4	19.08	6.2	6.452
Total	23.7	10.87	16.7	6.907	0.3	0.116
Veneto						
Manufacturing	14	3.305	−5.4	−1.23	−39.7	−9.19
Market services	42.8	19.06	30.2	11.29	38	12.77
Total	63.7	6.2	24.5	2.245	−12.9	−1.16
Friuli Venezia Giulia						
Manufacturing	11.1	9.72	3.2	2.554	−19.5	−15.2
Market services	9	10.71	7.7	8.28	11.8	11.72
Total	22.1	6.744	9.7	2.773	−1.3	−0.36
Liguria						
Manufacturing	4.2	3.279	−3.1	−2.34	−19.8	−15.3
Market services	8.5	4.67	15	7.874	12.4	6.034
Total	23.5	4.878	16.9	3.345	0.7	0.134
Emilia Romagna						
Manufacturing	18.4	4.775	5	1.239	−38.2	−9.35
Market services	41.6	17.99	27.6	10.11	28.6	9.517
Total	60.3	5.85	41.6	3.813	−17.2	−1.52
Toscana						
Manufacturing	14.6	4.185	−3.8	−1.05	−32.3	−8.98
Market services	30.3	15.03	20.5	8.84	33	13.07
Total	71.3	8.1	26.5	2.785	−1.7	−0.17

Appendix to Chapter 5 355

Table 5A.1 Employment changes by sector, 1970–84 – *continued*

	Employment (thousands)					
	1976–70		1980–76		1984–80	
	Absolute change	Percentage change	Absolute change	Percentage change	Absolute change	Percentage change
Umbria						
Manufacturing	11.9	20.1	2.8	3.938	−10.8	−14.6
Market services	6.2	18.84	5.6	14.32	3.7	8.277
Total	19.4	10.84	21.4	10.79	−17.9	−8.14
Marche						
Manufacturing	25.4	27.85	4	3.431	−18.9	−15.7
Market services	10.5	18.95	5.1	7.739	3.3	4.648
Total	39.7	13.63	15.8	4.775	−18.5	−5.34
Lazio						
Manufacturing	33.4	16.32	15.1	6.345	−3.5	−1.38
Market services	61.4	16.52	58.1	13.42	62.8	12.79
Total	114.8	9.631	91.6	7.009	36.1	2.582
Abruzzi						
Manufacturing	17.1	32.45	−1.4	−2.01	−2.9	−4.24
Market services	6.9	17.69	8.1	17.65	5.6	10.37
Total	26.4	12.97	12	5.22	14.1	5.829
Molise						
Manufacturing	3.8	42.7	0.7	5.512	−1.2	0.00
Market services	1.7	24.64	1.5	17.44	1.7	16.83
Total	9	19.19	5.7	10.2	0.2	0.325
Campania						
Manufacturing	39.6	16.28	7.4	2.616	−17.9	−6.17
Market services	24.6	12.39	38	17.03	22.3	8.538
Total	57.3	5.729	74.8	7.073	10.2	0.901
Puglia						
Manufacturing	31.2	20.77	6.9	3.804	−10.8	−5.74
Market services	20.4	17.66	20.3	14.94	32.9	21.06
Total	134.8	16.71	9.2	0.977	−26.8	−2.82
Basilicata						
Manufacturing	3	16.22	0.4	1.86	−1.5	−6.85
Market services	2.4	19.67	3.5	23.97	2.8	15.47
Total	19.3	17.99	9.6	7.583	3	2.203
Calabria						
Manufacturing	2.7	7.692	0.3	0.794	−2	−5.25
Market services	8	16.43	10.4	18.34	6.9	10.28
Total	−2.5	−0.63	9.7	2.451	−2.2	−0.54
Sicilia						
Manufacturing	9.2	6.162	4	2.524	−12.7	−7.82
Market services	20.1	11.62	30.8	15.95	35.6	15.9
Total	51.1	5.493	37	3.771	−12.7	−1.25
Sardegna						
Manufacturing	8.7	21.64	4	8.18	−3.9	−7.37
Market services	8.5	16.57	11.2	18.73	6.1	8.592
Total	23	8.403	16.3	5.494	7.4	2.364
Italia						
Manufacturing	249.2	5.352	−11.2	−0.23	−531	−10.8
Market services	447.6	15.22	415.3	12.26	479.1	12.59
Total	966	7.242	551.2	3.853	−121.9	−0.82

Table 5A.2 Growth rate of value added, 1970–6–80–4

	1970–6	1976–80	1980–4
Piemonte			
Manufacturing	18.9125832	8.20667727	−9.3767447
Market services	20.7684417	13.984556	3.35681864
Total	17.8473032	9.40736066	−3.232784
Valle d'Aosta			
Manufacturing	14.8325359	38.3333333	−9.4879518
Market services	73.582296	−6.6932271	−0.6831768
Total	40.754717	6.12153709	1.72631579
Lombardia			
Manufacturing	15.2671899	16.6072098	1.46680823
Market services	20.4593041	18.3490819	5.49045342
Total	15.2824368	14.8613081	1.94710544
Trentino-Alto Adige			
Manufacturing	43.40261	23.9973037	−10.437619
Market services	36.3771124	25.8913043	10.2572958
Total	27.2002813	19.4582642	0.23908684
Veneto			
Manufacturing	29.3107396	20.5950062	0.19089574
Market services	33.7236534	16.8741421	7.22906204
Total	23.8616473	14.2752162	1.47410424
Friuli Venezia Giulia			
Manufacturing	36.0990207	20.8674795	−10.980652
Market services	20.7136788	18.6655405	6.60735469
Total	19.4211018	15.6372166	−0.9361835
Liguria			
Manufacturing	19.9213865	10.1162098	−8.1450413
Market services	12.3744645	11.7430161	7.51118568
Total	8.8347733	8.85875865	2.50395276
Emilia Romagna			
Manufacturing	34.49169	24.2229712	−3.6690926
Market services	32.3653857	17.686211	8.08528055
Total	26.6861914	16.6120351	1.70431849
Toscana			
Manufacturing	31.0746472	19.7703833	−2.4045491
Market services	18.5892116	16.340098	7.05907626
Total	18.9113801	13.9389244	1.86423391
Umbria			
Manufacturing	39.8029135	26.0190009	−12.183852
Market services	31.1183733	17.4466907	6.67295449
Total	29.1194305	18.6727922	−0.0258087
Marche			
Manufacturing	48.1205674	26.1192243	−9.6051632
Market services	27.3555642	15.009895	4.27531436
Total	24.4831939	14.5274767	0.9961813
Lazio			
Manufacturing	36.6611486	18.5985837	10.6301069
Market services	21.5344963	11.4304155	9.29802786
Total	15.8346957	11.5153604	6.55011552
Abruzzi			
Manufacturing	57.8780178	21.3279678	−7.6616915
Market services	34.9860552	17.6538108	6.61463415
Total	29.1104227	14.354272	2.40557554

Appendix to Chapter 5

Table 5A.2 Growth rate of value added, 1970–6–80–4 – continued

	1970–6	1976–80	1980–4
Molise			
Manufacturing	106.334842	27.6315789	−7.2164948
Market services	42.2058824	14.3743537	2.98372514
Total	26.202709	18.0977054	5.54685052
Campania			
Manufacturing	23.6586007	18.7387004	−1.5047022
Market services	23.1432403	16.7245761	8.51252888
Total	17.2875514	14.6806827	5.37168089
Puglia			
Manufacturing	41.8519274	17.8576564	−10.821032
Market services	22.4169742	18.479444	5.2155477
Total	20.309856	20.9851847	1.75415935
Basilicata			
Manufacturing	23.8666667	41.0118407	−6.870229
Market services	27.2959184	16.3660655	3.27210103
Total	22.2648752	20.7417582	0.3738014
Calabria			
Manufacturing	22.0338983	4.67171717	8.62484922
Market services	27.775378	16.1595673	4.00174622
Total	12.2903062	17.9712631	−1.6118282
Sicilia			
Manufacturing	23.2814527	22.4443275	−4.854647
Market services	19.990458	16.94405	7.57237226
Total	15.9373034	14.074604	2.8287809
Sardegna			
Manufacturing	56.7811935	23.029604	−7.84375
Market services	22.1189189	17.1919263	−0.2417284
Total	17.2978256	11.6989516	2.43038289
Italia			
Manufacturing	25.0015072	17.5508826	−2.7202232
Market services	23.1935189	16.108047	6.53775883
Total	18.5610396	14.1700826	1.8576093

Table 5A.3 Productivity change, 1970–6–80–4

Productivity		Percentage change	
	1970–6	1976–80	1980–4
Piemonte			
Manufacturing	19.769	9.0695	11.442
Market services	4.2118	1.8083	−9.965
Total	10.579	6.2264	2.0691
Valle d'Aosta			
Manufacturing	6.2201	23.881	10.26
Market services	39.621	−21.96	−19.08
Total	23.367	−3.007	−1.453
Lombardia			
Manufacturing	15.642	20.627	14.129
Market services	3.5063	7.2651	−8.561
Total	10.214	12.084	2.3758
Trentino-Alto Adige			
Manufacturing	22.856	23.814	4.8762
Market services	23.196	5.7173	3.575
Total	14.733	11.741	0.1229
Veneto			
Manufacturing	25.174	22.102	10.325
Market services	12.32	5.0139	−4.912
Total	16.63	11.766	2.6612
Friuli Venezia Giulia			
Manufacturing	24.042	17.858	4.9448
Market services	9.0317	9.5918	−4.575
Total	11.876	12.517	−0.577
Liguria			
Manufacturing	16.114	12.758	8.4795
Market services	7.3604	3.5866	1.3931
Total	3.7732	5.3358	2.3667
Emilia Romagna			
Manufacturing	28.362	22.703	6.263
Market services	12.188	6.8771	−1.308
Total	19.685	12.329	3.2725
Toscana			
Manufacturing	25.81	21.036	7.2238
Market services	3.0944	6.8909	−5.32
Total	10.002	10.852	2.0416
Umbria			
Manufacturing	16.404	21.244	2.8465
Market services	10.327	2.733	−1.482
Total	16.494	7.1187	8.8377
Marche			
Manufacturing	15.854	21.936	7.1939
Market services	7.0637	6.7486	−0.356
Total	9.5482	9.3082	6.6891
Lazio			
Manufacturing	17.483	11.523	12.181
Market services	4.3007	−1.752	−3.094
Total	5.6588	4.2107	3.8687
Abruzzi			
Manufacturing	19.2	23.811	−3.573
Market services	14.694	0.0057	−3.403
Total	14.284	8.6815	−3.235

Table 5A.3 Productivity change, 1970–6–80–4 – *continued*

Productivity		Percentage change	
	1970–6	1976–80	1980–4
Molise			
Manufacturing	44.597	20.964	1.9098
Market services	14.095	−2.612	−11.85
Total	5.8838	7.1698	5.2053
Campania			
Manufacturing	6.349	15.712	4.9676
Market services	9.571	−0.257	−0.023
Total	10.932	7.1049	4.4309
Puglia			
Manufacturing	17.454	13.539	−5.395
Market services	4.0409	3.0817	−13.09
Total	3.0844	19.814	4.7058
Basilicata			
Manufacturing	6.5829	38.436	−0.022
Market services	6.3706	−6.136	−10.56
Total	3.6258	12.231	−1.789
Calabria			
Manufacturing	13.317	3.8475	14.643
Market services	9.7471	−1.844	−5.696
Total	13.008	15.149	1.070
Sicilia			
Manufacturing	16.126	19.43	3.2117
Market services	7.5005	0.8571	−7.185
Total	9.9	9.9297	4.1274
Sardegna			
Manufacturing	28.888	13.727	−0.509
Market services	4.7609	−1.295	−8.134
Total	8.205	5.882	0.0646
Italia			
Manufacturing	18.651	17.82	9.1184
Market services	6.9209	3.4317	−5.38
Total	10.555	9.934	2.7003

Table 5A.4 Ranking of Italian regions according to the level (A) and to the growth rates (B) of total productivity

(A)	1970 (a)	1975 (b)	1980 (c)	1984 (d)
Piemonte	5	6	6	8
Valle d'Aosta	2	1	1	3
Lombardia	3	3	2	1
Trentino-Alto Adige	11	12	11	11
Veneto	9	9	9	9
Friuli Venezia Giulia	6	5	5	6
Liguria	1	2	3	2
Emilia Romagna	8	7	4	4
Toscana	7	8	10	10
Umbria	12	11	8	5
Marche	16	15	12	12
Lazio	4	4	7	7
Abruzzi	17	17	15	15
Molise	20	20	20	18
Campania	14	14	16	16
Puglia	15	16	17	17
Basilicata	19	19	18	19
Calabria	18	18	19	20
Sicilia	13	13	14	14
Sardegna	10	10	13	13
Spearman's rank correlation index		(a/b) 0.922	(b/c) 0.951	(c/d) 0.980

(B)	1975/70 (e)	1980/75 (f)	1984/80 (g)	1984/70 (h)
Piemonte	19	9	16	17
Valle d'Aosta	8	19	18	16
Lombardia	18	10	10	15
Trentino-Alto Adige	16	6	15	10
Veneto	12	7	6	6
Friuli Venezia Giulia	9	8	17	8
Liguria	20	14	14	20
Emilia Romagna	3	5	9	4
Toscana	15	12	11	12
Umbria	4	1	2	2
Marche	5	2	8	3
Lazio	17	18	7	19
Abruzzi	6	3	12	5
Molise	2	11	1	1
Campania	10	16	3	13
Puglia	14	13	4	9
Basilicata	7	4	20	7
Calabria	1	17	19	11
Sicilia	11	15	5	14
Sardegna	13	20	13	18
Spearman's rank correlation index	(e/h) 0.742	(f/h) 0.748	(g/h) 0.338	

Source: Marelli, E. 1989. Crescita, produttivà e cambiamento strutturale nelle regioni italiane. *Rivista di Politica Economica* 5.

Table 5A.5 Shift–share analysis of productivity growth (differential, mix and reallocation effects, 1970–84)

	Mix	Reallocation	Differential	Total
Piemonte	1.391	−1.379	−8.925	−8.913
Valle d'Aosta	−6.743	3.254	−4.804	−8.293
Lombardia	0.416	−3.454	0.352	−2.686
Trentino-Alto Adige	−4.902	1.214	5.728	2.040
Veneto	0.301	0.551	9.078	9.930
Friuli Venezia Giulia	−6.810	2.626	8.998	4.814
Liguria	−7.739	0.048	−9.519	−17.210
Emilia Romagna	1.849	−1.187	15.162	15.824
Toscana	−3.367	1.560	3.020	1.213
Umbria	−0.599	−9.150	34.988	25.239
Marche	−2.171	8.525	12.068	18.422
Lazio	−11.486	2.907	−2.127	−10.706
Abruzzi	−0.698	6.601	8.745	14.648
Molise	0.213	13.321	16.749	30.283
Campania	−2.542	6.411	−3.122	0.747
Puglia	2.010	3.528	−2.039	3.499
Basilicata	1.757	5.010	−0.206	6.561
Calabria	−3.759	11.064	−5.902	1.403
Sicilia	−3.422	5.817	−4.350	−1.955
Sardegna	−2.965	2.902	−10.030	10.000
Centre and North	−2.085	−0.021	1.767	−0.339
Mezzogiorno	−1.754	6.015	−2.918	1.343

Source: Marelli, E. 1989. Crescita, produttivà e cambiamento strutturale nelle regioni italiane. *Rivista di Politica Economica* **5**.

Table 5A.6 Relative productivity and employment growth (firms with more than 20 employees)

	Relative productivity growth				Relative employment growth			
	1973–76	1976–79	1979–82	1982–86	1973–76	1976–79	1979–82	1982–86
Piemonte	1.044	0.945	0.982	1.082	0.961	1.003	0.952	0.904
Lombardia	0.989	1.023	1.022	0.984	0.97	0.97	0.991	0.969
Liguria	0.906	0.964	0.935	0.962	0.974	0.995	1.01	0.813
Veneto	1.028	1.032	1.036	1.037	0.988	1.016	0.991	1.117
Emilia Romagna	1.051	1.031	1.003	0.979	1.032	1.017	1.044	1.034
Toscana	1.033	1.024	1.044	0.978	0.981	0.978	1.008	1.019
Campania	0.946	1.005	0.964	0.933	1.033	1.04	1.031	0.981
Puglia	0.893	0.979	0.863	1.026	1.152	1.059	1.1	1.064
Sicilia	0.923	1.052	0.906	0.987	1.075	1.016	1.056	0.996
Abruzzi Molise	1.063	1.028	1.035	1.216	1.092	1.068	1.105	1.301

Appendix to Chapter 6

JUAN R. CUADRADO ROURA

Table 6A.1 Employment structure by sectors, 1960–87 (percentages)

	Agriculture and fisheries	Industry	Construction	Services
1960	42.25	23.02	6.70	28.19
1965	35.07	25.27	7.88	31.89
1970	29.41	26.16	8.47	35.95
1975	23.74	26.97	9.65	39.63
1980	19.27	26.95	8.98	44.80
1985	18.32	24.37	7.29	50.02
1987	15.15	24.16	8.14	52.55

Source: INE and FIES.

Table 6A.2 Participation of the production branches in the production value of industry (at 1964 prices)

1964	Percentage	1974	Percentage
Other food industries	15.2	Other food industries	15.2
Textiles	13.4	Construction	9.7
Basic metal industries	7.6	Basic metal industries	8.7
Canned foods and drink	7.6	Transport materials	8.0
Metal industries	4.7	Canned foods and drink	5.9
Wood	4.1	Cement, ceramics and glass	4.8
Non-electrical machinery	4.9	Electrical machinery	4.2
Cements, ceramics and glass	3.9	Metal industries	4.0
Transport materials	3.8	Non-electrical machinery	3.4
Electrical machinery	3.2	Electric energy	3.2
Paper	2.1	Basic chemicals	2.4
Publishing	2.0	Paper	2.3
Industrial chemistry	1.8	Industrial chemistry	2.3
Petroleum by-products	1.7	Plastics	2.3
Base chemicals	1.6	Synthetic and artificial fibres	2.2
Leather and footwear	1.6	Petroleum by-products	2.1
Shipbuilding	1.4	Shipbuilding	1.8
Extractive industries	1.2	Leather and footwear	1.5
Plastics	1.2	Publishing	1.5
Synthetic and artificial fibres	0.9	Cork	1.4
Pharmaceuticals	0.9	Pharmaceuticals	1.5
Mineral fuels	0.9	Fertilizers	0.9
Cork	0.9	Extractive industries	0.8
Fertilizers	0.8	Other manufactures	0.8
Coke and gas	0.6	Mineral fuels	0.5
Other manufactures	0.5	Coke and gas	0.4

Source: Author's calculations from MINER (Ministry of Industry and Energy) figures.

Table 6A.3 Structural changes by regions (in percentage: GRP = 100 percent)

Autonomous Communities	Agriculture and fisheries			Industry			Construction			Services		
	1973	1979	1985	1973	1979	1985	1973	1979	1985	1973	1979	1985
Andalucía	19.7	12.9	13.6	22.8	20.0	18.0	8.2	8.2	6.8	49.3	58.9	61.6
Aragón	16.8	11.2	8.8	32.6	28.9	32.0	6.1	6.7	5.2	44.5	53.2	54.0
Asturias	7.7	6.1	4.4	50.0	42.7	39.8	5.3	5.1	4.3	37.0	46.1	51.5
Baleares	7.2	4.3	2.6	43.1	12.6	11.0	8.9	7.0	6.5	70.8	76.1	79.9
Canarias	11.3	8.5	5.1	14.3	11.0	10.9	13.0	10.0	9.6	61.4	70.5	74.4
Cantabria	11.4	8.1	6.2	41.1	33.3	30.7	4.9	6.0	4.8	42.6	52.6	58.2
Castilla–La Mancha	31.1	20.6	16.9	22.7	22.4	23.9	8.1	10.0	8.8	38.1	47.0	50.4
Castilla y León	21.7	13.2	12.6	26.9	28.5	27.2	6.2	6.7	6.3	54.2	51.6	53.9
Cataluña	5.6	3.3	2.5	42.0	35.8	34.1	6.4	6.0	4.3	46.0	54.9	59.1
Comunidad Valenciana	11.4	6.9	5.1	32.5	29.4	28.4	7.6	7.1	5.7	48.5	56.6	60.8
Extremadura	29.3	17.6	16.9	17.5	17.8	16.2	8.0	9.0	8.2	45.2	55.6	58.8
Galicia	19.7	14.3	11.4	25.5	23.8	23.8	8.2	9.1	7.4	46.6	52.8	57.4
Madrid	1.0	0.6	0.3	26.6	21.9	19.7	6.9	5.9	4.2	65.5	71.6	75.8
Murcia	14.6	11.1	11.9	28.6	27.5	23.8	7.8	8.4	7.0	49.0	53.0	57.3
Navarra	16.7	10.8	7.3	36.3	37.2	35.3	5.5	6.3	5.5	41.5	45.7	51.9
País Vasco	4.7	3.7	2.3	49.7	44.4	44.0	5.2	4.0	3.8	40.4	47.9	49.9
Rioja, La	23.3	16.1	12.4	26.9	28.5	29.7	6.6	7.0	5.4	43.2	48.4	52.5
España	11.6	7.5	6.4	31.9	28.0	26.5	7.1	6.9	5.6	49.4	57.6	61.5

Source: Author's calculations based on figures from *Renta Nacional y Distrib. Prov.*, B.B.

Table 6A.4 ZUR: Investment and new jobs (at 31 December 1987)

ZUR	AP (1)	AI (2) (MPTAS)	IC (3) (MPTAS)	IA/AP (MPTAS)	PTS (4) NJC	NWC (5)	(IC/AI) 100	(NWC/NJC) 100	(AI/NJC) (MPTAS)
Asturias (11 May 1985) (*)	94	17,696	7,253	188.2	1,581	689	40.98	43.58	11.19
Barcelona (19 Dec. 1985)	150	57,934	29,764	386.2	4,411	2,518	51.37	57.08	13.13
Cádiz (8 Oct. 1985)	33	28,720	8,842	870.3	2,041	283	30.78	13.86	14.07
Galicia-Ferrol (24 July 1985)	32	12,777	4,804	399.3	868	240	37.60	27.65	14.72
Galicia-Vigo (24 July 1985)	80	14,375	4,852	179.7	1,735	684	33.75	39.42	8.28
Madrid (18 Sept. 1985)	69	70,364	34,043	1,019.8	3,698	1,676	48.38	45.32	19.02
Nervión (5 Nov. 1985) (**)	74	28,720	16,180	388.1	1,593	753	56.33	47.26	18.02
Total ZUR	532	230,586	105,738	433.4	15,927	6,843	45.85	42.96	14.47
ZPLI Sagunto (23 Dec. 1983)	71	31,852	25,800	448.6	2,381	2,300	80.99	96.59	13.38
Total	603	262,438	131,538	435.2	18,308	9,143	50.12	49.93	14.33

Source: Secretaría General Técnica, Ministerio de Industria y Energía.
ZUR = Urgent Reindustrialization Area.
ZPLI = Preferent Industrial Location Area.
(*) Date of starting.
(**) Without Sener Turbo Propulsión Project.
(1) Approved Projects.
(2) Approved Investments.
(3) Investments carried out.
(4) Number of jobs.
(5) New workers contracted.
MPTAS: Million Ptas.

Appendix to Chapter 8

PAUL CHESHIRE

The regions of the European Community

Eurostat, the official statistical office of the European Community, divides the territory of the member nations into regions for the purposes of reporting statistical data. This system of regional classification is, in turn, used by the European Commission for policy implementation. It is known, officially, as the 'Nomenclature of Territorial Units for Statistics', or more usually by its French acronym, NUTS. Regions are defined at three interrelated levels, in principle each level nesting with the one above it: Level 1 territorial units are usually made up of a whole number of Level 2 units and Level 2 units a whole number of Level 3 units. Ireland, the Grand Duchy of Luxembourg and Denmark are regarded as both Level 1 and 2 territorial units.

Table 8A.1 Correspondence between NUTS level and national administrative divisions

	Level 1	Level 2	Level 3
Belgium	régions	provinces	arrondissements
Denmark	—	—	Amter
Federal Republic of Germany	Länder	Regierungsbezirke	Kreise
Greece	Level 2 groupings	development regions	nomoi
Spain	Level 2 groupings	communidades autonomas	provincias
France	ZEAT	régions	départements
Ireland	—	—	planning regions
Italy	Level 2 groupings	regioni	provincie
Luxembourg	—	—	—
The Netherlands	landsdelen	provincies	COROP-Regio's
Portugal	Level 2 groupings	Level 3 groupings	groupings of concelhos
United Kingdom	standard regions	Level 3 groupings	counties, local authority regions

The problem with this system is that, as with much else in Europe, it represents an amalgam of very different historical traditions. Since each NUTS Level is composed of nationally defined spatial units and these units had different origins, were defined using different methodologies, at different dates and for different purposes, it is not surprising that the resulting regional system often provides a less than satisfactory basis for comparative analysis. The smallest practical units in the Federal Republic of Germany, the *Kreise*, are very large compared to, say, the French communes. There are 36,000 or more communes in France. The *Kreise* are, however, far smaller than the French *départements* or the British counties. The regional system in France and Italy has been stable for a very long time which makes it possible to analyse long-time series of data. Belgium, Denmark, the Federal Republic and the United Kingdom all redefined their basic units between 1970 and 1980, making it impossible to derive time series data in the Federal Republic, almost impossible in the United Kingdom (small area statistics are available as special tabulations allowing adjustment) and difficult in Belgium and Denmark. In short the European Community's regional system is an uneasy compromise. As a result both time series analysis and across country comparisons are extremely difficult and quite misleading without extensive interpretation or adjustment.

Notes

Notes to Chapter 1

1 Braudel, F. 1986. *The perspective of the world: civilization and capitalism 15th–18th century.* Vol. 3. New York: Harper & Row. pp. 21–44.
2 Morris, C. R. 1989. The coming global boom. *The Atlantic* **264**, (4). p. 56.
3 MacIver, R. M. 1957. Foreword. In K. Polanyi. *The great transformation.* Boston: Beacon Press, pp. ix–x.
4 Peter Hall. Structural transformation in the regions of the United Kingdom. This volume, Ch. 2, p. 58.
5 Ibid., pp. 41–2.
6 Ibid., pp. 45–6.
7 Ibid., pp. 46–9.
8 Ibid.
9 Ibid., p. 49.
10 Ibid., pp. 57–8.
11 Ibid., pp. 68–9.
12 Franz-Josef Bade and Klaus Kunzmann. Deindustrialization and regional development in the Federal Republic of Germany. This volume, Ch. 3, p. 72 and Fig. 3.1.
13 Ibid., p. 71.
14 Ibid., p. 98.
15 Jean-Paul de Gaudemar and Rémy Prud'homme. Spatial impacts of deindustrialization in France. This volume, Ch. 4, pp. 105–6.
16 Ibid., pp. 118–23.
17 Ibid., pp. 123–5.
18 Ibid., pp. 125–8.
19 Roberto P. Camagni. Regional deindustrialization and revitalization processes in Italy. This volume, Ch. 5, pp. 134–47.
20 Ibid., pp. 147–8.
21 Ibid., pp. 148–63.
22 Juan R. Cuadrado Roura. Structural change in the Spanish economy: their regional effects. This volume, Ch. 6.
23 Ibid., pp. 190–98.
24 Folke Snickars. Regional perspective of the deindustrialization of Sweden. This volume, Ch. 7, pp. 198–9.
25 Ibid., pp. 206–8.
26 Ibid., p. 232.
27 Ibid., pp. 211–15.
28 Ibid., pp. 208, 216.
29 Ibid., p. 218.
30 Ibid., pp. 215–18.

31 Ibid., pp. 223–4.
32 Ibid., pp. 224–8.
33 Paul Cheshire. Problems of deindustrialization in the European Community. This volume, Ch. 8, p. 241. See also ibid., note 5 for further detail on the adjustments necessary when component data were not available.
34 Ibid., pp. 248–57.
35 Ibid., pp. 249–50.
36 Ibid., pp. 264–5.
37 Ibid., p. 264.
38 Ibid., pp. 239–48.
39 Ibid., pp. 262–3.
40 Ibid., p. 253.
41 Ibid., pp. 266–7.
42 Paul Cheshire, Roberto P. Camagni, Jean-Paul de Gaudemar, and Juan R. Cuadro Roura. 1957 to 1992: Moving toward a Europe of regions and regional policy. This volume, Ch. 9.
43 Ibid., p. 269.
44 Ibid., p. 277.
45 Ibid., pp. 268–78.
46 Ibid., pp. 281–2.
47 Ibid., pp. 284–5.
48 Ibid., pp. 286–8.
49 Ibid., p. 295.
50 Roberto P. Camagni, Paul Cheshire, Jean Paul de Gaudemar, Peter Hall, Lloyd Rodwin and Folke Snickars. Europe's regional-urban futures: conclusions, inferences and surmises. This volume, Ch. 10.
51 Ibid., p. 308.
52 Ibid.
53 Ibid., p. 310.
54 Letter from Paul Cheshire to Lloyd Rodwin, 29 December, 1989.
55 This volume, Ch. 10, p. 309.
56 Ibid., pp. 312–15.
57 Ibid.
58 Ibid.
59 Ibid.
60 Cf. this volume, Ch. 5; also Chs. 2, 3 & 6.
61 See the chapters cited above. Cf. also Rodwin, Lloyd. 1970. *Nations and cities*. Boston: Houghton Mifflin. Chs. I, V, VI, VII. However, Paul Cheshire has pointed out in a letter to me (29 December 1989): It is true to describe Northern England as Britain's poor region, but I am not quite comfortable with the way you lump Britain's north with the poor south of Italy or Spain. The difference in per capita incomes within the UK is not that great. Excluding cities since they distort things, in 1985 Cleveland and Durham – the poorest English level 2 region – had a GDP per capita of 94.5% of the EC and Berkshire, Buckinghamshire and Oxford – the richest level 2 region – was 102.7. In Italy the range for level 2 regions was from 119 in Lombardia to 54.5 in Calabria. Also, Britain's "north" was rich until the First World War; Calabria has probably always been poor and Naples has been poor for 150 years or so. The main differences between regions in Britain – reflecting the fact that our regional problems are primarily not related to agriculture – are in unemployment.
62 This volume, Ch. 2, pp. 54–8, 67.

63 This volume, Ch. 4, p. 144.
64 This volume, Chs. 2–7.
65 This volume, Chs. 2, 3, 5, 6.
66 McClelland, David C. 1961. *The achieving society*. Princeton, N.J.: Van Nostrand. Chs. 1–3, 6–7.
67 J. C. Schumpeter, 'The creative response in economic history', *Journal of Economic History*, November 1947, vol. 7, no. 2, pp. 149–59.
68 Harrison, Bennett. *The big firms are coming out of the corner.* (unpublished paper).
69 Gras, N. S. 1922. *An introduction to economic history.* New York: Harper. Ch. 5.
70 Paul Cheshire has a reservation on this language. He says: 'I do not think that "inadequate and deteriorating public services"' is necessarily related. I think it mainly reflects an ideological trend – although there is a closer relationship in the US where more such services are funded at state and local level out of local revenues. The norm in Europe is for much more of local spending to be funded out of nationally raised revenues which breaks the link; and we have the example of Sweden, and to a lesser extent Germany and France, where there is deindustrialization which maintained public spending. The ' exceptions" are the US and UK, where in my view the explanation is primarily ideological'. (Letter to Lloyd Rodwin, 29 December, 1989.
71 Harrison, B. & B. Bluestone. 1988. *The great U turn.* New York: Basic Books; Thurow, L. 1987. The surge in inequality. *Scientific American* **255** (5).
72 Schultz, Charles Z., Barry Bluestone, Benjamin M. Friedman, Robert Z. Lawrence and Bennett Harrison. 1985. Who pays for economic change?. *Harpers.* February, pp. 37, 38–9, 42; Lawrence, Robert A. 1984. *Can America compete?* Washington, D.C.: Brookings Institution. Chs. 2–4.
73 On this score, we share the conviction of William James:

> Pretend what we may, the whole man within us is at work when we form our philosophical opinions. Intellect, will, taste, and passion co-operate just as they do in practical affairs; and lucky it is if the passion be not something as pretty as love of personal conquest over the philosopher across the way. The absurd abstraction of intellect verbally formulating all its evidence, and carefully estimating the probability thereof by a vulgar fraction, by the size of whose denominator and numerator alone it is swayed, is ideally as inept as it is actually impossible. It is almost incredible that men who are themselves working philosophers should pretend that any philosophy can be, or even has been, constructed without the help of personal preference, belief, or divination. How have they succeeded in so stultifying their sense for the living facts of human nature as not to perceive that every philosopher, or man of science either, whose initiative counts for anything in the evolution of thought, has taken his stand on a sort of dumb conviction that the truth must lie in one direction rather than another, and a sort of preliminary assurance that his notion can be made to work; and has borne his best fruit in trying to make it work? (James, William. 1897. *The will to believe and other essays.* New York: Longmans & Green. pp. 92–3.)

74 Peter Hall, *op. cit.*, pp. 49–53.
75 Franz Josef Bade and Klaus R. Kunzmann, *op. cit.*, pp. 4–6, 77–8.
76 Roberto P. Camagni, Abstract of an earlier version of the chapter published in this volume.

77 Thurow, Lester. 1989. 'Regional transformation and the service activities'. In Lloyd Rodwin and Hidehiko Sazanami (eds.). *Deindustrialization and regional economic transformation*. Boston: Unwin Hyman. pp. 3–25.
78 Holmes, O. Wendell. Lochner v. New York, 198 U.S. 45 (1905). Dissenting Opinion of Justice Holmes.
79 Other studies expressing parallel views include Prud'homme, Rémy. 1989. L'industrie a la rescousse. *Le Monde*. 12 January. pp. 17–18; also T. J. Noyelle, 'The rise of advanced services', *Journal of the American Planning Association*, 1983; and I. Levenson, 'The service economy in economic development: perspectives from the United States' (New York, Hudson Strategic Group, Inc., 1985).
80 Harrison. op. cit. See also Piore, M. J. & C. F. Sabel. 1984 *The second industrial divide*. New York: Basic Books, pp. 3–48.
81 Harrison, op. cit.
82 Harrison and Bluestone, B., 1988. op. cit.
83 Rodwin. 1989. op. cit. p. 20.
84 This volume, Ch. 8, pp. 262–3.
85 Ibid.
86 Rodwin. 1989. op. cit. p. 20.
87 Greenhouse, S. 1989. Europeans united to compete with Japan and the U.S. *The New York Times*, 21 August, 1989 (Vol CXXXVIII, No. 47969, pp. 1 and D8.)
88 This volume, Ch. 9. p. 281.
89 Alonso, W. 1990. Regional issues and European integration. Paper prepared for the International Congress on Regional Policy in the Europe of the Nineties, Madrid, 30 May – 3 June, 1989, p. 3.
90 Supra, p. 13. In the future, the most serious problem of migration flows may be from Eastern Europe, especially the German Democratic Republic.
91 Pike, John. A policy analysis for the Washington Based Federation of American Scientists. Cited in Greenhouse. 1989. op. cit.

Notes to Chapter 2

1 Martin, R. L. 1986a. Thatcherism and Britain's industrial landscape. In R. L. Martin & R. Rowthorn (eds.). (1986) *The Geography of De-industrialisation*. London: Macmillan, 240. Rowthorn, R. E. 1986. *De-industrialisation in Britain*. In Martin & Rowthorn (eds.). op. cit. pp. 2, 4–5, 7.
2 Rhodes, J. 1986. Regional dimensions of industrial decline. In Martin & Rowthorn (eds.). op. cit. pp. 142–3.
3 Martin. 1986a. op. cit. p. 261; Rhodes 1986. op. cit. p. 148; Martin, R. L. 1988. The policital economy of Britain's north–south divide. *Transactions of the Institute of British Geographers* **13** (n.s.) 396.
4 Ibid., p. 396.
5 Rhodes. 1986. op. cit. pp. 163–7.
6 Goddard, J. B. & A. E. Gillespie. 1987. Advanced telecommunications and regional economic development. In B. Robson (ed.). (1987) *Managing the City: the aims and impacts of Urban Policy*. London: Croom Helm. p. 87.
7 Martin. 1988. op. cit. p. 399.
8 Ibid.
9 Champion, A. & A. Green. 1987. *Local prosperity and the north–south divide:*

a report on winners and losers in 1980s Britain. Warwick: Institute of Employment Research, University of Warwick. p. 15.
10 Ibid. p. 16.
11 Hall, P., M. Breheny, R. McQuaid & D. Hart. 1987. *Western sunrise: the genesis and growth of Britain's major high tech corridor*. London: Allen & Unwin.
12 Cooke, P. 1987. Britain's new spatial paradigm: technology, locality and society in Transition. *Environment and Planning A* **19** 1293; Martin. 1988. op. cit. pp. 393–4.
13 Cooke, 1987. op. cit. pp. 1295–9; Scott, A. J. 1988. Flexible production systems and regional development: the rise of new industrial spaces in North America and Western Europe. *International Journal of Urban and Regional Research* **12** 179–80, 182, 186.
14 Begg, I. & G. C. Cameron. 1988. High technology location and the urban areas of Great Britain. *Urban Studies* **25** 365.
15 Ibid., pp. 368, 375.
16 Ibid., p. 365.
17 Hall, Breheny, McQuaid, & Hart. 1987. op. cit. p. 27.
18 Sayer, A. & K. Morgan. 1986. The electronics industry and regional development in Great Britain, In A Amin & J. B. Goddard (eds.). (1986) *Technological Change, Industrial Restructuring and Regional Development*. London: Allen & Unwin. p. 166.
19 Ibid., pp. 167–8.
20 Ibid., pp. 169–70.
21 Marshall, J. N. et al. 1988. *Services and uneven development*. Oxford: Oxford University Press. p. 27.
22 Ibid., p. 41.
23 Ibid., p. 63.
24 Gillespie, A. E. & A. E. Green, The Changing Geography of Producer Services Employment in Britain. *Regional Studies* **21** 1988, 404–5; Marshall et al., 1988. pp. 76, 79–80, 82.
25 Daniels, P. 1986. Producer services and the post-industrial space economy. In Martin & Rowthorn (eds.). op. cit. p. 306.
26 Gillespie & Green. op. cit. pp. 406, 409.
27 Marshall, et al. 1988. op. cit. p. 73.
28 Daniels. 1986. op. cit. p. 300.
29 Marshall et al. 1988. op. cit. p. 175.
30 Ibid., pp. 177–9, 198–200.
31 Ibid., pp. 306–9.
32 Ibid., pp. 310–11.
33 Howells, J. 1987. Developments in the location, technology and industrial organization of computer services: some trends and research issues. *Regional Studies* **21** 497–8.
34 Goddard & Gillespie. 1987. op. cit. p. 86.
35 Ibid., p. 96.
36 Ibid., p. 101–3.
37 Morris, J. L. 1988. Producer services and the regions: the case of large accountancy firms. *Environment and Planning A* **20** 751.
38 Goddard, J. B., A. Thwaites & D. Gibbs. 1986. The regional dimension to technological change in Great Britain. In Amin & Goddard (eds.). op. cit. p. 146.
39 Ibid., p. 151.

40 Ibid.
41 Harris, R. I. D. 1988. Technological change and regional development in the UK: evidence from the SPRU database on innovations. *Regional Studies* **22** 365.
42 Ibid. p. 369.
43 Storey, D. & M. Johnson. 1987. Regional variations in entrepreneurship in the UK. *Scottish Journal of Political Economy* **34** 172.
44 Ibid., pp. 167, 171.
45 Storey, D. J. 1986. The economics of smaller businesses: some implications for regional economic development. In Amin & Goddard (eds.). op. cit. p. 227; Storey & Johnson. 1987. op. cit. p. 172.
46 Massey, D. 1984. *Spatial Divisions of Labour: Spatial Structures and the Geography of Production*. London: Macmillan.
47 Goddard & Gillespie. 1987. op. cit. p. 89.
48 Ibid.
49 Hudson, R. 1986. Producing an industrial wasteland: capital, labour and the state in north-east England. In Martin & Rowthorn (eds.). op. cit. p. 196.
50 Marshall, M. 1986. *Long waves of regional development*. London: Macmillan. p. 220.
51 Ibid. p. 207.
52 Hall, P. 1989. *London 2001*. London: Unwin Hyman. pp. 51–2.
53 Buck, N., I. Gordon & K. Young. 1986. *The London Employment Problem*. Oxford: Clarendon Press. pp. 66–7.
54 Hall, 1989. op. cit. p. 55.
55 Hall, Breheny, McQuaid & Hart. op. cit. 1987. Chs. 7 & 8.
56 Scott, A. J. 1986. High Technology Industry and Territorial Development: the Rise of the Orange County Complex, 1955–1984. *Urban Geography* **7**. pp. 3–45.
57 Boddy, M., J. Lovering & K. Bassett. 1986. *Sunbelt City? A study of economic change in Britain's M4 corridor*. Oxford: Clarendon Press. p. 7.
58 Ibid., p. 10.
59 Ibid., p. 14.
60 Ibid., p. 20.
61 Ibid., pp. 207–9.
62 Ibid., p. 211.
63 Buck, N., & I. Gordon, 1987. The Beneficiaries of Employment Growth: An Analysis of the Experience of disadvantaged groups in expanding labour markets. In Hansner, V. A. (ed.). *Critical issues in Urban Economic Development*. Vol. 2. Oxford: Clarendon Press. pp. 111–12, 194.
64 Grafton, D. J. & N. Bolton. 1987. Counter-urbanisation and the rural periphery: some evidence from north Devon. In Robson (ed.). op. cit. pp. 200–3.
65 Spencer, K. et al. 1986. *Crisis in the industrial heartland: a study of the West Midlands*. Oxford: Clarendon Press. pp. 56–7.
66 Ibid., pp. 57–64.
67 Ibid., p. 66.
68 Ibid., pp. 17, 29.
69 Ibid., p. 69.
70 Ibid., p. 110.
71 Martin. 1986a. op. cit. p. 266.
72 Spencer et al. 1986. op. cit. pp. 90–1.
73 Ibid., pp. 123–4, 127.

74 Ibid.
75 Lever, W. & C. Moore (eds.). 1986. *The city in transition: policies and agencies for the economic regeneration of Clydeside.* Oxford: Oxford University Press. P. 1.
76 Ibid., p. 2.
77 Ibid., p. 3.
78 Ibid., p. 22.
79 Ibid., pp. 23, 28.
80 Ibid., pp. 10–11.
81 Ibid., p. 18.
82 Ibid., pp. 18–19.
83 Ibid., pp. 10–11, 18–19.
84 Ibid., p. 19.
85 Martin. 1986a. op. cit. p. 283.
86 Martin, R. L. 1986b. In what sense a jobs boom? Employment recovery, government policy and the regions. *Regional Studies* **20** 466.
87 Ibid., pp. 466–7.
88 Martin. 1986a. op. cit. pp. 280–1.
89 Ibid. p. 281.
90 Insofar as there were significant regional differences, they often cut across the grain of the traditional ones. At least as important as the old north–south divide was a new east–west one, between the regions and cities that looked westward to the old Atlantic economy (the North West: Glasgow, Liverpool) and those that looked east to the new economic opportunities within the European Community (East Anglia: Harwich, Felixstowe).

Notes to Chapter 3

1 Rodwin, Lloyd. 1989. Deindustrialization and regional economic transformation. In Lloyd Rodwin & Hidehiko Sazanami (eds.). *Deindustrialization and regional economic transformation. The experience of the United States.* Boston: Unwin Hyman. pp. 3–25.
2 Läpple, D. 1986. Süd-Nord-Gefälle. In Friedrichs, J., Häussermann, H., Siebel, W. 1986. Süd-Nord-Gefälle in der Bundesrepublik? Opladen. Sinz, M., Strubelt, W. 1986. Zur Diskussion über das wirtschaftliche Süd-Nord-Gefälle unter Berücksichtigung entwicklungsgeschichtlicher Aspekte, in: Friedrichs, Häussermann and Siebel, (eds.). op. cit. 1986.
3 Stanback, Thomas M., Jr. & Thierry J. Noyelle. 1988. Productivity changes and employment growth in the services. In Colloque International "Productivité et valorisation". Paris. 1–3 June 1988. In Thurow, Lester, C. 1989. Regional transformation and the service activities. In Rodwin & Sazanami. (eds.). op. cit. pp. 179–98.
4 Bluestone, Barry & Bennett Harrison. 1986. *The great American jobs machine. The proliferation of low wage employment in the U.S. Economy.* Washington: Joint Economic Committee.
5 Krupp, Hans-Jürgen. 1986. Der Strukturwandel zu den Dienstleistungen und Perspektiven der Beschäftigungsstruktur. *Mitteilungen für Arbeitsmarkt- und Berufsforschung* **1** 145–58.
6 Fels, Gerhard & K. -W. Schatz. 1974. Sektorale Entwicklung und Wachstumsaussichten der Westdeutschen Wirtschaft bis 1980. *Die Weltwirtschaft* **1**.

7   Ochel, Wolfgang & Paul Schreyer. 1988. *Beschäftigungsentwicklung im Bereich der privaten Dienstleistungen USA – Bundesrepublik Deutschland im Vergleich*, Berlin (Schriftenreihe des Ifo-Instituts für Wirtschaftsforschung, Nr. 123).
8   Gershuny, Jonathan I. & I. Miles. 1983. *The new service economy. The transformation of employment in industrial societies.* London: Frances Pinter; Gershuny, Jonathan, I. 1984. The future of service employment. IIM/LMP-discussion paper 84/7, Wissenschaftszentrum Berlin.
9   Clark, Colin. 1957 (3rd edn.). *The conditions of economic progress.* London. Fisher, A. G. B. 1939. Production, primary, secondary, tertiary. *Economic Journal.* **15**.
10 Bell, Daniel. 1973. *The coming of postindustrial society.* New York.
11 Porat, Mare Uri. 1976. *The information society.* Stanford.
12 Thurow. 1989. op. cit.
13 Müller, Jürgen. 1983. *Sektorale Struktur und Entwicklung der industriellen Beschäftigung in den Regionen der BRD.* Berlin: Beiträge zur angewandten Wirtschaftsforschung. Bd. 12; Bade, Franz-Josef 1987. *Regionale Beschäftigungsentwicklung und produktionsorientierte Dienstleistrungen.* Berlin: Deutsches Institut für Wirtschaftsforschung. Sonderheft 143.
14 Petzina, D. 1986. Wirtschaftliche Ungleichgewichte im Deutschland. *Der Bürger im Staat* **4** 267–74.
15 Institut für Medienforschung und Urbanistik. 1988. *Stuttgart – Problemregion der 90er Jahre?* IMU Studie 7.
16 Mikat, Paul. 1989. *Bericht der Kommission Montanregionen des landes Nordrhein-Westfalen.* Düsseldorf: Hg. Minister für Wirtschaft, Mittelstand und Technologie des Landes Nordrhein-Westfalen.
17 Kunzmann, Klaus R. 1988. Military production and regional development in the Federal Republic of Germany. In Michael Breheny. 1988. *Defence Expenditure and Regional Development.* London, Oxford: Alexandrine Press.
18 Sachverständigenrat. 1988. *Jahresgutachten 1988/89 des Sachverständigenrates zur Begutachtung der gesamtwirtschaftlichen Entwicklung.* Bonn.
19 Jacoby, Herbert. 1987. Sind die Löhne im Ruhrgebiet zu hoch? In *Mitteilungen des Arbeitskreises für Raumplanung* (Dortmund) **37**.
20 Kunzmann, Klaus R. 1986. Structural problems of an old industrial area: the case of the Ruhr district. In Walter H. Goldberg (ed.). *Ailing Steel.* Aldershot: Gower.
21 Kommunalverband Ruhrgebiet. 1987. *Kultur und Wirtschaft im Ruhrgebiet, Ergebnisse einer Unternehmensbefragung.* Essen.
22 Hennings, Gerd, Klaus R. Kunzmann & Klaus R. Kunzmann. 1988. Pittsburgh, nothing is more successful than success. Unpublished working paper. Institut für Raumplanung, Universitat Dortmund.
23 Ministerium für Wirtschaft, Mittelstand und Technologie. 1987. *Wirtschaftsförderungsprogramm Baden-Württemberg.* Stuttgart.
24 Ludwig, E. 1989. Erweiterung der Universität auf dem Eselsberg hat Modellcharakter. *Die Welt.* October 6, 1989.
25 Späth, L. 1989 (2nd edn.). *Wende in die Zukunft.* Hamburg: Rowohlt.

Notes to Chapter 4

1. Donnellier, J. C. et al. 1987. *L'appareil productif régional de 1975 à 1984. Une analyse de la valeur ajoutée.* Paris: INSEE (Coll. de l'INSEE R64); Jayet, H. (ed.). 1988. *L'espace économique français.* Paris: INSEE; A. Lopez. 1987. Un scénario pour l'emploi régional. *Economie et Statistiques.* December 1987; Uhrich, R. 1987. *La France Inverse? Les régions en mutation.* Paris: Economica.
2. Zindeau, B. 1987. Politique régionale et création d'emploi industriels: enseignement d'un modèle multi-régiona, *Revue d'Econoimie Régional et Urbaine.* **5** 753–67; Mathieu, E. 1983. *Aides régionales et structures industrielles. Sept années de primes de développement régional (1974–80).* Paris: Ministère de l'Industrie, Direction Générale de l'Industrie, Service d'Etude des Stratégies et des Statistiques Industrielles **29**; Gillouard, A. 1985. *Les primes régionales à la création d'entreprises et d'emplois: mesure et analyse de leur effet incitatif.* Rennes: CREFE. Charvet, E. 1986. Aide au développement régional et localisations industrielles dans le Grand Sud-Ouest, 1967–82. *Revue d'Economie Régionale et Urbaine* **1986–2** 219–45.
3. Zindeau. 1987. op. cit.
4. Charvet, 1986. op. cit.
5. Davezies, L. 1989. *La redictribution interdépartementale des revenus induite par le budget de l'Etat 1984.* Créteil: L'OEIL, Université de Paris XII.
6. Northeast is defined as the following five contiguous regions: Nord–Pas-de-Calais, Picardie, Champagne–Ardennes, Lorraine, and Franche-Comté. Alsace, which belongs geographically to the Northeast, has been excluded, because this region's economic behaviour has been completely different from that of the other five regions of the North East. South is defined as the following five contiguous regions: Aquitaine, Midi–Pyrénées, Languedoc–Roussillon, Provence–Côte-d'Azur, and Corse. The Paris region consists of only one region, Ile-de-France.
7. Graduate or post-graduate degree means a post-high school qualification (i.e. higher than the *baccalauréat*). The figures quoted are from INSEE. *Statistiques et indicateurs des régions françaises 1988.* p. 399.
8. Figures for migrations are only available for the 1975–82 period. Net migrations were positive for most French regions. They were negative for (a) the Paris region; (b) four out of the five regions of what we defined as the Northeast (Nord–Pas-de-Calais, Lorraine, Champagne–Ardennes, and Franche-Comté); and (c) two additional regions, Haute–Normandie and Basse–Normandie. See INSEE. *Statistiques et indicateurs des régions françaises 1988.* p. 166.
9. See INSEE. *Statistiques et indicateurs des régions françaises.* 1988. pp. 201, 214. The figures are as follows for 1983:

	Researchers (thousands)	Labour force (thousands)	Researchers as percentage of labour force
France	63.3	21,472	2.9
France excluding Paris region	26.6	16,910	1.5
Paris region	36.2	4,562	7.9
Northeast	2.9	3,831	0.7
South	10.3	3,946	2.6

10   There are five regions in our South, but data for Corse, which do not weigh much anyway, were not available for the shift–share analysis.
11   "Personal and industrial services" is an imperfect translation for *services marchands* (literally: traded services), a French Statistical Office concept that refers to services minus transportation, telecommunications, trade, finance and insurance, and non-traded services.
12   The rates of change of industrial employment for the Paris region and the rest of France for different periods were as follows (in percent per year):

	1986/82	1975/68	1982/75	1987/82
Paris region	−0.8	−0.3	−2.0	−2.5
Rest of France	0.7	1.6	−1.2	−2.1

See IAURIF. *Bilan de l'évolution des emploi en Ile de France et en province de 1954 à 1987*. Paris: IAURIF. p. 10.
13   The exact figures are 9.6 for value added and 9.7 for employment, and they refer to "automobile and land transportation equipment," including therefore bicycles, motorcycles, and railroad cars; see INSEE. 1988, *Les comptes de l'industrie – la situation de l'industrie française en 1987*. Paris: INSEE. p. 121.
14   Thurow, L. 1989. Regional transformation and the service activities. In L. Rodwin & H. Sazanami (ed.). *Deindustrialization and regional economic transformation. The experience of the United States*. Boston: Unwin Hyman. pp. 179–98.

Notes to Chapter 5

1   Camagni, R. 1980. Modèles de réstructuration économiques dans les régions européenes pendant les années '70. In P. Aydalot (ed.). *Crise et espace*. Paris: Economica.
2   Camagni, R. & R. Cappelin. 1985. *Sectoral productivity and regional policy*. Bruxelles: Commission of the European Communities. Document 92–825–5535–6.
3   Ibid.
4   Knight, R. V. 1984. The advanced industrial metropolis: a new type of world city. Paper presented at the Seminar on The Future of the Metropolis. October.
5   Tornqvist, G. 1984. Creativity and regional development. Paper presented at the Seminar on Rédeploiement industriel et aménagement de l'espace. Montreal. September; Anderson A. 1985. Creativity and regional development. *Papers and Proceedings of the Regional Science Association* **56**.
6   Verdoorn, P. J. 1951. On an empirical law governing the productivity of labour. *Econometrica* **19**; Verdoorn, P. J. Verdoorn's law in retrospect: a comment. *Economic Journal* **90**; Kaldor, N. 1970. The case for regional policies. *Scottish Journal of Political Economy* **3**; Kaldor, N. 1975. Economic growth and the Verdoorn law. A comment on Mr. Rowthorn's article. *Economic Journal* **85**; Dixon, R. J. & A. P. Thirlwall. 1975. A model of regional growth differences on Kaldorian lines. *Oxford Economic Papers* **27**.
7   A proof of this comes from the results of a shift–share analysis on regional development in Europe in recent years. If the demand view of regional growth were correct, the mix effect should account for most of relative regional growth (the "total shift"). However, almost all recent experiments show that

the mix effect no longer plays a role in the relative performance of the regions (with the exception of a few that are highly specialized in declining sectors, such as Lorraine and Wallonia in Europe) while the "differential effect" or "competitiveness effect", linked to the innovativeness and dynamism of the local fabric explains it almost entirely. See Camagni, R. & R. Cappelin. 1981. Policies for full employment and more efficient utilisation of resources and new trends in European regional development. *Lo spettatore internazionale* 2; Camagni & Capellin. 1985. op. cit.

Only in the 1950s and 1960s, when the differences in regional productive structures were more evident and the relative growth rates of the different sectors more differentiated, was the mix effect significant. In the 1960s it was sufficient, for example, to counterbalance the effects of a diminishing sectoral competitiveness in the core areas. See Camagni, R. & R. Capellin. 1980. Struttura economica regionale e integrazione economica europea. *Economica e Politica Industriale* 27.

In the case of the Italian regions, Verdoorn's law was carefully tested by Soro through an econometric analysis. (See Soro, B. 1986. Crescita della produttività, dell'occupazione e della produzione manifatturiera nell'esperienza regionale italiana. In R. Camagni & L. Malfi (eds.). 1986. *Innovazione e sviluppo nelle region: mature*. Milan: F. Angel.) Let us start from a production function of the type:

$$(Y/N)_t = A Y_t^v e^{pt}$$

where Y and N are value added and employment and p and v are respectively the "autonomous" growth rate of productivity and the income elasticity of productivity (the Verdoorn coefficient, hypothesized close to 1/2); taking the log derivatives with respect to time:

$$y - n = vy + p$$

we get finally the expression that can be estimated econometrically:

$$n = (1 - v) y - p$$

where the small letters indicate the growth rates of the respective variables. In the period 1973–81 this relationship was statistically significant, at the national level, in 11 industrial sectors out of 13 (but not for the manufacturing sector as a whole) and only in 11 regions out of 18 in the manufacturing sector. However, even in the cases where the relationship was statistically significant, the autonomous growth of productivity was proved to be of overwhelming importance particularly in respect to product growth, which accounted for about one-third of total productivity growth.

8 Camagni, R. 1980. Il mutamento strutturale nell'industria di una regiona europea. *Economia e Politica Industriale* 26; Camagni & Cappelin. 1985. op. cit.

9 In analytical terms, where P means productivity, as usual, and i the single sectors (no subscript indicating all sectors together):

$$(P^1/P^0)_r - (P^1/P^0)_n = \quad \text{(total shift)}$$
$$= \Sigma_i (Y^0_{ir}/Y^0_r) \left[(P^1_i/P^0_r - P^1_{ir}/P^0_{ir}) + (P^1_{ir}/P^0_{ir} - P^1_{in}/P^0_{in}) + (P^1_{in}/P^0_{in} - P^1_n/P^0_n) \right]$$
$$\text{(reallocation effect)} \quad + \quad \text{(differential effect)} \quad + \quad \text{(mix effect)}$$

It is easy to understand that the sum of the three effects gives the "total shift;" and it is also easy to forecast that the new reallocation effect may prove important in periods of rapid structural changes.

10 Osservatorio Economico territoriale dell'Area Metropolitana Milanese (OETAMM). 1988. Strutttura economica e dinamica recente dell'area metropolitana milanese. Documento 2.
11 Camagni, R. & R. Rabellotti. 1988. Technology, innovation and industrial structure in the textile industry in Italy. Paper presented at Conference on The Application of New Technologies in Existing Industry. University of Newcastle-upon-Tyne. March.
12 Camagni. 1984. op. cit.; Camagni, R. & R. Capello. 1988. Italian success stories of local development: theoretical conditions and practical experience. In W. Stöhr (ed.). *Success stories of local development*. London: Cassell Tycooly (United Nations University International Project).
13 Camagni, R. 1985. Spatial diffusion of pervasive process innovation. *Papers and Proceedings of the Regional Science Assocation* **58**.
14 Camagni, R. 1988a. Spatial implications of technological diffusion and economic restructuring in Europe with special reference to the Italian case. In *Urban development and impacts of technological, economic and sociodemographic changes*. Paris: OECD (Report of an Expert Meeting, June).
15 OETAMM. 1988. op. cit.
16 Freeman, C. & C. Peres. 1986. The diffusion of technical innovations and changes of techno-economic paradigm. Paper presented at the International Conference on Innovation Diffusion, Venice (to be published by Oxford University Press in a volume edited by F. Arcangeli et al.).
17 I.Re.R. – Progetto Milano. 1988. *La trasformazione economica della città*. Milano: Franco Angeli.
18 Camagni. 1988a. op. cit.; Camagni, R. 1988b. Functional integration and locational shifts in the new technology industry. In P. Aydalot & D. Keeble (eds.). 1988. *High technology industry and innovative environments*. London: Routledge.
19 Pichierri, A. 1989a. *Strategie contro il declino in aree di antica industrializzazione*. Torino: Rosenberg & Sellier.
20 Pichierri, A. 1989b. Crisi e ristrutturazione dell'industria siderurgica. In R. Catanzaro & V. Nanetti (eds.). *Politica in Italia*. Bolgna: Il Mulino.
21 Pichierri, A. 1986. Diagnosis and strategy in the decline of the European steel industry. Berlin: IIMV-IIM Wissenschaftszentrum (Discussion paper **22**).
22 OETAMM. 1988. op. cit.

Notes to Chapter 6

1 The idea of deindustrialization has been heavily supported by the evolution of employment in the many branches of industrial production and by the commercial balance of several European countries. However, the interpretation of this concept has given place to well-known controversy. Tertiarization is another highly questioned concept and the object of interesting debates. It is enough to compare, for example, the works of Gemmel, N. Economic development and structural change: the role of the service sector. *Journal of Development Studies* **19**; Noyelle, T. J. 1983. The rise of advanced services.

A. P. A. *Journal*; Gershuny, J. I. & I. Miles. 1983. *The transformation of employment in industrial societies*. F. Pinter, London: Frances Pinter. See also Cuadrado, J. R. & C. Del Río. 1989. Structural change and evolution of the service sector in the OECD. *The Service Industries Journal* **9** (3). pp. 439–68.

2 See Myro, R. 1989. La industria. Expansión, crisis y reconversión. In J. García. *España. Economía*, Madrid: Espasa-Calpe. Ch. 5. The concepts of weak, medium and strong demand have been used to compare and analyse some European economies. (See *Economie Européene*. 1985. **25**).

3 In 1974, US income per capita was already only 3.04 times that of Spain, the United Kingdom's was 1.57 bigger and Italy's 1.37 bigger. The improvement with respect to 1969 was, therefore, important.

4 General Franco died in November 1975. In December 1978 the new Constitution was passed. In the intermediate period, diverse political events occupied the scene, among which were the consolidation of political parties and the satisfaction of the claims for regional autonomy that gave way to a new quasi-federal state.

5 The Moncloa Pacts (December 1977) constitute the expression of this agreement, including a wide program of other structural and reforming measures arising from this.

6 These last figures, according to some recent studies, are overvalued, since a large volume of subterranean employment existed, the mean average of which could reduce the 1985 unemployment rate by 1.5 per cent.

7 See Cuadrado Roura, J. 1987. *Los desequilibrios regionales y el Estado de las Autonomías*, Barcelona: Edit. Orbis; and Cuadrado Roura, J. 1988. Tendencias económico-regionales antes y después de la crisis en España. *Papeles de Economía Española* **34** 16–61.

8 The analysis has been made with data in constant pesetas (1980). Eleven branches were differentiated within industry in each region and an identical analysis was developed with figures at provincial level.

9 See Cuadrado Roura, J. & J. Aurioles. 1989. *La localización industrial en España*. Madrid: Edit. FIES, Estudios. See also Cuadrado Roura, J. 1989. Facteurs de localisation industrielle. Nouvelles tendences *Revue d'Economie Régionale et Urbaine* **3** 471–90.

10 See: Klaasen, L. H. and W. Molle. 1986. *Industrial mobility and migration in the European Community*. Aldershot: Gower; Keeble, D. E. and E. Wever. 1986. *New firms and regional development in Europe*. London: Croom Helm.

11 The intensity indicator of the placement of investments has been calculated from information and data on new investments and amplifications or renewed investments relative to the weight of each province and considering it with the weight of relative industrial movement of each province with the national total. See: Cuadrado J. and Aurioles, J: *La localizacion industrial en España*. Col. Estudios. FIES, Madrid. 1989.

12 Giraldez, E. 1988. Comportamiento inversor de los sectores de alta tecnología 1975–85. Tendencias espaciales. *Papeles de Economía Española* **34** 431–53.

13 See Vazquez, J. A. 1988. Regiones de tradición industrial en declive: la Cornisa Cantábrica. In J. L. García (ed.). *España. Economía*. Madrid: Edit. Espasa–Calpe. Ch. 20. Del Castillo, J. 1989. El País Vasco como región industrializada en declive. In SPRI: *Regiones europeas de antigua industrialización*. Bilbao.

14 Pedreño, A. 1988. Un eje de expansión económica: Cataluña–Mediterráneo. In J. L. García (ed.). *España. Economía*. Madrid: Edit. Espasa–Calpe. Ch. 21.

Notes to Chapter 7

1. See Snickars, F. et al. 1989. *Chances for North Bothnia*. Luleå: North Bothnian Learning Association (in Swedish).
2. See, e.g., Johansson, B. & U. Strömquist. 1988. *Technology diffusion and import substitution: the role of the Stockholm region in technical change*. Stockholm: The Office of Regional Planning and Urban Transportation, Stockholm County Council, and Regional Economic Department, Stockholm County Board (in Swedish).
3. See, e.g., Bluestone, B. & B. Harrison. 1982. *The deindustrialization of America*. New York: Basic Books.
4. See Andersson, Å. E. & U. Strömquist. 1988. *The Future of the K-Society*. Stockholm: Prisma (in Swedish).
5. See Ståhlberg, L. 1987. *The private service sector: tendencies and prospects*. Stockholm: Swedish Industrial Board (in Swedish).
6. See Snickars et al. 1989. op. cit.
7. The solidary wage policy means that wages should be equalized among establishments operating in any sector.
8. See Johansson, B. & U. Strömquist. 1982. *Productivity and profits in Swedish manufacturing industry*. Stockholm: Swedish International Board (in Swedish).
9. See Johansson, B. 1988. *The future labour market: Dynamic competition and regions*. Stockholm: Council of Job Security, Swedish Employers' Association.
10. See Andersson & Strömquist. 1988. op. cit.
11. See Jacobs, J. 1984. *Cities and the wealth of nations*. New York: Basic Books; Johansson. 1988. op. cit.
12. See Andersson, Å. E. 1987. *Creativity – the future of the metropolis*. Stockholm: Prisma (in Swedish).
13. See Ståhlberg. 1987. op. cit.
14. See Törnqvist, G. et al. 1986. *The Swedish economy in a geographical perspective*. Stockholm: Liber (in Swedish).
15. See also Snickars, F. 1989. Regional development and urban material infrastructure. *Journal of Economic and Social Geography* 12 (3) 23–37.
16. See Snickars. 1989. op. cit.

Notes to Chapter 8

1. Camagni, R. 1984. Modèles de réconstructuration économique de régions Européennes pendant les années "70". In P. Aydalot (ed.). *Crise et éspace*. Paris: Economica. See also Ch. 5 this volume.
2. Hall, P. G. & D. G. Hay. 1980. *Growth centres in the European urban system*. London: Heinemann Educational.
3. Cheshire, P. C. & D. G. Hay. 1989. *Urban problems in Western Europe: an economic analysis*. London: Unwin Hyman.
4. Cheshire, P. C. & N. Bevan. 1988. *Data consolidation and update for territorial units in the Community: Final report*. Dept. of Economics, University of Reading.
5. This defines the "ideal" system of definition. In reality data limitations

enforced departures from this ideal. In Greece, Portugal and Spain employment by place of work data were not available: in Ireland, Italy, Portugal, and Spain commuting data were not available. How these problems were overcome is detailed in Hall & Hay. 1980. op. cit; Cheshire & Hay. 1989. op. cit.

6   There are clearly disadvantages of this system since precise data, measured directly for FUR components, may be combined with estimated data. For example, a far wider range of variables was collected directly for 50 case study FURs and their associated Level 2 regions (the second tier in the nesting NUTS hierarchy) so actual observations exist for some FURs but only estimated values for others. Similarly because of European diversity, some data, for some countries are not available for the smallest regions or, in the case of some variables, not available at all.
7   Cheshire & Hay. 1989. op. cit.
8   See, for example, Evans, A. W. & D. Eversley (eds.). 1980. *The inner city: employment and industry.* London: Heinemann, especially R. Dennis. The decline of manufacturing employment in Greater London, 1966–74.
9   Camagni. 1984. op. cit.
10 Berkshire now has the highest per capita income of any non-metropolitan county in the United Kingdom. See *Economic Trends.* November 1988. London: H.M.S.O.
11 See Breheny, M. J., P. C. Cheshire & R. Langridge. The anatomy of job creation: industrial change in Britain's M4 corridor. *Built Environment* 9 (1) 1983; P. G. Hall, M. J. Breheny, R. McQuaid & D. A. Hart. 1987. *Western sunrise: the genesis and growth of Britain's major high tech corridor.* London: Unwin Hyman.
12 See Cheshire, P. C. & S. Sheppard. 1989. The British planning system and access to housing: some empirical estimates. *Urban Studies* **26** (5) 469–85.
13 See P. C. Cheshire. 1973. *Regional unemployment differences in Great Britain.* Cambridge: National Institute of Economic and Social Research. Cambridge University Press.
14 See, for example, Cheshire, P. C. 1979. Inner areas as spatial labour markets. *Urban Studies*; or Gordon, I. R. & D. Lamont. 1982. A model of labour market interdependence in the London region. *Environment and Planning* **A**.
15 See Rowthorn, R. E. & J. R. Wells. 1987. *De-industrialization and foreign trade.* Cambridge: Cambridge University Press. pp. 31–6.
16 See, for example, Metzler, L. 1950. A multiple region theory of income and trade. *Econometrica*; or Airov, J. 1963. The construction of interregional business cycle models. *Journal of Regional Science.*
17 See, for example, Brechling, F. 1967, Trends and cycles in British regional unemployment. *Oxford Economic Papers.*
18 Eversley, J. T., A. E. Gillespie & A. O'Neal. 1984. *Regional disparities in European industrial decline.* Paper to European Congress of Regional Science, Milan.
19 The Flemish and French forms for the bilingual city of Brussel–Bruxelles are used interchangeably.
20 Keeble, D., P. L. Owens & C. Thompson. 1981. *The influence of peripheral and central locations on the relative development of regions: Final Report.* Dept. of Geography, University of Cambridge; Keeble, D., P. L. Owens & C. Thompson. 1983. The urban–rural manufacturing shift in the European Community. *Urban Studies* 20 405–18; Keeble, D., J. Offord & S. Walker. 1988. *Peripheral regions in a Community of twelve member states.* Luxembourg: Office of Official Publications of the European Communities.

21 See note 20.
22 Rowthorn & Wells. 1987. op. cit.
23 *Encuesta de Población Activa.* Madrid: Instituto Nacional de Estadística (quarterly).
24 There remain doubts as to the interpretation of unemployment data. These stem from the diversity of European Community regions that has been stressed throughout this chapter. In the poor rural regions of the south under-employment may be as much, or more of a problem as unemployment although that may not be relevant for an analysis of urban regional deindustrialization for obvious reasons. In addition there is in some countries or regions, especially Spain and southern Italy, an important informal or black economy. Unemployment in such regions may overstate rather than understate their problems. There is also the question of cultural expectations and social structures. With female participation rates around 80 per cent in Denmark, even though the survey question may be the same, people's expectations of the norm may be so different from those in, say, northern Spain that their behaviour is also different and the reported unemployment rate differs as a result.
25 The source of this data and that for the variable 'coal' is the *Oxford Regional Economic Atlas: Western Europe.* 1971. Oxford University Press.
26 The arguments of Fothergill, S. and Gudgin, G. London: 1982. *Unequal growth: urban and regional employment change in the United Kingdom.* London: Heinemann support this formulation of the influence of coalfields on local economies.
27 Clark, C., Wilson, F. and Bradley, J. 1969. Industrial location and economic potential in Western Europe. *Regional Studies* **3**.
28 Keeble, Offord & Walker. 1988. op. cit.
29 This is a composite indicator of urban problems reflecting what might be thought of as an "output" measure rather than an "input" measure. It gives most weight to economic factors, but also reflects to some extent environmental and social factors. Weights to apply to the component indicators were estimated using discriminant analysis and were not arbitrarily assigned. Full details are given in Cheshire & Hay. 1989. op. cit. Ch. 3.
30 Camagni, 1984. op. cit.
31 Adams *et al.* 1988. Trends in the distribution of earnings, 1973–86, *Employment Gazette* **96** (2) document these changes for individual earnings. They show that after a period of approximate stability for as long as data had been available, the distribution of earnings became steadily more unequal from 1975.
32 See Buck. 1989. Social polarization in the inner city: an analysis of the impact of labour market and household change. Paper to the British Sociological Society's Annual Conference. This uses alternative sources to examine the changing distribution of household incomes between 1979 and 1985. His results are provisional but suggest the most important contributory factor was rising unemployment.
33 Rowthorn & Wells, 1987. op. cit.
34 Camagni. 1984. op. cit.
35 Cheshire & Hay. 1989. op. cit.
36 See Cheshire, P. C. 1990. London's development prospects. *Land Development Studies.* This summarizes the evidence and offers further analysis.
37 This date is selected as the watershed because it is the year in which the Department of the Environment published *Inner area studies; Liverpool,*

Birmingham and Lambeth, summaries of consultants' final reports. London: HMSO.

Notes to Chapter 9

1. Landaburu, C. 1989. Paper to conference on *The European Community: internal market and regional policy*. Brussels.
2. Commission of the European Communities. 1987. *Third periodic report from the Commission on the social and economic situation and development of the regions of the Community*. Brussels.
3. Commission of the European Communities. 1981. *Study of the regional impact of the Common Agriculture Policy*. Regional Policy Series 21. Brussels.
4. The three original European Communities were the European Economic Community (EEC), the European Coal and Steel Community (ECSC) and EURATOM. They were combined by the Merger Treaty of 1967 and came to be known as the European Communities. This name has tended to give way to the more obvious European Community.
5. The Local Employment Act of 1960 redefined the areas qualifying for Regional Assistance and extended the range and value of support. It led to a more than sevenfold increase in British spending on regional aid. This was followed by a substantial simplification of the criteria for qualification and a further increase in regional spending in 1963. In 1967 the Regional Employment Premium, a direct subsidy on labor costs in the Development Areas, was introduced.
6. The problems of much of Northern Ireland, the South West and northern Scotland were nearer those of backward rural regions.
7. *The Times* December 1st, 1974.
8. Commission of the European Communities. 1973. *Report on the regional problems of the enlarged Community* [the Thomson Report]. Brussels: CEC.
9. Ibid.
10. Speech to the European Parliament, July 11th, 1974; quoted in *The Times* July 12th, 1974.
11. Thomson Report. 1973. op. cit.
12. Speech to British House of Commons, May 8th, 1973.
13. Quoted in *The Times* October 1974.
14. The indicative range for each member state for 1984–87 was as follows:

Belgium	0.9	to	1.2
Denmark	0.5	to	0.7
Federal Republic of Germany	3.8	to	4.8
France	11.1	to	14.7
Greece	12.4	to	15.7
Ireland	5.6	to	6.8
Italy	31.9	to	42.6
Luxembourg	0.06	to	0.08
Netherlands	1.0	to	1.3
United Kingdom	21.4	to	28.6

Source: Council Regulation No. 1787/84 of June 24th, 1984.

15 Proposals for a European high-speed rail network were presented to the European Commissioner for Public Works and Transport in January 1989.
16 Tenth report from the Commission on the European Regional Development Fund, Com (85) 516 final Brussels 1985.
17 Initiated in Council Regulation (EC) No. 2052/88 of June 24th, 1988 and followed up with implementation regulations, Council Regulations (EC) No. 4253/88 and No. 4254/88 both of December 19th, 1988.
18 Council Regulation (EC) No. 2052/88, Article 2.
19 Council Regulation (EC) No. 2052/88, Article 12, para. 5.
20 See Alonso, W. 1989. Regional issues and European integration. Paper for congress on *Regional policy in the Europe of the nineties*. Madrid.
21 Council Regulations (EC) No. 2052/88, Article 8.
22 *Indicateurs statistiques pour la mise en oeuvre de la réforme des Fonds structurels*. Eurostat 1988.
23 Information P8, Brussels, March 8th, 1989.
24 Ibid.
25 Camagni, R. P. 1976. 'La struttura settoriale della bilancia commerciale del Mezzogiorno: una analisi quantitativa'. *Gionale degli Economisti*. July.
26 Masera R. 1980. *L'Unificazione monetaria e lo S.M.E.* Bolgna: Il Mulino.

Notes to Chapter 10

1 *European Economy* Sup. A. 2. February 1989. Table 6.
2 This is consistent with the observation made elsewhere: that, by the standards of even large federal countries, regional variation in Europe is greater.
3 On the concept of the innovative regional milieu, see Aydalot, P. & D. E. Keeble. 1988. *High technology industry and innovative environments: the European experience*. London: Routledge & Kegan Paul.
4 Cheshire, P. C. & D. G. Hay. 1989. *Urban problems in Western Europe: an economic analysis*. London: Unwin Hyman. The Functional Urban Region is generally smaller than the so-called Level 2 regions, distinguished by the European Community statisticians, which tend to correspond more to historic provinces.
5 Hall, P. 1989. *London 2001*. London: Unwin Hyman.
6 Hall, P. 1990. Moving information: a tale of four technologies. In: J. Brotchie, M. Batty, & P. Hall (eds.). *Information: technological change and spatial impacts*. London: Unwin Hyman.
7 Goddard, J. & A. Gillespie. 1988. Advanced telecommunications and regional economic development. In M. Giaoutzi & P. Nijkamp (eds.). *Informatics and regional development*. Aldershot: Avebury. pp. 121–46.
8 Kamada, M. 1980. Achievements and future problems of the shinkansen. In A. Straszak & R. Tuch (eds.). 1980. *The Shinkansen high-speed rail network of Japan: proceedings of an IIASA conference, June 27–30, 1977*. Oxford: Pergamon. pp. 41–56; Sanuki, T. 1980. The Shinkansen and the future image of Japan. In Straszak & Tuch (eds.). op. cit. pp. 227–51.

Notes to Postscript

1. Ng, C. Y., R. Hirono, Narongchai Akrasanee (eds.). 1988. *Industrial restructuring in ASEAN and Japan. An overview.* Singapore: ISEAS (Institute of Southeast Asian Studies).
2. Dore, R. P. with contributions by K. Taira. 1986. *Structural adjustment in Japan, 1970–82.* Employment, Adjustment and Industrialization 1, Geneva: International Labor Office.
3. Adachi, Fumihiko. 1989. Economic adjustment in Japan and its impact on Asian economies. Paper presented at the International Conference on Industrial Transformation and Regional Development in an Age of Global Interdependence. Nagoya/Tokyo: UNCRD.
4. Recent trends indicate an increase in inter-regional trade interdependence in East and Southeast Asia. This, coupled with the vastly increased levels of JDFIA and manufactures imports into Japan, is likely to make this part of the world a major growth pole, sustaining the global economy. See Watanabe, Toshio (ed.). 1989. *Ajia Sangyo Kakumei no Jidai* (Age of the Asian Industrial revolution). Tokyo: JETRO (Japan Export Trade Organization).
5. Adachi. 1989. op. cit.
6. Kobayashi, Hajime. 1989. Industrial estates and regional development in Japan: current issues and vision for the future. Paper presented at the International Seminar on Industrialization and Development Focus on ASEAN, Bangkok: UNCRD and NIDA (National Institute of Development Administration).
7. Ministry of International Trade and Industry (ed.). 1989. *Shin Kogyo Saihaichi Keikaku no Kaisetsu* (A Guide to the New Industrial Location Plan). Tokyo: Tsusho Sangyo Chousakai.
8. Sazanami, Hidehiko. 1988. Le trasformazioni della regione metropolitana di Tokyo dopo la Seconda Guerra Mondiale (The Transformation of the Tokyo Metropolitan Region after the Second World War). In *Urbanistica* **92** June. Milan: Istituto Nazionale di Urbanistica.
9. "Civil minimum:" A minimum standard of amenities that local governments endeavored to achieve in the wake of dyseconomies and adverse environmental effects of rapid industrial growth. First propounded by Metropolitan Tokyo's Governor, Ryokichi Minobe, the concern is also implicitly echoed in the Second NCDP (1969).

Notes on Contributors

Franz-Josef Bade, Professor of Economics at the Department of Spatial Planning, University of Dortmund, Federal Republic of Germany, has been a Visiting Professor in France (Aix-en-Provence) and in the United Kingdom (Glasgow). His research fields are industrial and regional location economics, and innovation and labour market policies. Two of his more important books are *Die mobilität von industriebetrieben* (The mobility of industrial firms) and *Die funktionale struktur der wirtschaft und ihre räumliche arbeitsteilung* (The functional structure of economy and its spatial division of labour). Recently, Dr. Bade has been inducted into the German Academy of Spatial Research. He is also a member of the Scientific Council of Economic Advisors on Structural Change for the Department (Ministry) of Economy of Nordrhein-Westfalen, and a member of the managing board of the German Section of the Regional Science Association.

Roberto Camagni is Professor of Economic Policy and Regional Economics at the University of Padua, and Professor of Economics at the Bocconi University, Milan. Since January 1990 he has been Chairman of the Italian Regional Science Association. He is also President of GREMI, an international association for the study of innovative environments, located in Sorbonne University, Paris; a member of various scientific Committees including the French Commissariat Général Au Plan, of DATAR, the French Agency for public territorial intervention; a member of the Regional Government of Lombardy; and Chairman of the Economic and Territorial Observatory of the Milan Metropolitan Area. He has also worked for EEC, OECD, and the Italian Minister of Industry in the fields of innovation diffusion and regional and urban development.

Paul Cheshire is Professor of Urban and Regional Economics at the University of Reading. A consultant to numerous government and international organizations, he has been a member of the Economic and Social Research Council's Environment and Planning Committee. Between 1983 and 1989 Paul Cheshire was a consultant for the European Commission and directed a large scale study of urban and regional development in the European Community. His present research interests encompass urban and regional systems, the economics of land use planning and the environmental economics of agriculture and agricultural policy. He is the author of *Regional unemployment differences* (1973), *Agriculture, the countryside and amenity: an economic critique* (with J. K. Bowers, 1983) and of *Urban problems in western Europe: an economic analysis* (with D. G. Hay, 1989) and numerous articles in professional journals and reports for government.

Juan R. Cuadrado Roura, author of *Los desequilbrios regionales y el Estado de las Autonomías* (1987) and *La localización industrial en España. Hechos y tendencias* (ed., 1989), has been Aggregate Professor, Universities of Barcelona and

Santiago, 1969–73 and Professor of Economics, University of Málaga, 1974–81, and Secretary for Transport, Tourism and Communications (Madrid 1981–5). He is presently Professor of Economics, University of Alcalá, Madrid, and Research Director of FIES (Fundación para la Investigación Económica y Social, Madrid). A member of various international organizations (Club de Roma, EEA, RSA, SID, etc.) he was one of the founders and later President of the Asociación Española de Ciencia Regional (Regional Science Spanish Association), 1973–83. He has worked with EEC from 1981 on questions related to employment, new technologies and regional problems in Europe, and he has served with OECD as Director of international working groups in the Dominican Republic, the Magreb Area and Colombia. He has published 12 books, about 70 articles and a number of reports and monographs, mainly on the Spanish economy, problems and tendencies of the service sector, and regional problems.

Jean-Paul de Gaudemar, the author of *La mobilisation générale*, and *L'ordre et la production*, has been rapporteur for regional planning of the 10th French national Plan. His many articles cover subjects such as labour market theory and industrial and regional development policies. Now Professor and Dean of the faculty of Economics at Aix-en-Marseille University (France), for several years he served as director of DATAR, the French national administration for regional policy. He is also chairman of the DATAR Scientific Committee, and of the OECD Working Party on Regional Development Policies, and he has served as consultant for various national and international public bodies. More recently, his work has been devoted to European and OECD economic policies from the viewpoint of structural adjustment, industrial and technological change, and regional policies.

Peter Hall is Director of the Institute of Urban and Regional Development and Professor of City and Regional Planning at the University of California at Berkeley. He has taught at the London School of Economics and at the University of Reading, where he was Dean of the Faculty of Urban and Regional Studies. A member of many official committees, he has given testimony to committees of Congress and of the California Legislature on the concept of the Enterprise Zone, which he has been credited with inventing and has received the Founder's Medal of the Royal Geographical Society for distinction in research. He is author of over twenty books on urban and regional planning and related topics, including *The world cities* (1966, 1977, 1983); *Planning and urban growth: an Anglo-American comparison* (with M. Clawson, 1973); *Urban and regional planning* (1975, 1982); *Europe 2000* (ed., 1977); *Great planning disasters* (1980); *Growth centres in the European urban systems* (with D. Hay, 1980); *The inner city in context* (ed., 1981); *Silicon landscapes* (with A. Markusen, 1985); *Can rail save the city?* (with C. Hass-Klau, 1985); *High-tech America* (with A. Markusen and A. Glasmeier, 1986); *The carrier wave* (with P. Preston, 1988); *Cities of tomorrow* (1988) and *London 2001* (1989). He has just completed a major study of the impact of defense spending on American regional development, to be published as *The rise of the gunbelt* (with A. Markusen, S. Campbell and S. Deitrick, 1989).

Klaus R. Kunzmann is Professor of Spatial Planning and has been Director of the Institut für Raumplanung, School of Spatial Planning, University of Dortmund (IRPUD) since 1974. He is an elected member of both the German Academy of Urban and Regional Planning and the Academy of Regional Research and Regional Planning. In 1984 he established SPRING, an English language post-graduate training programme, now jointly offered by the University of Dortmund, the

University of Science and Technology, Kumasi, Ghana, and the Asian Institute of Technology in Bangkok, Thailand. As a consultant to the 'Deutsche Gesellschaft für Technische Zusammenarbeit' (GTZ) he has been working in various Third World countries, including Thailand, Nepal, Yemen Arab Republic, Malawi, Brazil and Jordan. In 1986 he helped to establish the Foundation of the Association of European Schools of Planning (AESOP) and was elected in 1988 its first president. In 1990 he was a visiting professor at the Université Paris VIII. He has directed numerous research projects dealing with the restructuring of traditional industrial regions, spatial development in developing countries and European regional (spatial) planning policy among others for the German Ministry of Urban Planning, Housing and Regional Planning, the Council of Europe and the EEC. His publications include numerous articles and books on these themes, on spatial and environmental planning education and on planning in Germany in general.

Rémy Prud'homme is Professor at the University of Paris XII. His main areas of teaching and research are regional policies, local public finance, transportation policies and environmental policies. He is the head of a University based research group, L'OEIL, which is active in these areas. He has worked for the OECD as Deputy Director of the Environment Directorate for several years, and as a consultant for the World bank, Habitat, the EEC, and various governments. He was a member of many French government advisory committees, such as the committee on regional policy for the 8th Plan, of which he was the rapporteur. He has also taught twice as visiting professor at MIT. In addition to his contributions to many academic journals, he is the author or co-author of: *Regional policies in Turkey, Le financement des équipements publics de demain, L'avenir d'une centenaire: l'automobile, Le rôle des transports urbains dans le développement économique du Brésil, Le ménagement de la nature, Environmental policies in Japan, L'economie du Cambodge.* He is a board member of the International Institute of Public Finance and of the French Economic Association.

Lloyd Rodwin, Ford International Professor Emeritus at the Massachusetts Institute of Technology, was Head of the Faculty Policy Committee of the Joint Center for Urban Studies of MIT and Harvard University (1959-69), organiser and Director of the Guayana, Venezuela planning and research program of the Joint Center (1959-64) and Head of the Department of Urban Studies and Planning at MIT from 1969 to 1973. He founded the Special Program for Urban and Regional Studies of Developing Areas (SPURS) and served as its director from 1967 to 1988. President of the Regional Science Association in 1986-7, he has served on the Board of Editors of *Daedalus, the International Regional Science Review* and other professional journals. Also, he has been an adviser on problems of housing, urban policies and regional development to many governments and international agencies as well as non-governmental and private organizations. His previous books are *The British new towns policy*, 1956; *Housing and economic progress*, 1961; *Nations and cities*, 1970; *Cities and city planning*, 1981. He has edited *The future metropolis*, 1961; *Planning urban growth and regional development*, 1969; *Shelter, settlement and development*, 1986. He also co-edited (with Robert Hollister) *Cities of the mind: images and themes of the city in the social sciences*, 1984; and co-edited (with Hidehiko Sazanami) *Deindustrialization and regional economic transformation: the experience of the United States*, 1989.

Hidehiko Sazanami, Director of the United Nations Centre for Regional Development (UNCRD) in Nagoya, Japan, is former Professor of Urban Environmental

Planning at the Institute of Socio-Economic Planning, University of Tsukuba. He has served as Director of the Urban and Regional Planning Division at the Building Research Institute, Ministry of Construction, Government of Japan; Senior Lecturer at the Institute of Social Studies, the Hague, Netherlands; and in various kinds of central and local government advisory activities in Japan and in several Third World countries. In addition to many articles written for professional journals, his books include Metropolitan planning management, 1982 and Local social development planning, 1984. He has also co-edited (with Lloyd Rodwin) Deindustrialization and regional economic transformation: the experience of the United States, 1989.

Folke Snickars, currently the President of the European Regional Science Association, is Professor of Regional Planning at the Royal Institute of Technology in Stockholm. He has spent part of his professional career in government, in planning departments within the City of Stockholm and in an expert group for regional studies within the Swedish Ministry of Industry. He has also been on the staff of the International Institute for Applied Systems Analysis in Vienna for two years. Professor Snickars has been a joint editor, and contributing author, of several internationally published volumes in the field of regional science, including Spatial interaction theory and planning models 1978 (with Anders Karlqvist, Lars Lundqvist and Jörgen W. Weibull); Regional development modeling: theory and practice 1982 (with Murat Albegov and Åke E. Andersson); Multiregional Economic Models: Practice and Prospect 1983 (with Boris Issaev, Peter Nijkamp and Piet Rietveld); Economic faces of the building sector 1985 (with Börje Johansson and T. R. Laksmanan), and The economics of a regulated housing market 1987 (with Alex Anas, Göran Cars, Jan R. Gustafsson, Björn Hårsman and Ulf Jirlow). He has published numerous articles in international journals, and written several books and articles on regional economic planning for a Swedish audience, concerning infrastructure analysis, the building sector, regional economics, population and the regional consequences of long-term economic change. He has also consulted extensively with planning agencies in Sweden, Finland and Australia.

Acknowledgements

As in the case of our earlier study, *Deindustrialization and regional economic transformation: the experience of the United States*, the ten studies in this book were commissioned by the United Nations Centre for Regional Development. Abbreviated versions of the first eight chapters were presented in Nagoya and Tokyo on September 18–22, 1989, at the International Conference on Industrial Transformation and Regional Development: Regional Economies in a Borderless Age. Chapters 9 and 10, dealing with the evolution of the European Community regional policy, and with Europe's regional–urban futures, were prepared following the September meetings in Japan by two teams of the contributors. The Postscript, too, on the issues and prospects for urban and regional structural transformation in Japan, was prepared after these meetings.

The contributions and co-sponsorship of several organizations made these meetings and publication possible. They include the Organizing Committee for the International Conference on Comparative Regional Development Studies, the Prefectural Governments of Gifu and Aichi, the Nagoya City Government, the Japan Development Bank and the Japan Centre for Area Development Research. We appreciate their assistance very much.

We also extend our thanks to the discussants and participants who attended the conference for their helpful comments and suggestions; to the staff of UNCRD for their exceptional efficiency and gracious assistance; and to Adiyana Sharag-Eldin for all sorts of very capable assistance – research, editing and typing – in the preparation of this manuscript.

Finally, all of the contributors, particularly Roberto Camagni, Paul Cheshire, Peter Hall, Rémy Prud'homme and Folke Snickars, invested an immense amount of time and energy in critiquing the papers and in making constructive suggestions for the conduct of this enterprise.

Lloyd Rodwin
Hidehiko Sazanami

List of tables

		page
2.1	Informational occupations as a percentage of total employment	45
2.2	Major high-tech concentrations, 1975–81	48
2.3	Changes in producer service location by local labour market areas, 1971–81	50
2.4	Business and producer services: location quotients, 1981	51
2.5	Types of business service: percentages of national offices, 1981	51
2.6	Top ten United Kingdom accountancy firms: major locations	54
2.7	Employment effect of innovation	55
2.8	New firm formation rates, 1980–3	57
2.9	Employment changes by sectors, South East region, 1971–81	59
3.1	The final demand for services, 1980	76
3.2	Employment vs. production changes: United States and Federal Republic of Germany	78
3.3	The states of the Federal Republic of Germany	82
3A.1	The sectoral development of production	350
3A.2	The sectoral development of gross value added	351
3A.3	The sectoral development of employment	352
3A.4	The functional development of employment	353
4.1	Four concepts of industrialization, 1967–87	107
4.2	Changes in final demand, foreign trade and output: industrial goods and services, 1970–86	109
4.3	Value-added and employment growth in industry: France and Europe, selected periods	112
4.4	Selected economic indicators: Northeast, South, Ile-de-France, 1976–86	121
4.5	Growth of value added by sector, 1982–6: Paris and France	129
4.6	Office construction: France and Paris region, 1976–88	129
4.7	Employment in the automobile industry, 1976–86, by groups of regions	133
5A.1	Employment changes by sector, 1970–84	354
5A.2	Growth rate of value added, 1970–6–80–4	356
5A.3	Productivity change, 1970–6–80–4	358
5A.4	Ranking of Italian regions according to the level and growth rates of total productivity	360
5A.5	Shift–share analysis of productivity growth	361
5A.6	Relative productivity and employment growth	362

6.1	Growth of industrial activities, 1960–73	170
6.2	Foreign trade coverage in some European Community countries and Spain, 1975	171
6.3	The evolution of industrial demand	173
6.4	Employment trends in the Spanish service industries, 1976–86	176
6.5	The balance of trade of Spanish industry, 1970–86	176
6.6	Changes in the structure of Spanish industry, 1970–86	177
6.7	Spanish GDP growth, 1986–9	178
6.8	Regional growth of GRP and industry regional growth, 1973–85	183
6.9	Evolution of regional employment by sectors, 1973–9 and 1979–85	184
6.10	Shift–share figures by regions, 1973–85	187
6.11	Location of new industrial investments, 1980–5	189
6.12	GRP growth by autonomous communities, 1986–8	192
6A.1	Employment structure by sectors, 1960–87	363
6A.2	Participation of the production branches in the production value of industry	364
6A.3	Structural changes by regions	365
6A.4	ZUR: investment and new jobs	366
7.1	Traits and development conditions of industries operating under price and product competition	206
7.2	Employment in three parts of the Swedish economy, 1950–80 and a forecast to 1990	207
7.3	Development of population and employment in functional regions of Sweden, 1975–84 and forecast for 1985–95	216
7.4	Sectors in the Swedish economy ranked according to employment growth for the period, 1970–83	221
7.5	Share of R&D personnel of total employment in private service sectors in four parts of Sweden, 1970–84	223
7.6	Funds for education and R&D in five subject areas allocated to six national university regions in Sweden in 1984	231
8.1	Summary data on the experience of deindustrialization at national level, 1977–86	238
8.2	Manufacturing employment loss from selected FUR cores, FURs and Level 2 regions 1971–81, FUR unemployment 1985–7, and FUR per capita GDP 1985	245
8.3	Industrial production and employment for EC member states: cumulative change 1983–7	250
8.4	Dependent variable: change in mean U rate, 1977/81–1985/87	259
8.5	Dependent variable: change in urban problem index, 1971–87 for FUR	262
8A.1	Correspondence between NUTS levels and national administrative divisions	367
9.1	Structure of the European Community's budget	272
9.2	The significance of Objective 2 regions	294
9.3	Balance sheet of European Community policies	295
11.1	Principal targets of the former and new industrial location plans, and achievements as of 1985	330

11.2	Concentration of international business and other central functions in the Tokyo Metropolitan Region	334
11.3	Land prices and land use intensity in the 23 central wards of Metropolitan Tokyo, 1983 and 1988	335
11.4	Basic characteristics and conceptual features of designated Technopolises	344

List of figures

		page
1.1	Europe	5
1.2	Regions of the United Kingdom	7
1.3	States of the Federal Republic of Germany	10
1.4	Regions of France	12
1.5	Regions of Italy	14
1.6	Regions of Spain	15
1.7	Counties of Sweden	17
2.1	Employment changes by region, 1971–8, 1978–81, 1981–8	40
2.2	Gross Domestic Product changes by region, 1971–8, 1978–81, 1981–8	41
2.3	Standard regions of the United Kingdom	42
2.4	Ratios of services to production, 1971–8, 1978–81, 1981–4	43
2.5	Ratios of information to goods, 1971–8, 1978–81, 1981–4	44
2.6	High-technology industry, location quotients, 1981 and 1984	47
2.7	Top ten United Kingdom accountancy firms: regional distribution	54
2.8	Standardized innovation rates by region	56
2.9	Changes in London employment, 1951–84	60
2.10	Employment changes, 1979–83, 1983–5	66
2.11	Regional employment changes, 1979–83, 1983–5	67
3.1	Sectoral changes in employment 1973–86, Federal Republic of Germany and United States	72
3.2	The tertiarization of employment in functional and sectoral terms	74
3.3	The expansion of the service sector in terms of production, income and labour input	75
3.4	Sectoral gains and losses of employment	78
3.5	The contribution of producer services to the tertiarization of employment	79
3.6	The states (Länder) of the Federal Republic of Germany	81
3.7	The spatial structure of the Federal Republic of Germany	83
3.8	The hierarchical delineation of the spatial structure of the Federal Republic of Germany	84
3.9	Spatial trends in the change of employment	84
3.10	Change of employment in the agglomerations of the Federal Republic of Germany	85
3.11	Spatial trends in the change of sectoral employment	86
3.12	Sectoral changes of employment in the agglomerations	87
3.13	Spatial trends in the employment of selected service industries	88